LANDSCAPE
LOGIC

LANDSCAPE
LOGIC

1946   *1970*   *1989*   *2006*

EDITORS: TED LEFROY, ALLAN CURTIS, ANTHONY JAKEMAN & JAMES MCKEE

CSIRO

PUBLISHING

National Library of Australia Cataloguing-in-Publication entry

Landscape logic: integrating science for landscape management/edited by Ted Lefroy ... [et al.].

9780643103542 (pbk.)
9780643103559 (epdf)
9780643103566 (epub)

Includes bibliographical references and index.

Natural resources – Australia – Management.
Water quality management – Australia.
Environmental management – Australia.

Lefroy, E. C. (Edward C.)

363.700994

Published by
CSIRO Publishing
36 Gardiner Road, Clayton VIC 3168
Private Bag 10, Clayton South VIC 3169
Australia

Telephone: [+613] 9545 8555
Local call: 1300 788 000 (Australia only)
Fax: +61 3 9662 7555
Email: csiropublishing@csiro.au
Website: www.publishing.csiro.au

Front cover: Aerial photo montage courtesy of the Department of Sustainability and Environment. Influence diagram courtesy Jenifer Ticehurst, Carmel Pollino and Wendy Merritt. Photo of people by Jennifer Hemer.

Edited by Adrienne de Kretser, Righting Writing
Cover and text design by James Kelly
Original cover concept by Julia Dineen
Set in 10/12.5 Adobe Minion Pro and ITC Stone Sans
Typeset by Desktop Concepts Pty Ltd, Melbourne
Index by Bruce Gillespie
Printed by Ingram Lightning Source

CSIRO PUBLISHING publishes and distributes scientific, technical and health science books and journals from Australia to a worldwide audience and conducts these activities autonomously from the research activities of the Commonwealth Scientific and Industrial Research Organisation (CSIRO). The views expressed in this publication are those of the author(s) and do not necessarily represent those of, and should not be attributed to, the publisher or CSIRO.

Contents

Preface

The research project described in this book had at its core the aim of establishing links between past environmental management and the state of the environment. Aware of the size of this challenge, we set out in numbers with intent to practise science that was interdisciplinary and participatory. That is, we intended to bring together scientific experts in many fields to work alongside environmental managers with practical experience to help shape the questions, participate as far as possible in the research, enhance knowledge exchange and apply the findings.

Support for such a large collaborative venture was made possible by the establishment of the Australian Government's Commonwealth Environmental Research Facilities (CERF) program in 2005, administered by the then Department of Environment, Water, Heritage and the Arts. The CERF program arose partly in response to the growing commercialisation of the Cooperative Research Centres program, which had been the major vehicle for public good collaborative research. CERF provided a vehicle for public good research dedicated to the environment, and Landscape Logic was one of eight research hubs funded in its first round.

Through its four-year life, this group of researchers and environmental managers confronted the challenge of establishing causality and along the way learned about accommodating uncertainty and applying multiple lines of evidence to strengthen inferred relationships. But perhaps most of all we learned about people and how they work together. About understanding and valuing different people's perspectives of landscapes and capturing their knowledge of places, processes and people. And about how to represent this collective understanding as a basis for management and a starting point for further learning.

In particular, we acknowledge our partners from regional natural resource management organisations and state environment agencies who brought their knowledge and experience to the project. They include John Riddiford, Bill O'Kane, Megan McFarlane, Richard Ingram, Catherine Murdoch, Chris Norman, Rod McLennan, Mark Cotter, Pat Feehan, Kathleen Broderick, Vanessa Keogh, Kaylene Allan, Peter Ockendon, Tim Barlow, Wayne Tenant, Geoff Taylor, David McCormack, Aniela Grun, Sue Botting, James Shaddick, Geoff Robinson, Adam Hood, Andrew Baldwin, Scott Schilg, Ian Higgins, Peter McRostie, Rohan Hogan, Jenny Alexander, Catriona Grantham, Vanessa Elwell-Gavins, Alistair Kay, Peter Zund, Jim Blackney, Jane Roots, Carla Miles, Adrian Martin, Stuart Brownlea, Polly Buchhorn, Mel Kelly, Peter Voller, Rhys Stickler, Louise Gilfedder, Penny Wells, Allan Harradine, Martin Read and Kate Wilson.

The chair of our advisory board, John Williams, offered wise counsel and guidance over the course of the project as did the rotating membership of the board: Mike Lee, John Whittington, Hans Drielsma, Charlie Zammit, Sean Sullivan, Gavin Hanlon, Damien Wells, Kim Lowe, Jacky Tierney, Dave Johnson, Geoff Richardson, Ian Gaze, Kathleen Broderick, Margaret Johnson and Colin Steele.

Graham Harris and David Rissik chaired midterm reviews of our research projects, and we are grateful for their frank advice and that of panel members Kim Lowell, Alexander Held, John Cary, Ian Lunt, Phil Gibbons, Tony McAlister, Imogen Fullagar, Richard Stirzaker, Damien Wells, Geoff Robinson, Rod McLennan and Steve Harris.

Our thanks also to those whose job it was to keep this large and widely distributed group connected and functioning: Geoff Park, Greg Pinkard, Mignon Jolly, Jennifer Hemer, Liam Gash, Rebecca Kelly, Jenny Baulis, Corrine Jager and Felix Andrews.

Finally, our thanks to David Lyons who helped with the original proposal and suggested the project name, Imogen Fullagar who oversaw the reviewing and editing of the manuscript and Julia Dineen who prepared numerous figures and designed the cover.

Ted Lefroy, Allan Curtis, Anthony Jakeman and James McKee
June 2011

Contributors

Catherine Allan   Institute for Land, Water and Society, Charles Sturt University, PO Box 789, Albury NSW 2640. callan@csu.edu.au

Jason Beard   Institute for Marine and Antarctic Studies, University of Tasmania, Private Bag 49, Hobart TAS 7001. Jason.Beard@utas.edu.au

Ulrike Bende-Michl   CSIRO Land and Water, and Water for a Healthy Country Flagship, GPO Box 1666, Canberra ACT 2601. Ulrike.Bende-Michl@csiro.au

Shane Broad   Tasmanian Institute of Agricultural Research, PO Box 3523, Burnie TAS 7320. Shane.Broad@utas.edu.au

Bill Cotching   Tasmanian Institute of Agricultural Research, PO Box 3523, Burnie TAS 7320. Bill.Cotching@utas.edu.au

Christine Crawford   Institute for Marine and Antarctic Studies, University of Tasmania, Private Bag 49, Hobart TAS 7001. Christine.Crawford@utas.edu.au

Hamish Cresswell   CSIRO Land and Water, and Water for a Healthy Country Flagship, GPO Box 1666, Canberra ACT 2601. Hamish.Cresswell@csiro.au

Allan Curtis   Institute for Land, Water and Society, Charles Sturt University, PO Box 789, Albury NSW 2640. acurtis@csu.edu.au

Peter Davies   School of Zoology, University of Tasmania, Private Bag 5, Hobart TAS 7001. P.E.Davies@utas.edu.au

David Duncan   Arthur Rylah Institute for Environmental Research, Department of Sustainability and Environment, Heidelberg VIC 3084. David.Duncan@dse.vic.gov.au

Elizabeth Farmer   Geospatial Science, RMIT University, GPO Box 2476V, Melbourne VIC 3001. Elizabeth.Farmer@rmit.edu.au

Barry Gallagher   School of Zoology, University of Tasmania, Private Bag 5, Hobart TAS 7001. JohnG6@utas.edu.au

John Gibson   Institute for Marine and Antarctic Studies, University of Tasmania, Private Bag 49, Hobart TAS 7001. John.Gibson@utas.edu.au

Aniela Grun   NRM South, PO Box 425, South Hobart TAS 7004. agrun@nrmsouth.org.au

Peter Hairsine   CSIRO Land and Water, GPO Box 1666, Canberra ACT 2601. Peter.Hairsine@csiro.au

Nelli Horrigan   c/- School of Zoology, University of Tasmania, Private Bag 5, Hobart TAS 7001. Nelli.Horrigan@utas.edu.au

Anthony Jakeman   iCAM, Fenner School of Environment and Society, and National Centre for Groundwater Research and Training, Building 48, Australian National University, Canberra ACT 0200. Tony.Jakeman@anu.edu.au

Simon Jones   Geospatial Science, RMIT University, GPO Box 2476V, Melbourne VIC 3001. Simon.Jones@rmit.edu.au

Rebecca Kelly   isNRM Pty Ltd, PO Box 8017, Trevallyn TAS 7250. Rebecca@isnrm.com.au

Marit E Kragt   School of Agricultural and Resource Economics, University of Western Australia, 35 Stirling Highway, Crawley WA 6009. Marit.Kragt@uwa.edu.au

Garreth Kyle   Arthur Rylah Institute for Environmental Research, Department of Sustainability and Environment, Heidelberg VIC 3084. Garreth.Kyle@dse.vic.gov.au

Michael Lacey   Tasmanian Institute of Agricultural Research, GPO Box 3523, Burnie TAS 7320. Michael.Lacey@utas.edu.au

Alex Lechner   Geospatial Science, RMIT University, GPO Box 2476V, Melbourne VIC 3001. Alex.Lechner@rmit.edu.au

Ted Lefroy   Centre for Environment, University of Tasmania, Private Bag 141, Hobart TAS 7001. Ted.Lefroy@utas.edu.au

Shaun Lisson   CSIRO Ecosystem Sciences, Private Bag 54, Hobart TAS 7001. Shaun.Lisson@csiro.au

Regina Magierowski   School of Zoology, University of Tasmania, Private Bag 5, Hobart TAS 7001. Regina.Magierowski@utas.edu.au

Nicki Mazur   Institute for Land, Water and Society, Charles Sturt University, PO Box 789, Albury NSW 2640. NickiMazur@grapevine.net.au

Stephen McGowan   Institute for Marine and Antarctic Studies, University of Tasmania, Private Bag 49, Hobart TAS 7001.

James McKee   NRM North, PO Box 1224, Launceston TAS 7250. admin@nrmnorth.org.au

Rod McLennan   Goulburn Broken Catchment Management Authority, PO Box 1752, Shepparton VIC 3630. reception@gbcma.vic.gov.au

Wendy Merritt   iCAM, Fenner School of Environment and Society, and National Centre for Groundwater Research and Training, Building 48, Australian National University, Canberra ACT 0200. Wendy.Merritt@anu.edu.au

Wendy Minato   Institute for Land, Water and Society, Charles Sturt University, PO Box 789, Albury NSW 2640. Wendy.Minato@csiro.au

Naoko Miura   Geospatial Science, RMIT University, GPO Box 2476V, Melbourne VIC 3001.

Daniel G Neary   USGA Forest Service, Rocky Mountain Research Station, 2500 South Pine Knoll Drive, Flagstaff AZ 86001 USA. DNeary@fs.fed.us

Graeme Newell   Arthur Rylah Institute for Environmental Research, Department of Sustainability and Environment, Heidelberg VIC 3084. Graeme.Newell@dse.vic.gov.au

Tony Norton   School of Agricultural Science, University of Tasmania, GPO Box 3523, Burnie TAS 7320. Tony.Norton@utas.edu.au

Geoff Park   North Central Catchment Management Authority, PO Box 18, Huntly VIC 3551. Geoff.Park@nccma.vic.gov.au

Kevin C Petrone   CSIRO Land and Water, Private Bag 5, Wembley WA 6913. Kevin.Petrone@csiro.au

Greg Pinkard   Centre for Environment, University of Tasmania, Private Bag 55, Hobart TAS 7001. Greg.Pinkard@utas.edu.au

Carmel Pollino   CSIRO Land and Water, GPO Box 1666, Canberra ACT 2601. Carmel.Pollino@csiro.au

Digby Race   Institute for Land, Water and Society, Charles Sturt University, PO Box 789, Albury NSW 2640. drace@csu.edu.au

Steve Read   Forestry Tasmania, Box 207, Hobart TAS 7001. Steve.Read@forestrytas.com.au

Karin Reinke Geospatial Science, RMIT University, GPO Box 2476V, Melbourne VIC 3001. Karen.Reinke@rmit.edu.au

Maureen Rogers Institute for Land, Water and Society, Charles Sturt University, PO Box 789, Albury NSW 2640. maureen_rogers@clearmail.com

Jeff Ross Institute for Marine and Antarctic Studies, University of Tasmania, Private Bag 49, Hobart TAS 7001. Jeff.Ross@utas.edu.au

Libby Rumpff School of Botany, University of Melbourne, Parkville VIC 3010. LRumpff@unimelb.edu.au

Kathryn Sheffield Geospatial Science, RMIT University, GPO Box 2476V, Melbourne VIC 3001.

Philip J Smethurst CSIRO Ecosystem Sciences, Private Bag 12, Hobart TAS 7001. Philip.Smethurst@csiro.au

Jenifer Ticehurst iCAM, Fenner School of Environment and Society, and National Centre for Groundwater Research and Training, Building 48, Australian National University, Canberra ACT 0200. Jenifer.Ticehurst@anu.edu.au

Kirsten Verburg CSIRO Land and Water, and Water for a Healthy Country Flagship, GPO Box 1666, Canberra ACT 2601. Kirsten.Verburg@csiro.au

Peter Vesk School of Botany, University of Melbourne, Parkville VIC 3010. PVesk@unimelb.edu.au

Brendan Wintle School of Botany, University of Melbourne, Parkville VIC 3010. b.wintle@unimelb.edu.au

Introduction: improving the evidence base for natural resource management

Ted Lefroy, Allan Curtis, Anthony Jakeman and

James McKee

THE EFFECTIVENESS OF ENVIRONMENTAL MANAGEMENT

Throughout the world we face increasing evidence of degrading natural resources. At the same time there is a growing body of evidence that restoration efforts are not having their expected impacts, with low levels of return on investment in restoration partially attributed to a lack of clear and measurable goals and objectives (e.g. Bernhardt *et al.* 2005). The Australian experience is salutary, with a long history of investment and a high level of community engagement but limited evidence of improvement.

In 1989, the Australian government's State of Environment report announced the launch of the Decade of Landcare (DAFF 1995). This self-help approach to environmental management was based on a joint proposal from the Australian Conservation Foundation and the National Farmers' Federation and relied heavily on local community groups. In 1997, community environmental management was boosted by the establishment of the Natural Heritage Trust, funded from the sale of one-third of Telstra, the country's publicly owned telecommunications provider. Since that time, over $4.2 billion of public funds have been invested in environmental management, much of it through Landcare and other volunteer community groups (Hajkowicz 2008). As the Landcare movement developed momentum throughout the

Figure 1.1: Australia's 56 natural resource management regions. The six regions in the Landscape Logic research partnership are shaded: in Tasmania they are the Cradle Coast Natural Resource Management (north-west), Natural Resource Management North (north-east) and Natural Resource Management South; in northern Victoria, from west to east, they are the North Central Catchment Management Authority, Goulburn Broken Catchment Management Authority and North East Catchment Management Authority.

1990s the number of groups exceeded 2500, which stretched the capacity of governments at all levels (Campbell 1994; Curtis 2000; Paton *et al.* 2004). The Australian and state governments responded by establishing a network of 56 catchment management organisations across the country to act as peak regional bodies (Fig. 1.1). Their role has been to identify priorities for environmental management and oversee investment across portfolios that typically include water quality, vegetation extent and condition, invasive pests, soil health, sustainable agriculture and human capacity-building (Pannell *et al.* 2008).

Four audits of public environmental programs over the last 15 years have cast some doubt, however, on the effectiveness of this investment (ANAO 1997, 1998, 2001, 2008). The 1997 audit included the comment 'It is difficult to determine the extent to which programs are achieving their intended outcomes' and each successive audit has echoed that sentiment. While the audits have found evidence of thousands of kilometres of fencing, millions of trees planted, hundreds of hectares of wetlands treated and thousands of people trained, they have not been able to determine if this has made any material difference to the state of the environment.

There are several perfectly reasonable explanations of why this evaluation of success has been indeterminate, not just in Australia but globally. The scale of intervention of most of these projects has been too small and they have been too fragmented to make any measurable impact on processes occurring at landscape and larger scales. The long time lags between intervention and response mean it is often unrealistic to expect change for decades. Human intervention is frequently overwhelmed by forces outside our influence such as climate variability, fluctuating markets and extreme events. And we are frequently dealing with complex, non-linear processes that involve as yet undefined thresholds of change – all in a social context where there are many different land owners and managers, often with competing interests and values (Curtis *et al.* 1998).

But there is a limit to how long these explanations should satisfy the auditors or the funders. The increasing emphasis on evidence-based policy and scrutiny of public funding means that large public environmental programs may be at risk if evidence of their success cannot be produced.

One of the biggest challenges is the lack of long-term data capable of showing the state of the environment and the direction in which it is heading. Without that information, we cannot hope to discern if we are making any difference when we intervene. The choices are to set up well-designed interventions in the spirit of adaptive management and monitor them for several decades, or see what we can glean from existing datasets, natural experiments, expert opinion and modelling.

WHAT WE SET OUT TO DO

The Landscape Logic research partnership was set up to measure the effects of past human intervention on the environment using a combination of biophysical and social evidence, and the experience of land and water managers. The types of evidence at our disposal included historic aerial photos, long-term datasets from stream gauging stations, water quality and vegetation surveys, analysis of historic records, space-for-time experiments, and contemporary studies assembling landscape histories from old aerial photos, social surveys and interviews. There were also the historic records laid down in the landscape itself that can be interpreted through surrogates such as stable isotope analysis of soil, water and vegetation.

The primary objective was to see if we could detect change in environmental condition and determine the extent to which that change could be attributed to human intervention. The secondary objective was to use this knowledge of how landscapes have responded to past intervention to give environmental managers and their public and private investors more confidence in their plans, strategies and ability to report on progress. A necessary condition of both objectives was to work closely with the environmental managers for whom the new knowledge was intended. This helped to ensure that the research questions were appropriate, the interpretation of results made sense and the knowledge produced was relevant to their responsibilities of choosing where to invest, how to intervene and how to know if the intervention worked.

The research partnership consisted of four universities (University of Tasmania, Australian National University, RMIT University and Charles Sturt University), three state agencies (Department of Primary Industries, Parks, Water and Environment Tasmania, Forestry Tasmania, and the Department of Sustainability and Environment Victoria) and six catchment management organisations (Fig. 1.1) with funding from the federal Department of Sustainability, Environment, Water, Population and Communities.

The three catchment management organisations in Victoria (North Central Catchment Management Authority, Goulburn Broken Catchment Management Authority and North East Catchment Management Authority) were among the largest and longest-running in the country. These organisations were approaching 10 years of age, with legislative responsibilities for some aspects of land and water management, up to 40 staff and annual budgets in excess of $50 million. The three comparative Tasmanian organisations (Natural Resource Management South, Natural Resource Management North and Cradle Coast Natural Resource Management) were among the most recently established and smallest in the country. One of our challenges, therefore, was to develop decision support tools and other information that could be used by organisations with different levels of responsibility, capacity and experience.

The first question the researchers posed to the catchment management organisations was 'What is the biggest single investment area across your portfolio about which you are least certain?' In Tasmania the response was water quality. The State of Environment report (Tasmanian Planning Commission 2009) described the condition of Tasmania's rivers as 'steady', but the picture is uneven. Forty-four percent of the island is protected for conservation and only one-third of its river network is subject to disturbance, but Tasmania has the highest proportion of land under irrigation of any Australian state. Its agricultural sector contributes $1 billion a year to the state's economy, primarily through its horticulture ($250 million pa) and dairy industries ($200 million pa), and there is a new wave of interest in agricultural development generated by over a decade of drought on mainland Australia and planning constraints in other jurisdictions. This new interest in irrigation development has raised questions about safe limits of intensification. The question of interest to the Tasmanian natural resource management bodies was how to best manage areas under intensive farming to ensure protection of water quality. In the case of the three Victorian regions, vegetation condition was the main concern. The regions are in the heartland of Australia's Landcare movement and have seen over 20 years of concerted effort devoted to improving the extent and condition of native vegetation, but they did not have the means to determine the effectiveness of those efforts.

PROJECT STRUCTURE

The biophysical disciplines required within our research teams were determined by the two areas of interest identified by our partners: water quality and vegetation condition. The need to integrate knowledge from different disciplines with the experience of managers suggested the need for social sciences focused on understanding human behaviour and practices. It also required the need for participatory processes and an integration framework. Consequently, three different types of activity were considered necessary for the research to be both interdisciplinary and participatory (or transdisciplinary, defined as unrelated disciplines working with end users to solve problems, Tress *et al.* 2005). These were knowledge discovery, knowledge integration and knowledge broking. Ten teams were set up, with eight devoted to knowledge discovery, one to knowledge integration and one to knowledge broking.

KNOWLEDGE DISCOVERY

Water quality research was carried out by five teams. Four teams focused on the influences on water quality in agricultural catchments (see Chapters 2 to 5, on land use and land management, river health, estuarine health and riparian buffering) while the fifth team considered the dynamics of water movement across a whole catchment (Chapter 7, catchment spatial diagnosis). Vegetation extent and condition was examined across three scales (landscape, property and site) within one team (Chapters 10 and 12). In addition to these six teams, one knowledge discovery team was devoted to understanding the social contexts within which water quality and vegetation are managed (Chapters 6, 11, 15 and 16), including the social acceptability of recommended practices and the impacts of investment on human and social capital, while one team focused on the use of remotely sensed data in measuring change in vegetation extent and condition (Chapters 13 and 14, spatial analysis).

KNOWLEDGE INTEGRATION

The integration team had three vital roles within the partnership. It had to identify how we might integrate and share information of different types (quantitative and qualitative) and quality and from different sources (empirical, simulated and experiential); it had to train the research teams and research users in data application; and it had to develop decision support systems that incorporated the outputs of the knowledge discovery teams. Based on several criteria (e.g. Jakeman *et al.* 2007) the primary integration framework selected was Bayesian networks (Chapter 18). Following training of researchers and environmental managers there was wide acceptance of the first stage of Bayesian network modelling, using influence or causal loop conceptual diagrams as a systematic and inclusive approach to articulating hypotheses of ecosystem response to intervention. Subsequently, a range of different modelling approaches and decision frameworks emerged within different teams, tailored to the nature of the data and the intended use. These included process models (Chapters 2 and 5), decision trees (Chapters 4 and 7) and interactive maps (Chapter 12) as well as Bayesian network models (Chapters 3, 12 and 18). Chapter 8 describes a PhD thesis supervised within the integration team that combined biophysical modelling and economic valuation of water quality.

KNOWLEDGE BROKING

The role of the knowledge broking team was to maintain a two-way flow of information between the individual research teams, between researchers and our end user partners, and between the project as a whole and our wider group of stakeholders among state and national government, industry groups and NGOs (Chapters 19 and 20).

BOOK STRUCTURE

This book is divided into three parts. Part I (Chapters 2–9) describes the water quality research, with a concluding chapter summarising what was learned from the point of view of the research teams and environmental managers. Part II (Chapters 10–17) describes the vegetation change research, with a concluding chapter summarising what was learned, also from the point of view of the research teams and environmental managers. Part III (Chapters 18–21) contains chapters on three aspects of transdisciplinary research: integrating knowledge from disparate sources to aid decision-making (Chapter 18), the role of the knowledge broker in large research collaborations (Chapter 19), and a reflection on the project's success in meeting the needs and interests of researchers, end users and funders based on a survey carried out across the partnership (Chapter 20). The last chapter (Chapter 21) summarises the lessons learned about managing natural resources and carrying out transdisciplinary research.

REFERENCES

ANAO (Australian National Audit Office) (1997) 'Commonwealth natural resource management and environment programs'. Audit Report no. 36 1996-97. Australian National Audit Office, Canberra.

ANAO (Australian National Audit Office) (1998) 'Preliminary inquiries into the National Heritage Trust'. Audit Report no. 42 1997-98. Australian National Audit Office, Canberra.

ANAO (Australian National Audit Office) (2001) 'Performance information on Commonwealth financial assistance under the Natural Heritage Trust'. Audit Report no.43 2000-01. Australian National Audit Office, Canberra.

ANAO (Australian National Audit Office) (2008) 'Regional delivery model for the National Heritage Trust and the National Plan for Salinity and Water Quality'. Audit Report no. 21 2007- 08. Australian National Audit Office, Canberra.

Bernhardt ES, Palmer MA, Allan JD, Alexander G, Barnas K, Brooks S, Carr J, Clayton S, Dahm C, Follstad-Shah J, Galat D, Gloss S, Goodwin P, Hart D, Hassett B, Jenkinson R, Katz S, Kondolf GM, Lake PS, Lave R, Meyer J L, O'Donnell TK, Pagano L, Powell B and Sudduth E (2005) Synthesizing U.S. river restoration efforts. *Science* **308**, 636–637.

Campbell AC (1994) *Landcare: Communities Shaping the Land and the Future*. Allen and Unwin, Sydney.

Curtis A (2000) Landcare: approaching the limits of volunteer action. *Australian Journal of Environmental Management* **6**, 26–34.

Curtis A, Race D and Robertson A (1998) Lessons from recent evaluations of natural resource management programs in Australia. *Australian Journal of Environmental Management* **5**(2), 109–119.

DAFF (Dept of Agriculture, Forestry and Fisheries) (1995) The Decade of Landcare Plan: National Overview. Dept of Agriculture, Forestry and Fisheries, Canberra. <http://www.daff.gov.au/natural-resources/landcare/publications/decade-plan>.

Hajkowicz SA (2008) The evolution of Australia's natural resource management programs: towards improved targeting and evaluation of investments. *Land Use Policy* **26**, 471–478.

Jakeman AJ, Letcher RA and Chen S (2007) Integrated assessment of impacts of policy and water allocation change across social, economic and environmental dimensions. In *Managing Water for Australia: The Social and Institutional Challenges*. (Eds K Hussey and S Dovers) pp. 97–112. CSIRO Publishing, Melbourne.

Pannell DJ, Ridley A, Seymour E, Regan P and Gale G (2008) Regional natural resource management arrangements for Australian states: structures, legislation and relationships to government agencies. SIF3 Working Paper 0809, CRC for Plant-based Management of Dryland Salinity, Perth. <http://cyllene.uwa.edu.au/~dpannell/cmbs3.pdf>.

Paton S, Curtis A, McDonald G and Woods M (2004) Regional NRM – is it sustainable? *Australasian Journal of Environmental Management* **11**(4), 259–267.

Tasmanian Planning Commission (2009) State of the Environment Tasmania 2009. Tasmanian Planning Commission, Hobart. <http://soer.justice.tas.gov.au/2009/> (accessed 5 May 2011).

Tress B, Tress G and Fry G (2005). Integrative studies on rural landscapes: policy expectations and research practice. *Landscape and Urban Planning* **70**(1/2), 177–191.

Part I

MANAGING WATER QUALITY IN AGRICULTURAL CATCHMENTS

2

Modelling the influences of land use and land management on water quality

Bill Cotching, Shane Broad, Shaun Lisson and Rebecca Kelly

SUMMARY

This chapter describes the approach taken, the findings and the implications for land managers of research into the influences of land use and land management on water quality in Tasmanian rivers. The objective was to assist future environmental management decisions and investment in water quality protection. Our team approached this task through six activities that spanned four spatial scales (multiple catchment, single catchment, paddock and site) using a combination of historic gauging station data, aerial imagery and modelling. This work resulted in new information on characteristic nutrient generation rates for different land uses, modelled daily nutrient and turbidity concentrations, the effectiveness of current cropping practices in managing water and nutrient leaching, the impact of past riparian intervention on water quality and the major drivers of nutrient loads and turbidity in Tasmanian catchments. The core of this multi-scale approach, if not all the results, should be of wide relevance to different landscapes across Australia and elsewhere where there are insufficient data and information to be of direct use for managing water quality.

INTRODUCTION

The focus on water quality as the key resource for investigation and analysis was identified by the three participating Tasmanian natural resource management (NRM) regions due to the uncertainty concerning the relative impacts on water quality of land use, land management and previous landscape interventions. There was also a need to know that investment to improve water quality will have environmental outcomes. That is, investment should be based on knowledge rather than assumptions. Our end user partners made it clear that quantitative evidence of changes in resource condition was only one of the inputs to their investment decision-making processes. Catchment managers also acknowledged that community values, local history and 'gut feel' influenced their decisions on how and where to invest public funds to improve water quality (Richard Ingram, *pers. comm.*). To our end user partners, natural resource management was essentially a people-oriented process. To researchers, however, understanding the drivers of water quality was essentially a biophysical challenge. We recognised that seriously involving our end user partners in the research process would need to be a central component of project communication if their knowledge and perspective were to be integrated and the work was to have influence.

The research was predominantly retrospective, based on historic datasets, given the lack of specifically designed monitoring that related documented changes in land use and land management to water yield and quality. Our team made the decision at the outset to focus on the most widely available water quality indicators of nitrogen, phosphorus and sediment. For the purposes of this chapter, unless otherwise qualified, the term 'water quality' refers to the status of these indicators. The long-term water quality data underpinning our team's work was provided by the Tasmanian Department of Primary Industry, Parks, Water and Environment. Some activities also used data sourced from local groups and smaller organisations. Modelled information was generated and tested with the collection of new data. The results were integrated, along with outputs from the river health (Chapter 3) and estuary health (Chapter 4) research, into a decision support system (Chapter 18) built by the integration team.

LAND USE AND LAND MANAGEMENT TEAM ACTIVITIES

One of our industry partners with experience in research collaboration gave early advice to avoid the 'focus catchment trap', where a research group learns a great deal about one catchment but is unable to generalise beyond its watershed (John Riddiford, *pers. comm.*). We were encouraged to use a multiple catchment approach so that findings would be applicable to many land managers in different landscapes. This led the team to an initial activity examining the relationships between land use (e.g. dairying, cropping, forestry) and water quality in Tasmania's 34 gauged catchments (activities 1, 2 and 4). Subsequent activities examined the impacts of land management (different cultural practices within a land use) at subcatchment (activities 3 and 6), farm and paddock scales (activity 5). An additional activity was undertaken at small plot scale to improve understanding of how water movement through the soil profile affected the transport of contaminants (Hardie *et al.* 2011).

The six sequentially staged activities were:

1   estimating nutrient generation rates for major land uses in 34 gauged catchments;
2   hydrologic modelling of multiple catchments;
3   modelling daily total phosphorus concentrations at subcatchment scale;

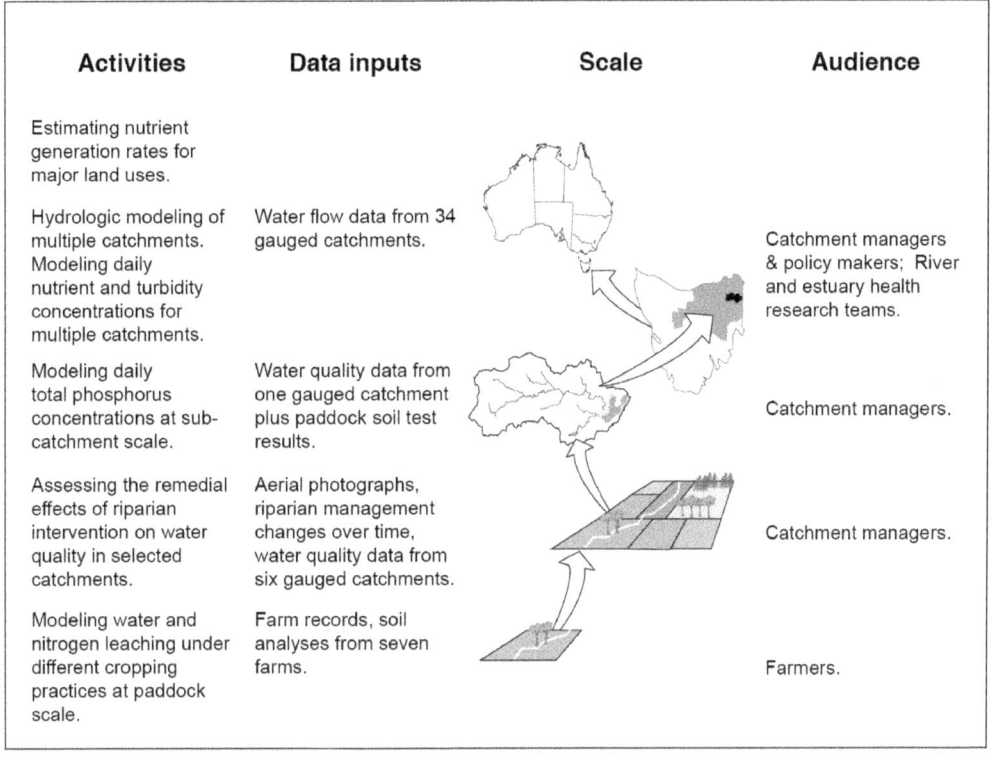

Figure 2.1: Research activities, data inputs, scales of consideration and intended users of research into the relationships between land use, land management and water quality.

4   modelling daily nutrient and turbidity concentrations for multiple catchments;

5   modelling water and nitrogen leaching under different cropping practices at farm and paddock scale;

6   assessing the remedial effects of riparian intervention on water quality in selected catchments.

These activities are summarised in Figure 2.1 and are considered in turn below.

NUTRIENT GENERATION RATES

The aim of this activity was to increase confidence in the nutrient generation rates used in modelling studies by deriving values from annual loads calculated from gauged data. Initially, we developed a conceptual model of the major drivers of nutrient delivery to surface waters (both natural and anthropogenic). This involved reaching agreement on which drivers could be influenced by management (Fig. 2.2). Rainfall, topography and soil type are factors over which we have little or no influence, while tillage practices, fertiliser application rates, grazing management and riparian zone width are anthropogenic factors that can be influenced through extension and incentives. Other human-induced factors such as land use, drainage and stream incision are not so directly amenable to influence.

Acknowledging that there are numerous influences driving nutrient delivery to rivers, we sourced data on actual river nutrient loads from Tasmania's 34 gauged catchments which have a mosaic of land use rather than a predominance of one land use. These data were related to

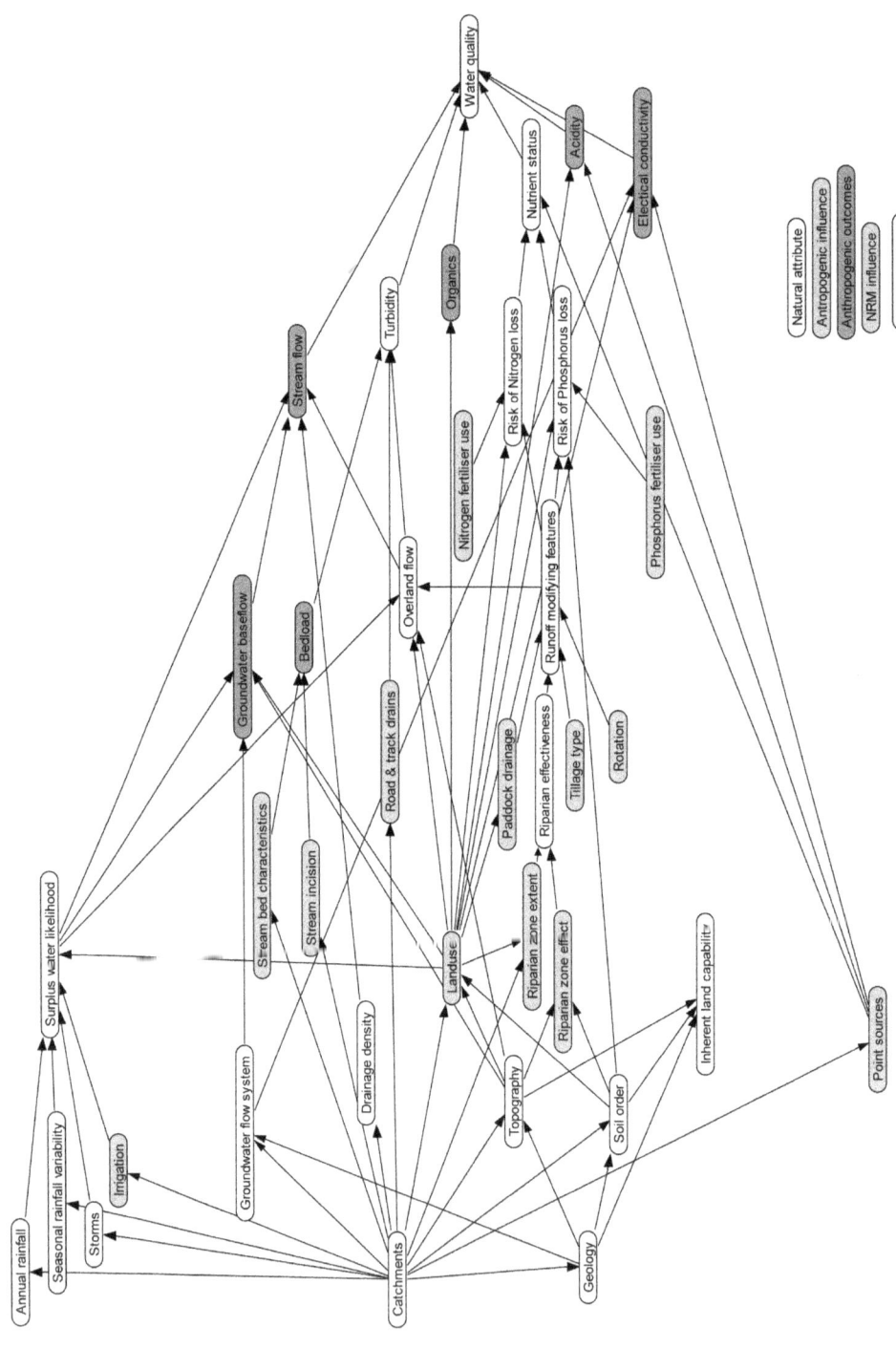

Figure 2.2: Influence diagram illustrating the factors thought by researchers and managers to influence delivery of nutrients to surface waters.

land use mapping to derive estimates of annual nutrient generation rates for the major land uses. A nutrient 'generation rate' is the mass of a pollutant generated from a given land use area during a specified time interval, usually a year, and expressed in kg/ha/yr (Marston *et al.* 1995). The selection of nutrient generation rates can be difficult as they vary with climate, hydrologic conditions, land use distribution and physiography. Accepted sources of nutrient generation rates include local or expert knowledge of the catchment, previous model applications (Letcher *et al.* 2002) and detailed field studies monitoring pollutant concentrations and stream flows for specific catchment conditions and land use composition.

We started modelling by using a nutrient balance model, the Catchment Management Support System (CMSS; Davis and Farley 1997), to predict total loads produced in a catchment. CMSS assumes that nutrient exports are dependent on land use, and that each land use has a specific nutrient generation rate (Young *et al.* 1995). The CMSS model does not explicitly account for the effects of different soil types, rainfall, slope or land management factors but rather relies on these factors being integrated into the defined land uses (Young *et al.* 1995). We then endeavoured to refine both the land use data and the methodology for calculating nutrient generation rates, which are used in models such as CMSS. Our inspection of the available land use data indicated that important errors and omissions existed, especially for dairy pastures which are typically significant sources of nutrient and sediment runoff (Barlow *et al.* 2005). Several months were spent updating the land use mapping data using the most recent aerial imagery for the catchments of interest. We were able to identify dairy pastures by their unique production characteristics such as extensive systems of laneways leading to a central dairy shed, and rotational grazing which creates a graduated series of light brown to dark green paddocks.

We then used a Bayesian modelling framework to estimate 'best fit' land use generation rates for the gauged data across the 34 catchments. This approach allowed all data across the 34 catchments to be simultaneously considered with proper propagation of uncertainty throughout the model (Broad and Corkrey 2011). We were then able to calculate end-of-catchment estimates for phosphorus and nitrogen loads (Fig. 2.3). The results show that in Tasmanian agricultural catchments, the highest catchment nutrient loads come from intensive land uses which typically occur in flatter landscapes with high rainfall/runoff (north-west). East coast catchments with low rainfall/runoff and less intense land use have the lowest nutrient loads. All catchments suffer from a lack of flood sampling data and so it is likely that the total nutrient loads are underestimated in every catchment, but this is likely to be exacerbated in high rainfall/runoff catchments.

Using this method, we found that intensive land use, in particular cropping and dairy production, was the major driver of sediment and nutrient delivery to surface waters at the catchment scale. We infer from this that intensification of land use, currently under way in Tasmania, is likely to result in greater surface water nutrient and sediment loads. Dairy pastures showed total phosphorus and total nitrogen generation rates of 10–12 kg/ha/yr and 20–30 kg/ha/yr respectively (Fig. 2.4). These are at the higher end of published values (Davis and Farley 1997). The mean values for grazing modified pastures, production forestry, irrigated cropping and remnant forest from our study were lower than published values and showed less variation between catchments. The modelling approach has provided a constrained estimate of nutrient export from most agricultural catchments in Tasmania and generated an envelope of nutrient export coefficients for the dominant land uses. We suggest that management interventions should focus on reducing nutrient sources and transport at the landscape scale rather than solely relying on abatement in riparian zones to affect nutrients and sediment in surface

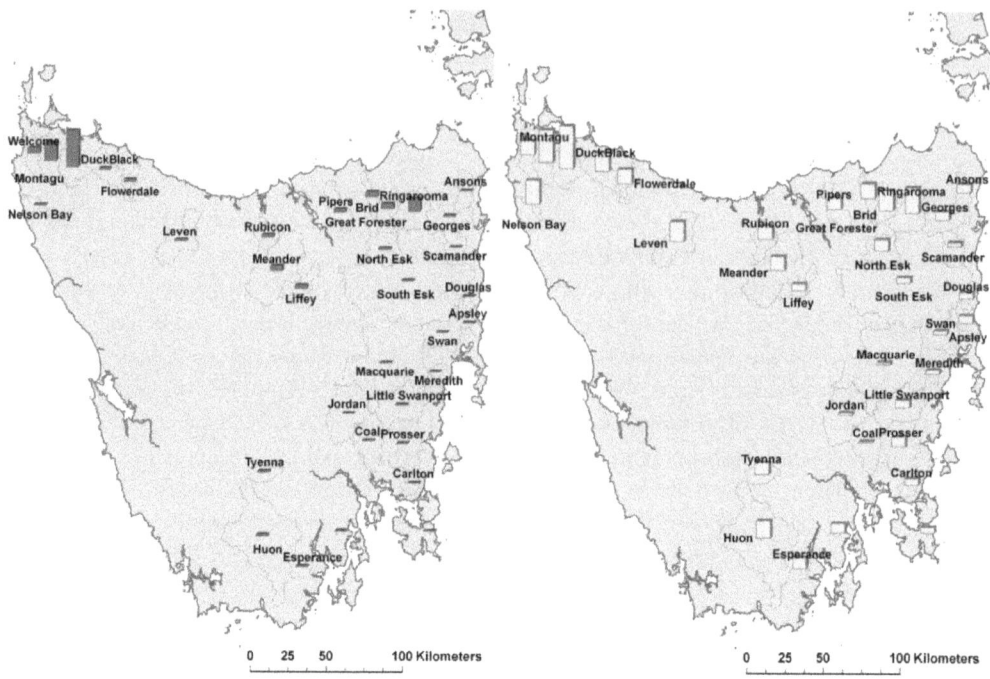

Figure 2.3: Estimated relative total annual catchment loads for total phosphorus (left) and total nitrogen (right).

waters. This means managing rates and timing of fertiliser application across most farms in a catchment and across whole farms.

HYDROLOGIC MODELLING

One function of our team's focus on water quality was to provide the river health team (Chapter 3) and estuary health team (Chapter 4) with catchment nutrient data on a daily basis as input data for their ecological response models. This required developing a daily time-step rainfall/runoff model from the available Tasmanian flow data for an ensemble of Tasmanian catchments. The catchments selected for study represented a wide range of size (8–3300 km²), annual rainfall (580–1800 mm), annual evaporation (730–1100 mm) and flow (2700–2.5 million ML per year).

The Australian Water Balance Model (AWBM) was used for the purpose, being one of the most widely used rainfall/runoff models in Australia (Boughton 2005). AWBM is a six-parameter lumped conceptual rainfall/runoff model that can be used for determining rainfall/runoff relationships in river catchments under changing climate scenarios (e.g. Chiew *et al.* 2010). We estimated the parameter values for each catchment using modelled climate and measured stream flow data by a nonlinear optimisation method (Broad and Corkrey, *in press* a).

PHOSPHORUS CONCENTRATIONS AT SUBCATCHMENT SCALE

An opportunity to test the nutrient generation rates from activity 1 was provided by the availability of a paddock-scale dataset of soil nutrient levels in the Duck River catchment (James and Cotching 2010). We combined available subcatchment-scale water quality data with the

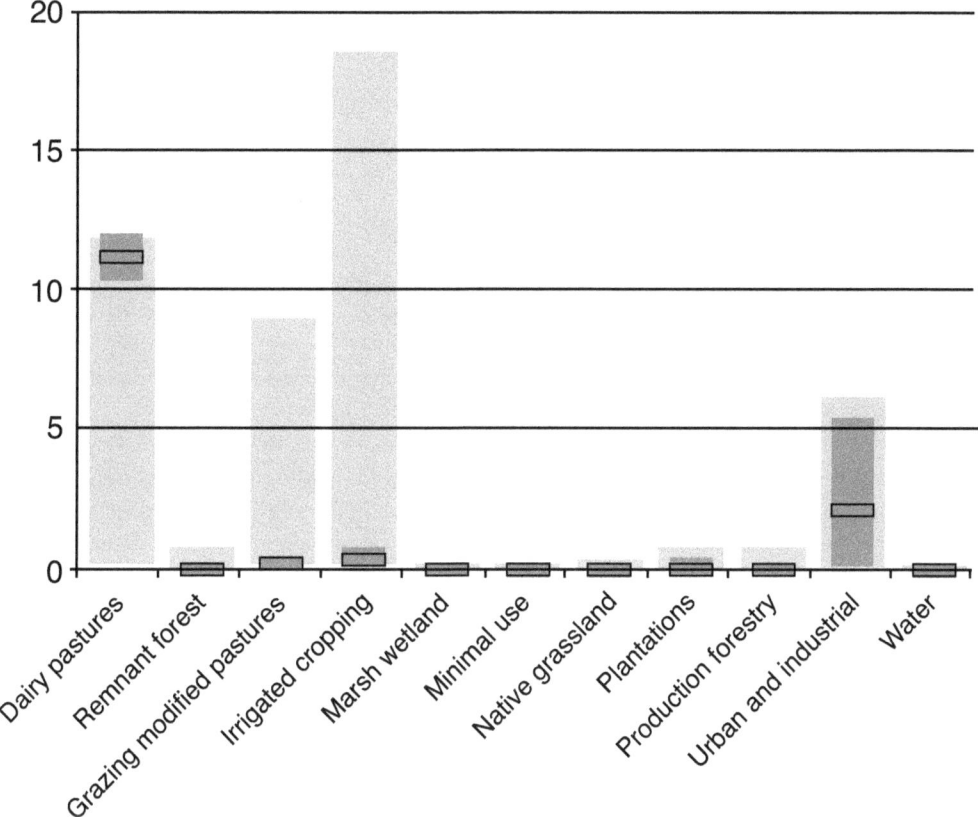

Figure 2.4: Tasmanian phosphorus generation rates by land use. Mean = black bar; Tasmanian range = dark grey; published range = column.

soil nutrient data by using the WaterCAST (Argent *et al.* 2004) hydrologic model to determine the extent to which paddock-scale management was related to nutrient generation at subcatchment scale. Such subcatchment-scale data is rarely available due to the time and cost associated with data collection.

Our results indicated that dairy pastures in different parts of the catchment exhibited large variation in total phosphorus (TP) losses, resulting in TP concentrations of 0.2–5.0 mg/L (Broad and Cotching 2009), with the major nutrient enrichment mechanisms being nutrient supply and proximity to drainage. The variation in TP concentrations between subcatchments is not apparent in the overall land use generation rates or catchment annual load results.

The high average levels of soil Olsen P (Olsen *et al.* 1954) found in the paddock-scale survey (mean 37.1; std dev 18.7) led us to believe that managing the rate and timing of fertiliser and effluent applications would provide more effective opportunities to improve water quality than altering the drainage system. We are aware that inherently poor drainage is a significant factor limiting sustainable farming in this landscape, derived from draining swamp land, but the installation of man-made drains has provided a large degree of connection in the landscape that provides pathways for the easy and quick egress of applied fertiliser nutrients to local rivers and streams (see Chapter 7 for a detailed analysis of this landscape).

Table 2.1: Suggested confidence with which modelled data can be used for the 34 Tasmanian catchments studied

Model use	Turbidity	Total phosphorus	Total nitrogen
Use with confidence (NS* > 0.5)	16	2	8
Use with caution (NS* = 0.1–0.5)	16	10	13
Do not use (NS* < 0.1)	2	22	13

* Nash Suttclife criterion.

DAILY NUTRIENT AND TURBIDITY CONCENTRATION

Turbidity and nutrient concentrations control many ecological functions in rivers and estuaries, including light penetration and phytoplankton production. Activity 4 resulted in a model of daily turbidity, total phosphorus and total nitrogen delivery for 34 Tasmanian catchments as input data for the river and estuarine and health activities (Chapters 3 and 4). To accommodate the different scales of river and estuary sampling, the model needed the ability to extrapolate results from subcatchment- to whole estuary-scale. This was achieved by recognising land use as a primary driver of nutrient and sediment losses (Baginska *et al.* 2003) and developing a power function relationship between land use and flow for turbidity, TN or TP using a Bayesian approach (Broad and Corkrey, *in press* b).

Model runs confirmed that differences between catchments were too great to enable generalisation across all catchments, with different parameters in the power function relationships applying in each catchment. By ranking model performance, we were able to demonstrate that modelled daily nutrient and turbidity loads could be used with confidence in some catchments, with caution in others and, in some cases, not at all (Table 2.1; Broad *et al.* 2010). The daily nutrient data generated were ultimately integrated into the overall Tasmanian Aquatic Condition Decision Support System (Chapter 18) through the river health and estuary health components but the different levels of confidence between catchments could not be accommodated.

WATER AND NUTRIENT LEACHING IN MIXED CROPPING

Despite the significance of vegetable production to Tasmania and widespread recognition of soil erosion as a major management issue (Cotching *et al.* 2002), gauging station data from predominately cropped catchments in Tasmania remains a major gap in the state's water quality and flow data. Consequently, cropping could only be analysed as a small component of catchments dominated by other land uses, with nutrient generation rates likely to be subject to significant error. In order to fill this knowledge gap, we initiated an activity using APSIM (Agricultural Production Systems Simulator; Keating *et al.* 2003) to explore the impacts of different intensive cropping practices on production and leaching of water and nitrogen.

This modelling was based on the crop rotations, irrigation practices and fertiliser usage of five intensive cropping farmers in the Panatana catchment of north-west Tasmania. Our simulation modelling estimated that mean nitrogen losses ranged from 1 kg/ha to 16 kg/ha with seasonal peaks of 4 kg/ha to 36 kg/ha. Potatoes were found to be the 'leakiest' crop, with an average of 29 kg N/ha leached below the root zone each year (with seasonal peaks up to 140 kg N/ha), four times more than other crops. Our simulation results also suggested that relatively simple changes in management (deficit-based irrigation, irrigation scheduling, and reduced and split nitrogen fertiliser application) have the potential to reduce the risk of off-site nutrient pollution (Table 2.2) and produce significant cost savings without any reduction in yield (Lisson

Table 2.2: Comparison of modelled water and nitrogen leaching and potato yield based on current management practices used on two farms in the Panatana catchment, Tasmania

	Farm 3			Farm 1
Scenario	Current management	+ Deficit-based irrigation scheduling	+Deficit-based irrigation + reduced N	Current management
Basal N (kg/ha)	215	215	100	154
Topdressing (kg/ha)	5 x 115 urea	5 x 115 urea	5 x 80 urea	5 x 125 urea
Irrigation (mm)	260	174	172	198
Irrigation interval (days)	11	5	5	5 (deficit-based)
Amount per event (mm)	26	13	13	13
Drainage (mm)	135	52	51	57
N leached (kgN/ha)	53	6	3	1
Tuber yield (T/ha)	56	56	56	63

and Cotching 2011). These simulated results were consistent with the practices of one farmer operating within the studied catchment, which was used as one of the parameterised farms in the modelling. Despite the significant cost savings, experience suggests that adoption of these practices is unlikely to occur without substantial extension effort, including local demonstration and mentoring, as many farmers are often resistant to change unless there is direct local evidence that illustrates that similar results can be achieved in their particular situation.

RIPARIAN VEGETATION AND WATER QUALITY

Our team's final activity was to determine if there were any detectable differences in the nutrient loads in catchments where there was significant riparian intervention in the form of fencing and revegetation. We recognised that riparian zone management can be one of the most effective means of reducing non-point source pollution in managed landscapes (Phillips 1989). This is because managed riparian zones function as barriers or filters to nutrient and sediment inputs from disturbance associated with agriculture and forestry. For many years, funding riparian restoration and management has been the single most important area of investment undertaken by our end user partners to achieve improvement in water quality in rivers and estuaries. Despite this commitment, there is a lack of local data to indicate the impact of investment in riparian zone management on water quality, hence our focus on this intervention.

For this activity, we selected four catchments that had been subject to substantial investment in riparian management consisting of some combination of removing introduced willows (*Salix fragilis*), revegetation with endemic species, and streamside fencing to manage stock access to the streamside. The dataset included one small (31 km²) catchment where almost the entire stream length had been fenced for stock exclusion and a limited water quality dataset (N, P but not turbidity) for 12 subcatchments over seven years after intervention. The same four catchments, plus two others, were also the focus of research by the social research team to determine the drivers of land management intervention (Chapter 6). We detected no consistent response to intervention in water quality indicators in any of the four catchments. As well as no significant change in nutrient loads, aerial photo analysis did not show any measurable change in the extent or condition of riparian vegetation. This finding is in agreement with that of the social research team that, in 50% of instances where willows had been removed, they had not been replaced by native vegetation. The social research team found that, for a substantial proportion of riparian land owners, replanting and fencing to control stock access after willow removal made control of pest plants and animals and provision of water for stock more difficult.

Confounding factors in each of the four catchments such as unsealed road crossings being conduits for sediment, and the absence of good-quality long-term data, meant that there was little chance of detecting water quality change or attributing it to riparian intervention. Possible explanations for the lack of evidence linking riparian zone rehabilitation to changes in water quality include (Broad *et al.* 2010):

- lack of adequate monitoring design;
- the small scale of intervention;
- insufficient time since intervention to detect a response;
- residual nutrients and sediment being released from the stream bed;
- potential significance of road crossings or other structures as point source inputs of sediment and nutrients.

The lack of evidence linking riparian zone rehabilitation to changes in water quality in this activity does not imply that investment in riparian management has been futile. The removal of willows has inevitably reduced the amount of shade over river reaches, which was found by the river health team to result in increased algal growth (i.e. poorer water quality, Chapter 3). Growth of replanted native species is likely to take decades to produce similar levels of shade where landholders do replant, which the social research team found occurred in only 50% of cases. Investing in riparian zone management to minimise direct stock access to streams, channelised flow or runoff from roads and tracks draining to streams is critical to minimising nutrients and sediment in streams as these can circumvent buffers. However, it is likely that riparian buffers cannot completely compensate for the source of nutrients and sediments generated by intensive land use. There are also qualitative benefits of riparian zones which include provision of landscape connectivity for terrestrial wildlife and aquatic species, livestock and crop shelter, forage sources and as an aid in farm certification. In the absence of well-designed long-term studies that can detect change in water quality parameters, our studies suggest that there is little evidence to warrant a significantly greater focus on riparian zone management than on paddock- and landscape-scale management to influence sources of nutrients and sediment.

DISCUSSION AND CONCLUSIONS

Our initial view of the relationships between land use, land management and water quality were based on knowledge generated in other locations, on untested observations and/or on anecdotes, all of which enabled us to develop an initial influence diagram (Fig. 2.2). This method was the common language adopted by Landscape Logic. It was used initially as a means of communicating our hypotheses and ultimately, in some cases (see Chapter 18), as precursors to quantitative Bayesian network models. In terms of the simplicity cycle (Fig. 21.1, p. 287), Figure 2.2 represents this project's initial state of naïve simplicity and the results of activities 1 to 6 represent incremental steps towards the ultimate goal of informed simplicity. The team was able to combine research at a range of scales from plot to multiple catchments in order to derive improved understanding of the major drivers of nutrient loads and turbidity in

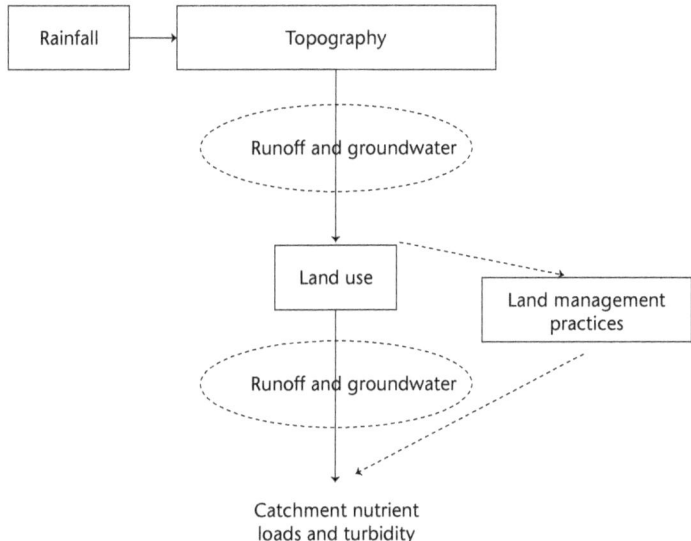

Figure 2.5: Simplified conceptual model of the major drivers of catchment nutrient loads.

Tasmanian catchments (Fig. 2.5). This conceptual model indicates that catchment nutrient loads are driven largely by rainfall, topography and land use, but rainfall and topography both influence land use, and land management practices act only as a moderator.

Before we can attribute changes in resource condition to specific NRM programs it is necessary to measure change and attribute it to particular human influences. We found that in some cases there was not enough data of sufficient quality to clearly demonstrate any change, let alone distinguish between human or natural causes; for example, our attempt to quantify the effect of riparian intervention on water quality at the subcatchment scale. Even where monitoring data did exist, results were often inconclusive for several reasons: the scale of intervention was too small to be recorded in end-of-catchment water quality data as larger-scale processes drowned out the smaller signal, insufficient time had elapsed between intervention and response, or the type of data collected was not designed to test appropriate hypotheses or to eliminate other drivers of change. Major knowledge gaps that still exist include nutrient concentrations and river flow data from catchments dominated by cropping as a land use, and well-designed monitoring studies of specific management interventions on water quality. The datasets required to close the knowledge gaps are site and river reach studies of intervention effects plus longitudinal (both spatial and temporal) measurement of water quality at catchment scales.

Consistently collected water quality data made available to this project was essential to our success and was found to have considerable value in providing greater understanding of landscape function. During the process of our work, it became apparent that the value of Tasmania's long-term water quality data could be enhanced through strategic event-based (flood) data collection and within-catchment sampling, as demonstrated in Chapter 7. The available water quality data were collected at monthly intervals, which is inappropriate for building understanding and quantitative decision support systems. However, using data from multiple catchments allowed us to develop broad-scale relationships between land use, land management and water quality for these landscapes.

Catchments with a high proportion of irrigated cropping are currently underrepresented in the available monitoring data, a gap which will become increasingly important in landscape management as the state's irrigation sector expands. Also, catchments with a longer period of data had much better-performing models, as did catchments with at least one flood sampling dataset. Some of the data sourced from local groups and smaller organisations were not sufficiently rigorous, either spatially or temporally, lacking accurate locations of land management practices and inconsistent data on timing of interventions.

The modelling undertaken has been integrated into decision support systems at both annual and daily time scales for use by our NRM partners and other stakeholders including the Tasmanian government, consultants and non-government organisations. We found that the major driver of sediment and nutrient delivery to surface waters at the catchment scale is intensive land use, particularly the most intensive land uses of cropping and dairy production. We consider that our catchment-scale nutrient loads may be underestimates, due to the paucity of flood sampling in the available data. Given the heavy reliance on riparian intervention but the lack of data to demonstrate its effectiveness, as found in our studies, it would appear sensible to focus effort on reducing nutrient sources and transport at the landscape scale rather than solely relying on abatement in riparian zones. While land management practices can reduce the total nutrient and sediment delivery to rivers, as demonstrated by the subcatchment study in the Duck catchment, even where best management practices are adopted, nutrients and sediments resulting from intensification of land use will still be delivered at higher than natural rates.

Aspects of the research process we found of particular value were having the opportunity to develop detailed research goals and plans in collaboration with end users after funding had been secured, having a process to integrate results from different components into decision support systems, and regular (six-monthly) meetings of researchers and external partners to discuss findings and their implications as the work progressed.

The principal beneficiaries of our research are the partner regional NRMs who are making investment decisions about the management of natural resources and are responsible for meeting resource condition targets. Our partners from the NRM organisations identified that knowledge of landscape processes plays an important role in understanding which practices are likely to minimise off-site impacts. While this knowledge alone is not sufficient to overcome conflicts between the public pressure to conserve resources and their use in generating social and economic value, it removes some of the assumptions and uncertainty from those debates.

REFERENCES

Argent RM, Podger GM, Grayson RB, Fowler K and Murray N (2004) E2 modelling software user guide. eWater CRC, Australia. <http://www.toolkit.net.au>.

Baginska B, Pritchard T and Krogh M (2003) Roles of land use resolution and unit area load rates in assessment of diffuse nutrient emissions. *Journal of Environmental Management* **69**, 39–46.

Barlow K, Nash D and Grayson RB (2005) Phosphorus export at the paddock, farm-section, and whole farm scale on an irrigated dairy farm in south-eastern Australia. *Australian Journal of Agricultural Research* **56**, 1–9.

Boughton W (2005) Catchment water balance modelling in Australia 1960–2004. *Agricultural Water Management* **71**(2), 91–116.

Broad ST and Corkrey R (2011) Estimating annual generation rates of total P and total N for different land uses in Tasmania, Australia. *Journal of Environmental Management* **92**, 1609–1617.

Broad ST and Corkrey R (in press a) Nonlinear methods outperform standard algorithms for the optimisation of a rainfall-runoff model across an ensemble of catchments. *Environmental Modelling and Software.*

Broad ST and Corkrey R (in press b) Modelling turbidity across an ensemble of catchments using a bayesian statistic approach. *Environmental Modelling and Software.*

Broad ST and Cotching WE (2009) Assessing the spatial variation of dairy farm total phosphorus losses in the Duck river, NW Tasmania. *18th World IMAC/MODSIM Congress.* 13–17 July 2009, Cairns, Australia. <http://mssanz.org.au/modsim09>.

Broad S, Corkrey R, Cotching W and Coad J (2010) 'Estimating nutrient loads and turbidity for Tasmanian catchments'. Technical Report no. 18. Landscape Logic, Hobart.

Chiew FHS, Kirono DGC, Kent DM, Frost AJ, Charles SP, Timbal B, Nguyen KC and Fu G (2010) Comparison of runoff modelled using rainfall from different downscaling methods for historical and future climates. *Journal of Hydrology* **387**(1–2), 10–23.

Cotching WE, Hawkins K, Sparrow LA, McCorkell BE and Rowley W (2002) Crop yields and soil properties on eroded slopes of red ferrosols in north-west Tasmania. *Australian Journal of Soil Research* **40**, 625–642.

Davis JR and Farley TFN (1997) CMSS: policy analysis software for catchment managers. *Environmental Modelling andSoftware* **12**, 197–210.

Hardie MA, Cotching WE, Doyle RB, Holz G and Lisson S (2011) Effect of antecedent soil moisture on preferential flow in a texture-contrast soil. *Journal of Hydrology* **398**, 191–201.

James RA and Cotching WE (2010) 'Adoption of nutrient budgeting for sustainable dairy farms and healthy rivers technical report'. Tasmanian Institute of Agricultural Research, Burnie, Tasmania.

Keating BA, Carberry PC, Hammer GL, Probert ME, Robertson MJ, Holzworth D, Huth NI, Hargreaves JNG, Meinke H, Hochman Z, McLean G, Verburg K, Snow V, Dimes JP, Silburn M, Wang E, Brown S, Bristow KL, Asseng S, Chapman S, McCown RL, Freebairn DM and Smith CJ (2003) An overview of APSIM, a model designed for farming systems simulation. *European Journal of Agronomy* **18**, 267–288.

Letcher RA, Jakeman AJ, Calfas M, Linforth S, Baginska B and Lawrence I (2002) A comparison of catchment water quality models and direct estimation techniques. *Environmental Modelling and Software* **17**, 77–85.

Lisson S and Cotching WE (2011) Modelling the fate of water and nitrogen in the mixed vegetable farming systems of northern Tasmania. *Agricultural Systems* **104**, 600–608.

Marston F, Young W and Davis R (1995) *Nutrient Data Book.* 2nd edn. CSIRO Division of Water Resources, Canberra.

Olsen SR, Cole CV, Watanbe FS and Dean LA (1954) *Estimation of Available Phosphorus in Soil by Extraction with Sodium Bicarbonate.* US Government Printing Office, Washington.

Phillips JD (1989) An evaluation of factors determining the effectiveness of water quality buffer zones. *Journal of Hydrology* **107**, 133–145.

Young WJ, Farley TF and Davis JR (1995) Nutrient management at the catchment scale using a decision support system. *Water Science and Technology* **32**, 277–282.

Measuring and modelling the impacts of land use on ecological river condition

Peter Davies, Regina Magierowski, Steve Read and

Nelli Horrigan

SUMMARY

This project aimed to understand the influence of land use and land management on the ecological condition ('health') of Tasmanian rivers at local and catchment scales. The five project phases were developing a conceptual model linking land use and other drivers to ecosystem responses, finding evidence to support model parameterisation, constructing a simplified river health model for communication purposes; developing a Bayesian Belief Network (BBN) from the full conceptual model using both expert elicitation and local evidence, and evaluation of the BBN and modelling of management scenarios. A significant finding was a disturbance threshold associated with the area of a catchment upstream of a sampling point classed as 'grazing land', characterised by a change in algal-driven river metabolism and river macroinvertebrate communities. Understanding the spatial scales (site, reach, catchment) at which key drivers of ecological responses operated (e.g. riparian condition, sediment input) was also important. Project outcomes were incorporated into a decision support system containing the BBN, and into fact-sheets describing river health impacts of land use and management scenarios. Key lessons included the importance of a conceptual framework throughout the project, identifying relative inferential strength when integrating evidence for drivers, using both correlative and experimental data to identify relative roles of nutrients and fine sediment in driving benthic biological responses to land use, the need to manage drivers of river health at both catchment and local scales, defining management 'levers' connecting project design to deliverables, and maintaining an objective, adaptive scientific process when developing evidence-based management tools.

INTRODUCTION

The river health component of Landscape Logic was initiated in 2007, with a focus on the links between river ecosystem state and function ('river health'), and land use and management. The challenge was to identify those links based primarily on Tasmanian evidence and to embed them in a decision support framework for use by local managers.

This chapter describes the stages of project development, some of the lessons for land managers from each stage, and key lessons about the process of integrative science both within the project and in relation to other projects within the larger Landscape Logic project.

The project had five key phases:

1 developing a conceptual model linking land use and other drivers to ecosystem responses;
2 finding evidence for links within that conceptual model through:
   a mining spatially explicit data on land use, catchment and river characteristics, and river biota from 103 sites;
   b field surveys across gradients of grazing land use area and forest management history, involving 27 and 41 subcatchments respectively;
   c training neural networks developed from a stream mesocosm experiment to discriminate between nutrients and fine sediment as drivers of ecological responses;
3 constructing a simplified river health model for teaching and communication;
4 developing a Bayesian Belief Network (BBN) from the full conceptual model using both expert elicitation and local evidence;
5 evaluating the BBN and modelling management scenarios.

CONCEPTUAL MODEL DEVELOPMENT

Conceptual models which illustrate key structural components and system drivers assist in identifying the contexts and scope of processes that affect ecosystems (Karr 1991). They can also guide thinking across disciplinary boundaries (Allen and Hoekstra 1992). Conceptual models tend to be initially large and complex. Smaller, more focused models can be utilised to develop and refine specific questions concerning drivers and responses, and are more suited to decision support.

Barber (1994) described three essential steps in developing conceptual ecosystem models. First, it is necessary to identify the structural components of the resource and the interactions among these components, including interactions external to the model (inputs and outputs) and stressors that influence the resource's ecological operation and sustainability. Second, it is important to consider the relevant temporal and spatial dynamics of the resource at relevant multiple scales. Last, it is necessary to identify how major stressors can be expected to impact on the resource's structure and function.

A conceptual model (Fig. 3.1) was developed initially to guide our thinking and interaction, and build our team understanding. The model illustrated what we believed were the main drivers of river ecosystem condition, with a focus on responses of benthic macroinvertebrates and algae. This initial model involved identifying all possible significant links between land use and river health responses. It also captured a range of catchment contexts relevant to Tasmania, relevant intermediate (proximal) drivers of biological responses, and outputs useful for and interpretable by catchment and river managers. Importantly, the conceptual model was biophysically based with drivers, condition states and outcomes structured into a realistic representation of a major subset of Tasmanian catchments. This allowed communication with and acceptance by scientists and managers, and allowed subsequent direct parameterisation using experimental data from these catchments.

The rationale for the choice of biological endpoints for the model involved several concepts. Macroinvertebrates and algae form key compositional and functional components of Australian stream ecosystems (Boulton and Brock 1999). A well-established culture of bioassessment exists in river management at national, state and regional levels that includes macroinvertebrates and, to a lesser extent, algae (e.g. Krasnicki *et al.* 2001; NRM South 2009a). Macroinvertebrates are seen as key indicators of river health (Schofield and Davies 1996), while conditions stimulating 'nuisance' algal growth are of management concern (Biggs 2000; Dodds and Welch 2000; Dodds *et al.* 2002). There is a growing understanding of the links between land use, water quality and algal response through the food chain to invertebrates and higher-order predators such as fish and platypus (e.g. Bojsen and Barriga 2002; Allan 2004).

A framework for reporting on river condition, the Tasmanian River Condition Index (TRCI), has been recently developed, with a standardised scoring and reporting system for aquatic biota (NRM South 2009a, b). Our conceptual model outputs included components of this scoring system, specifically the TRCI macroinvertebrate and algal components and overall aquatic life condition score. This allowed the outcomes to be placed into a routine river assessment context for managers.

The conceptual model was derived from several iterative group meetings, with initial scoping followed by justification for each link. Opposing pressures to add complexity and to simplify were carefully balanced. An important early step was to define the geographic bounds of the model and explicitly omit sets of catchments to which the model would not apply, for reasons such as unusual geology, mine drainage or the presence of substantial river diversions (dams). The temporal scale of the model was fixed at c. five to 10 years, acknowledging that several links within it operated at shorter time-steps, but no iterative time-steps were built into the model. While the overall model operated at spatial scales from subcatchment to catchment (c. 100–2000 km²), it was recognised that different processes functioned at different spatial scales, with outputs being more relevant at smaller (site or reach) scales while drivers operate at scales between reach and catchment.

While complex, with over 30 nodes and over 100 links, the model proved flexible and useful throughout the project, undoubtedly because of its biophysical basis, and it formed the basis of the subsequent BBN design.

EVIDENCE

Development of decision support for management requires a high level of evidentiary support in order to defend and demonstrate evaluations of likely outcomes for environmental

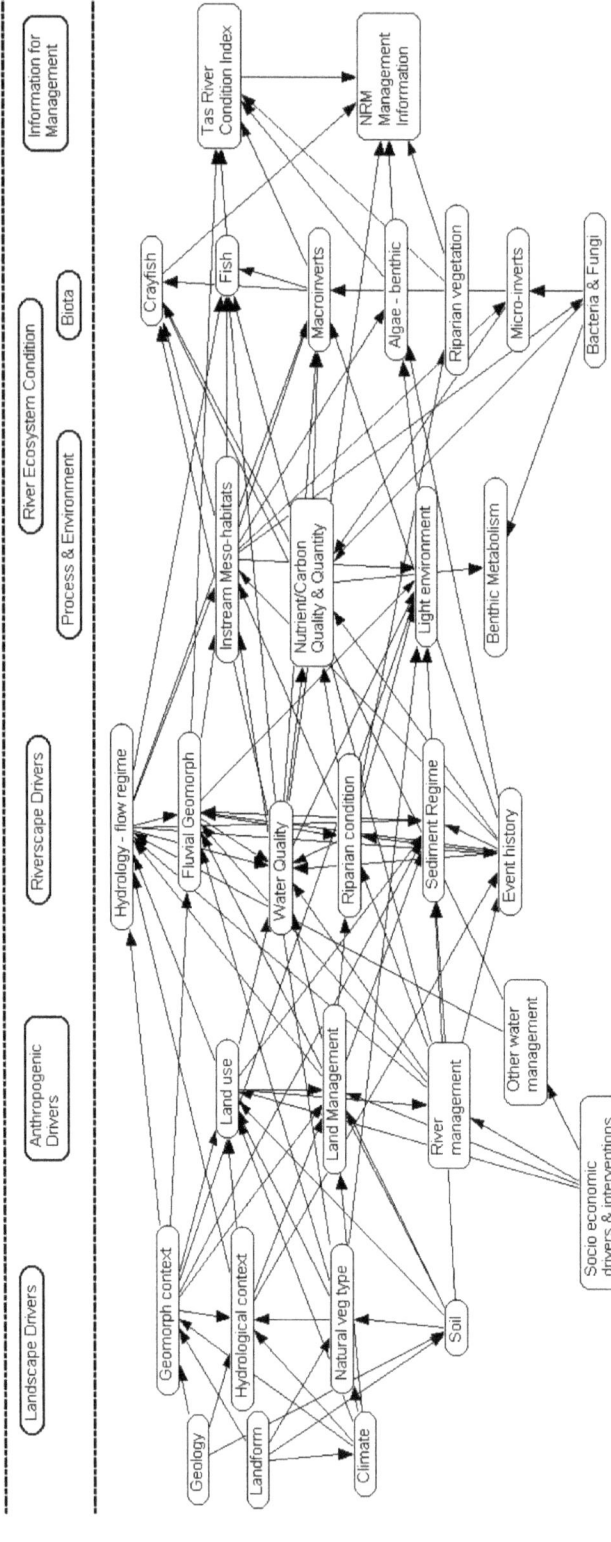

Figure 3.1: The initial conceptual model depicting what we believed were the important linkages between land use and macroinvertebrate and algal condition. The model included important landscape and riverscape drivers (left and centre left), river ecosystem responses (centre right), and management objectives or end points (right).

decision-making (Matthies *et al.* 2007). Concepts of causal association and inference have been explored for a considerable time in medicine (Hill 1965; Susser 1977, 1991) and several are directly applicable here. A range of criteria can be used to evaluate and assess inferential strength and to support causal inference (Downes *et al.* 2002; Norris *et al.* 2005; Downes 2010).

We evaluated the evidence for key links in our conceptual model on the basis that inferential strength increases:

- as one moves from relationships based on correlations or models of state measures, to relationships based on correlations or models of state and process measures, to relationships based on experimental manipulations;
- when direct measures of response (rather than surrogates) are utilised;
- with conceptual relevance, i.e. consistency with the conceptual model.

A subset of links and nodes in our conceptual model were then selected as a focus for seeking further evidentiary support. Evidence was primarily sought that had direct regional relevance, ideally being drawn from locally available data or similar environmental contexts. Three main lines of evidentiary support were pursued: mining existing data from Tasmanian catchments, conducting field surveys along designed gradients in Tasmania, and developing diagnostic information (process measures and neural networks) that allowed us to interpret several key cause-and-effect links. Each of the sets of results were directly applied to parameterising the BBN derived from our conceptual model, and used to shape a much simpler diagrammatic model of influences used for purposes of explanation and communication.

EXISTING DATA FROM TASMANIAN CATCHMENTS

Three extensive datasets were available for analysis:

- data on stream biota and habitat assessment from the National River Health program (NRHP) in Tasmania (Davies 2000), with key data collected in the late 1990s to 2003;
- Bureau of Rural Sciences (BRS) land use data (ANRDL 2003, 2008);
- a large set of Tasmanian catchment and stream feature data derived from the Conservation of Freshwater Ecosystem Values (CFEV) framework (DPIW 2005).

All three datasets were contemporary with each other, were spatially comprehensive and were derived, assigned and attributed spatially at the same mapping scale (the 1:25 000 scale relevant to local planning decisions).

A number of key correlations were identified in these datasets, focusing on the relationships between grazing land use, riparian vegetation condition and measures of macroinvertebrate community structure (Horrigan *et al.*, unpub. data). The proportion of catchment area under grazing land use was a dominant correlate with a range of stream responses. We were able to examine these relationships at three different scales: stream reach/site scale, local scale and whole-of-subcatchment scale. The relationships between grazing land use and riparian condition with macroinvertebrate variables were substantially more significant at catchment scale than at reach or site scale. We also explored thresholds, as break-points in relationships, using regression tree analysis. Key thresholds included several at 40–60% of catchment area under grazing land use, above which benthic macroinvertebrate community condition showed substantive declines (Fig. 3.2; Horrigan *et al.*, unpub. data). The value of these thresholds depended on the index of macroinvertebrate condition, and may be expected to vary between catchments.

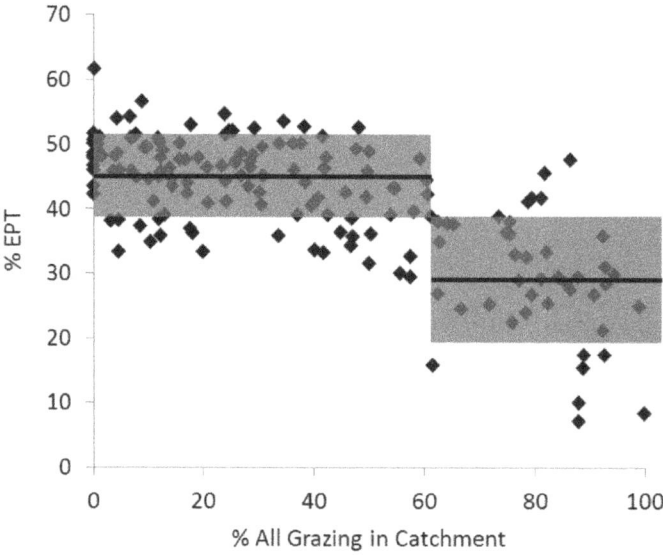

Figure 3.2: Relationship between percentage of benthic macroinvertebrate abundance in the EPT taxonomic grouping (aquatic insects of the orders Ephemeroptera, Plecoptera and Trichoptera) and percentage of catchment area under grazing land for Tasmanian streams. A threshold (change-point) exists at 62%. Bold lines indicate mean values differentiated by the change-point analysis (Horrigan *et al.*, unpub. data). Grey bands indicate standard deviation.

Key issues in this data-mining approach were:

- inconsistent information on data quality;
- the need to establish the conceptual, spatial and temporal relevance of datasets for the particular purpose of this work;
- the potential for pseudoreplication, given that some sites were located within the same catchment, and that sites experienced different intensities of data-collection;
- the potential for confounding of environmental variables;
- poor inferential strength (only correlative relationships can be established).

Notwithstanding all these issues, the extensive nature of the NRHP and CFEV datasets allowed statements about the correlative relationships between land use and stream biota to be made with confidence in the context of the characteristics of individual catchments.

DESIGNED FIELD SURVEYS

We designed and conducted two separate field surveys at stream sites from multiple catchments across a range of grazing land use areas (Magierowski *et al.*, unpub. data) and forestry management history (Davies *et al.*, unpub. data). An intensive phase of data analysis accompanied site selection for these surveys to optimise the site distribution across the land use gradients, remove sites affected by other impacts (flow diversion, mining etc.) and ensure consistency of geomorphological and climatic contexts. For the grazing gradient survey, 27 independent (non-overlapping) catchments were selected across northern Tasmania, ranging across a wide range of proportion of catchment area under grazing. For the forestry gradient survey, 41 independent catchments were selected across a range of forest management histories and intensities but with no substantive agricultural land use.

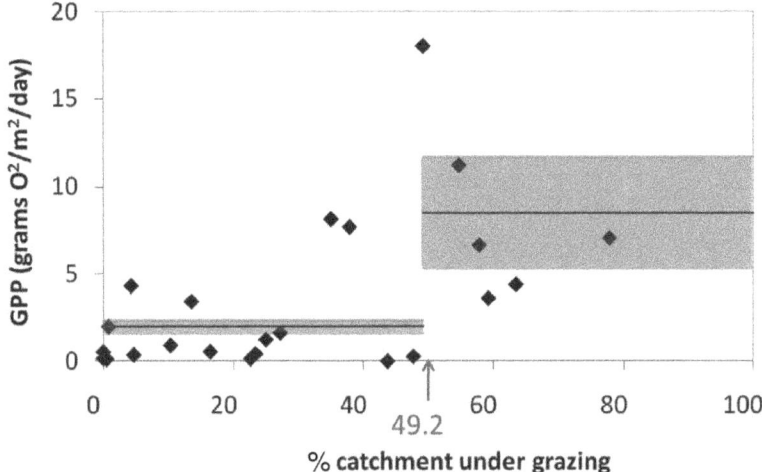

Figure 3.3: Relationship between the rate of gross primary production, estimated using a single-station open-channel technique, and the percentage of catchment area under grazing land for Tasmanian streams. A threshold (change-point) exists at 49%. Bold lines indicate mean values differentiated by the change-point analysis (Magierowski *et al.*, unpub. data). Grey bands indicate standard deviation.

Sites were sampled for stream macroinvertebrates, algae and a range of physical and riparian features. A subset of sites was also assessed for algal growth rates and algal nutrient limitation (using nutrient-diffusing substrates; Hauer and Lamberti 2007), stream metabolism (using an open-channel method; Grace and Imberger 2006) and carbon and nitrogen stable isotope composition in the macroinvertebrates and their food sources. Land use information consisted of data on production and plantation forests supplied by Forestry Tasmania and Private Forests Tasmania and an update of the Bureau of Rural Sciences (BRS) land use data layer by Landscape Logic (DPIPWE 2009). Data on stream total phosphorus and nitrogen loads were provided by Cotching *et al.* (Chapter 2, this volume).

A diversity of correlations was explored (Magierowski *et al.*, unpub. data). Thresholds were observed in the relationships between the proportion of catchment area under grazing land use and benthic macroinvertebrate community condition, total phosphorus load and gross primary production at 44%, 40% and 49% respectively (Fig. 3.3; Magierowski *et al.*, unpub. data).

Key issues in this gradient survey approach were:

- obtaining sufficient replicate sites for analysis due to confounding impacts;
- limitations of GIS data in representing true on-ground conditions;
- the need for catchment independence (selected catchments did not and should not overlap);
- management of confounding drivers in statistical analysis;
- the inadequacy of available nutrient and sediment concentration and budget data.

NEURAL NETWORKS FOR ANALYSIS OF STREAM MESOCOSM EXPERIMENT DATA

Data were available from two large stream mesocosm experiments designed to examine the relationships between light levels, nutrient concentrations, fine sediment loads and benthic biological responses (Magierowski *et al.*, unpub. data). Sets of four replicate flow-through stream mesocosms were each treated to multiple combinations of low and high shading, nutrient concentrations (soluble nitrogen and phosphorus) and fine sediment (sand-clay) additions over a period of three months. Macroinvertebrate abundance and community composition, and the biomass and cover of benthic algae, were assessed at the end of each treatment period. The streams had been previously colonised via source water from Tasmanian streams of similar geomorphological and hydrological character to those in the gradient survey study above.

These biological data were used as inputs to train neural networks to derive the probability of a gradient survey study sample being classified as a high or low nutrient or high or low fine-sediment case. The four or five best-performing neural networks (with a >85% successful classification rate, n=4 for the forestry survey sites and n=5 for the grazing survey sites) were selected for both the nutrient and fine-sediment scenarios as diagnostic tools to apply to the gradient field survey data. Relationships were then developed between the neural network output probabilities for the different nutrient and fine-sediment states and the percent area of catchment under grazing or intensive forestry (clearfell, burn and sow) in the catchment gradient surveys. The probabilities were used to derive linearly interpolated scores for nutrient status (with 1.0 representing low nutrient status and 2.0 representing high nutrient status) and fine-sediment status (with 1.0 representing low fine-sediment status and 2.0 representing high fine-sediment status). The primary correlation for both land use gradients (grazing and forestry) was with the fine-sediment status score (Fig. 3.4). The correlation with the nutrient status score was significant but weaker for grazing, and completely absent for intensive forestry.

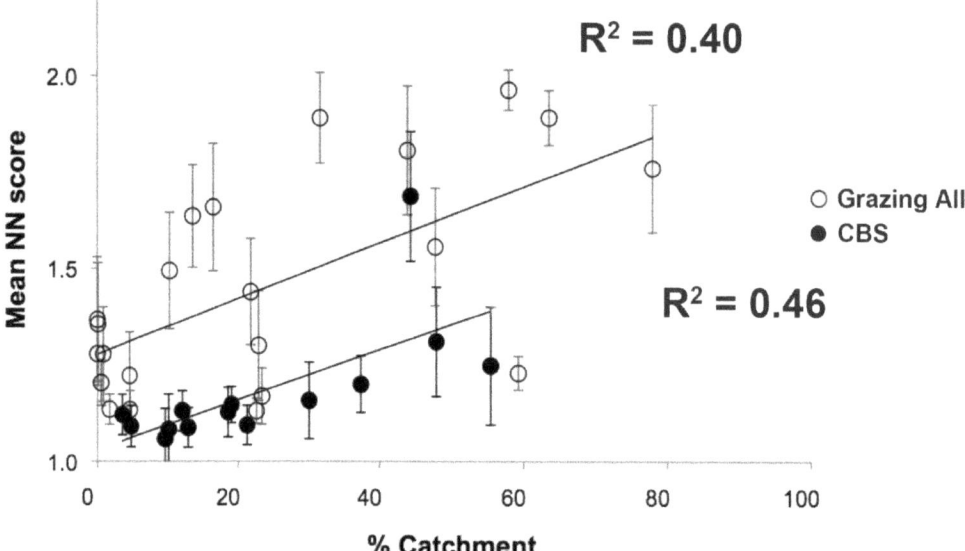

Figure 3.4: Mean neural network (NN) status score (± standard error) for fine-sediment data for field survey sites. Plotted values are derived from the best-performing neural networks developed from a stream mesocosm experiment examining the effects of sediments, nutrients and light on macroinvertebrate and algal condition. The results show a significant positive correlation between the area percentage of grazing ('GrazingALL', ○) and intensive forestry ('CBS': clearfell, burn and sow, ●) land uses and status score representing the predicted chance of a site experiencing high fine-sediment loads ($P_{1,20}$ = 0.002, $P_{1,12}$ = 0.007 respectively).

Key issues with this experimental approach relate to how well the mesocosms are representative of real streams. Careful consideration of the experimental design is required to ensure that the experiment occurs within the same ecosystem context (i.e. similar river systems with similar biota and habitat conditions) and under similar impact levels (in this study, similar nutrient and sediment conditions to those experienced at the gradient sites). Even under ideal conditions, the spatial and temporal scale of the mesocosms experiment may not be representative of real streams. However, a benefit of this approach is that the diagnostic potential and hence inferential strength is strong, especially when used in combination with compatible survey data on state and process measures.

Overall, this combination of experimental approaches provided strong causal evidence that fine sediments were the dominant driver of biological responses to these two land uses, and that nutrient concentration changes were playing a minor role in grazed catchments and none in catchments subject to intensive forestry.

CONSTRUCTING A SIMPLE RIVER HEALTH DIAGRAM

Gathering these different forms of evidence allowed us to distinguish the relative importance of the major drivers of river ecological responses in our study catchments, and to identify dominant drivers and responses without losing the range of scales or local context. We therefore shaped a much simpler diagrammatic model of these influences to capture this new synthetic understanding (Fig. 3.5), and used it to explain these simple concepts to natural resource managers and to experts in related agricultural and estuarine fields.

This simpler model showed how site-level benthic macroinvertebrate biota is driven by the catchment-wide influences of land use and riparian vegetation condition, while site algal levels are driven by catchment-level nutrient regimes and local riparian vegetation conditions (the latter through its influence on shading). Ratings of the strength of the interaction and the evidentiary strength are indicated for each link. The result is a model that demonstrates, in a simple and understandable format, the mixture of drivers and scales that are important for river health.

DESIGNING A BAYESIAN BELIEF NETWORK

The biophysically based conceptual model developed in Phase 1, refined through the conceptual understanding developed in Phase 2 and represented in Phase 3, formed the basis for a Bayesian Belief Network (Fig. 3.6). This included the major drivers of change in river ecological condition as well as a number of network nodes establishing environmental context (e.g. hydrological and fluvial-geomorphic context, and river section slope). The resulting BBN (River BN; Magierowski et al. 2010) allows the user to model the consequences of different land-management scenarios on the ecological condition of Tasmanian rivers at a chosen scale and river location (reach, section or site). The River BN converts expected macroinvertebrate and algal responses into the TRCI macroinvertebrate, algal and overall aquatic condition scores, allowing the outcomes to be placed into a routine river assessment context for managers. Key learnings from the process of developing the River BN are outlined below.

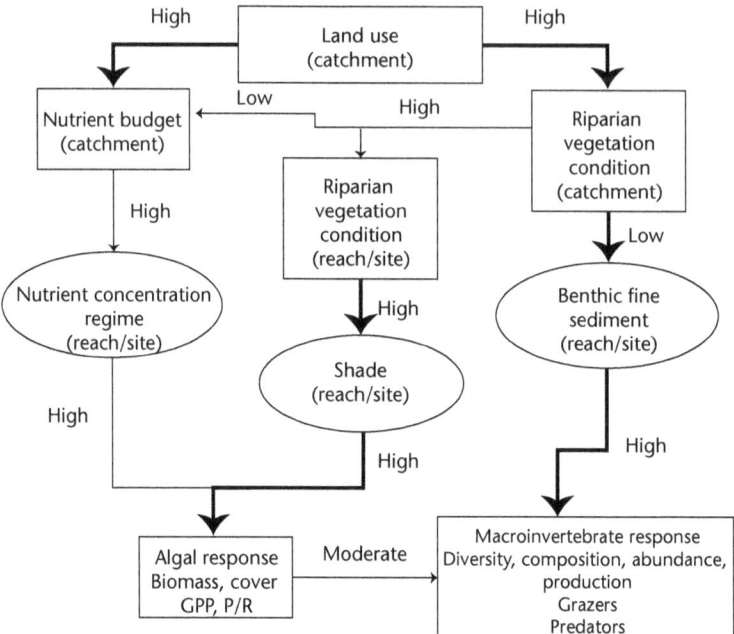

Figure 3.5: Dominant drivers and scales of macroinvertebrate and algal response to land use in mid- to lower-catchment Tasmanian rivers. Arrows represent the strength of interactions within Tasmania (bold = stronger, light = weaker). Arrow labels represent the strength of evidence for that interaction within Tasmania ('high' = high-quality quantitative data available from more than one study, 'moderate' = moderate amounts of good-quality quantitative data available, 'low' = limited quantitative data available, largely because these interactions were difficult to test on a large scale). Adapted from Allan *et al.* 2011.

While the basic architecture of a biophysical BBN for rivers can be reliably generated through a process of expert elicitation (as in the initial model of Figure 3.1), careful consideration and analysis are required to derive credible state descriptions of nodes, thresholds and probabilities. The amount of data required to build a moderately complex network can be very large, thus key challenges in BBN development relate to finding sufficient or critical minimum sets of data for parameterising the network. It is also important to communicate in the BBN documentation how these data have been utilised in this way.

It is unlikely that sufficient data will be available to parameterise an entire network of this sort, and expert elicitation is therefore necessary in parameterising a biophysically based BBN. A noted advantage of BBNs is the ability to mix information from a variety of sources (expert elicitation to fully parameterised data or modelled relationships) in 'mixed models' (Allan *et al.* 2011). A complicating factor arises when dealing with processes operating at large scales, as these are difficult to examine in isolation: factors are often confounded and an ideal BBN would separate these processes into individual nodes. Expert elicitation is therefore a necessary step in parameterising a biophysically based BBN, but may be less reliably utilised for this purpose than for development of the network architecture because expertise must be highly resolved and locally relevant.

Given that a BBN can be parameterised using a variety of data sources and expert opinion, it is important to ensure that the network structure and architecture are adequately documented.

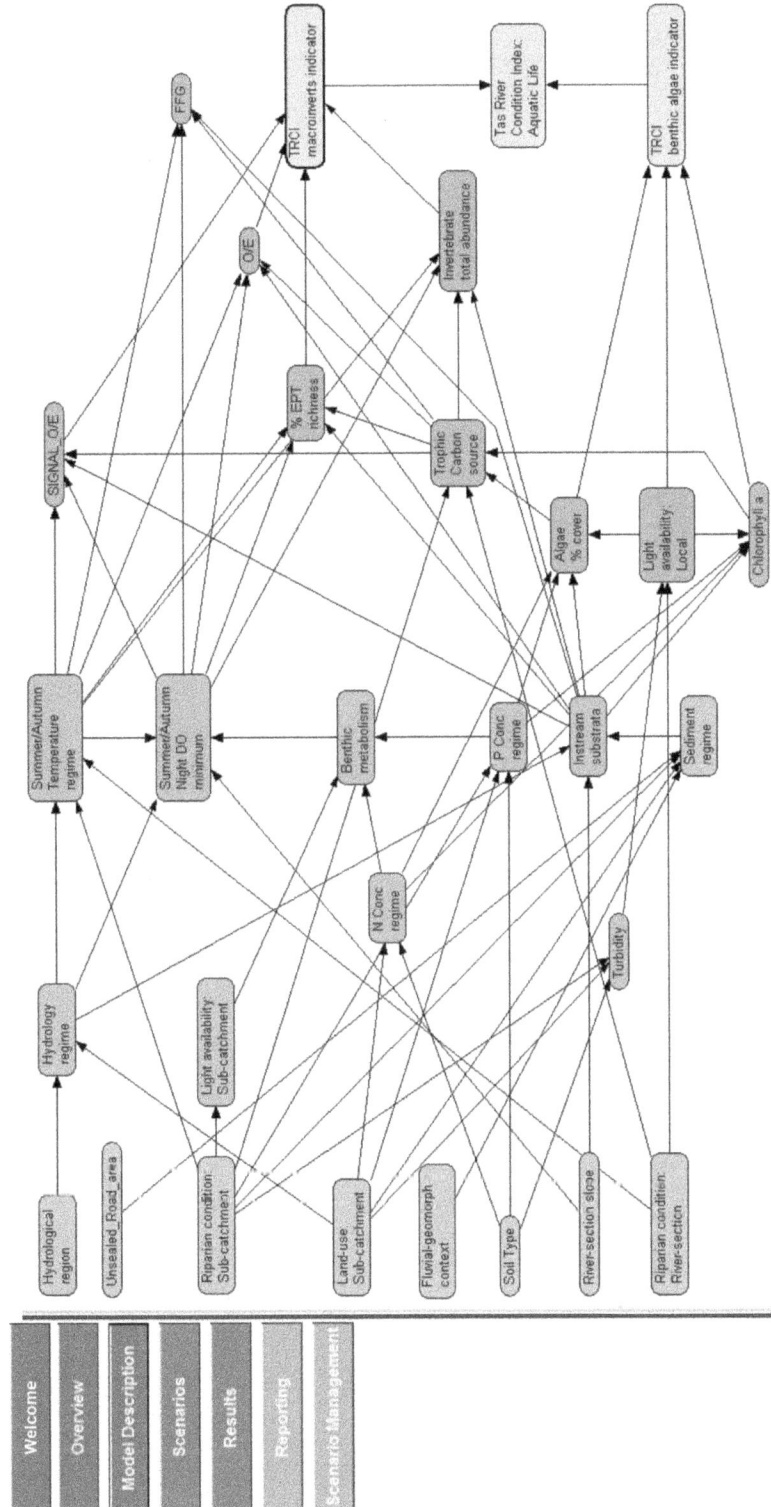

Figure 3.6: Basic architecture of the Tasmanian River Condition BBN. The model can be downloaded from www.landscapelogicproducts.org.au (node definitions in Magierowski *et al.* 2010).

Documentation should include data and model limitations, and detailed information on how state thresholds and probabilities are generated. Standards for network documentation are required to ensure that the limitation and biases of a BBN are adequately described.

EVALUATING THE BBN AND MODELLING MANAGEMENT SCENARIOS

The final phase of the project was to evaluate the BBN's performance, use the functioning BBN to test a wide variety of management scenarios and summarise key lessons and messages from the empirical data and the model outputs. Fact-sheets were then developed to communicate the results of management scenarios evaluated with the River BN (www.landscapelogicprod-ucts.org.au). When possible, scenario outcomes were linked to the results from the empirical data analyses: model outputs were congruent with empirical data, but both were used to construct more widely credible communications tools.

Four main scientific or technical messages emerged within the focus of the study and were captured in the fact-sheets:

- at the catchment scale, fine sediment sourced from erosion is a stronger driver of river ecological responses than light or nutrients;
- at the catchment scale, native forest riparian vegetation is a strong mediator of land use-driven fine-sediment impacts;
- sediment and light and nutrients all interact strongly to drive responses at reach and site scales;
- shade from riparian vegetation is a strong mediator of nutrient effects at the reach scale.

CONCLUSIONS

This project aimed to understand the influence of land use and land management on the ecological condition (health) of Tasmanian rivers at local (site and reach) and catchment scales, with the project team bringing a wide range of previous scientific and land management experience to a commonly agreed problem. Positive aspects of this integrative and participatory research included effective interaction and discussion among the multi-disciplinary team, a common and tractable regional focus, good access to relevant data with known provenance, and active and dynamic project hub and leadership engagement.

Important lessons from this collaborative experience are outlined below.

PROJECT DESIGN

The project was designed to integrate with a number of concurrent Landscape Logic projects. It is important to develop and maintain a core conceptual framework across all projects, and to design integration from the beginning. Greater cross-team conceptual exploration of potential ecosystem drivers at the outset of Landscape Logic would have assisted in identifying how the different projects could integrate better. Project design and resourcing should thus include all cross-project integration. On the other hand, there are periods in the research process when project teams need to work separately, to allow them to focus on specific problems. While time spent on detailed experimental design and data-analysis and time spent on integration are both necessary, these aspects of a large collaborative project cannot always happen simultaneously.

STAKEHOLDER ENGAGEMENT

To be useful to land managers, management tools must be regionally relevant, evidence-based, objective and adaptive.

Stakeholder engagement should occur during the early stages of research design and at both program and project levels. Project design and deliverables can then be developed to connect with particular management needs and levers. Equally, however, the structure of the material delivered needs to emerge from the biophysical models developed, whether represented in complex or simpler forms (contrast Figures 3.1, 3.5 and 3.6). Attempts to build models that link management levers directly to desired outputs, without connecting these through the requisite complexity of a model of biophysical reality, will fail (see Fig. 21.1, p. 287).

EXPERIMENTAL DESIGN

Open-ended data-mining is a high-risk activity and needs great care in interpretation. Nevertheless, it can reveal correlations for subsequent experimental analysis that could not be discovered by other means, and thus was an important first step in this work.

Managing river health requires management at catchment and local scales. A sustained focus on inferential strength for evidence linking cause and effect is required to ensure different scales of management are complementary in achieving desired NRM outcomes (Downes 2002). It is important to identify relative inferential strength when integrating evidence from multiple drivers. In surveys at large spatial scales where inferential strength is generally weak (typical in landscape ecology), we found a combination of correlative and diagnostic or experimental data could be used to markedly improve inference and implication of causality.

REFERENCES

Allan JD (2004) Landscapes and riverscapes: the influence of land use on stream ecosystems. *Annual Review of Ecology and Systematics* **35**, 257–284.

Allan JD, Yuan LL, Black P, Stockton T, Davies PE, Magierowski RH and Read SM (2011) Investigating the relationships between environmental stressors and stream condition using Bayesian Belief Networks. *Freshwater Biology*. doi: 10.1111/j.1365-2427.2011.02683.x.

Allen TFH and Hoekstra TW (1992) *Toward a Unified Ecology*. Columbia University Press, New York.

ANRDL (2003, 2008) Australian Natural Resources Data Library. <http://adl.brs.gov.au>.

Barber MC (Ed) (1994) *Environmental Monitoring and Assessment Program Indicator Development Strategy*. EPA/620/R-94/022. Environmental Protection Agency, Office of Research and Development, Environmental Research Laboratory, Athens, GA.

Biggs BJF (2000) Eutrophication of streams and rivers: dissolved nutrient-chlorophyll relationships for benthic algae. *Journal of the North American Benthological Society* **19**, 17–31.

Bojsen BH and Barriga R (2002) Effects of deforestation on fish community structure in Ecuadorian Amazon streams. *Freshwater Biology* **47**, 2246–2260.

Boulton AJ and Brock MA (1999) *Australian Freshwater Ecology: Processes and Management.* Gleneagles Publishing, Adelaide.

Davies PE (2000) Development of a national river bioassessment system (AUSRIVAS) in Australia. In *Assessing the Biological Quality of Freshwaters RIVPACS and Other Techniques.* (Eds JF Wright, DW Sutcliffe and MT Furse) pp. 113–124. Freshwater Biological Association, Cumbria.

Dodds WK and Welch EB (2000) Establishing nutrient criteria in streams. *Journal of the North American Benthological Society* **19**, 186–196.

Dodds WK, Smith VH and Lohman K (2002) Nitrogen and phosphorus relationships to benthic algal biomass in temperate streams. *Canadian Journal of Fisheries and Aquatic Sciences* **59**, 865–874.

Downes BJ (2010) Back to the future: little-used tools and principles of scientific inference can help disentangle effects of multiple stressors on freshwater ecosystems. *Freshwater Biology* **55**(S1), 60–79.

Downes BJ, Barmuta LA, Fairweather PG, Faith DP, Keough MJ, Lake PS, Mapstone BD and Quinn GP (2002) *Monitoring Ecological Impacts: Concepts and Practice in Flowing Waters.* Cambridge University Press, New York.

DPIPWE (2009) *Tasmanian Land Use Layer.* Tasmanian Dept of Primary Industries, Parks, Water and Environment, Hobart.

DPIW (2005) *CFEV Database, v1.0.* Conservation of Freshwater Ecosystem Values Project. Water Resources Division, Tasmanian Dept of Primary Industries and Water, Hobart.

Grace MR and Imberger SJ (2006) *Stream Metabolism: Performing & Interpreting Measurements.* Water Studies Centre, Monash University, Murray-Darling Basin Commission and NSW Dept of Environment and Climate Change. <http://www.sci.monash.edu.au/wsc/docs/tech-manual-v3.pdf>.

Hauer FR and Lamberti GA (Eds) (2007) *Methods in Stream Ecology.* 2nd edn. Academic Press (Elsevier), London.

Hill AB (1965) The environment and disease: association or causation? *Proceedings of the Royal Society of Medicine* **58**, 295–300.

Karr JR (1991) Biological integrity: a long-neglected aspect of water resources management. *Ecological Applications* **1**, 66–84.

Krasnicki T, Pinto R and Read M (2001) 'Australia wide assessment of river health. Tasmanian program final report'. Technical Report no. WRA 01/2001. Tasmanian Dept of Primary Industries, Water and Environment, Hobart.

Magierowski RH, Davies PE and Read SM (2010) 'The Tasmanian River Condition Bayesian Network'. Technical Report no. 26. Landscape Logic, Hobart.

Matthies M, Giupponi C and Ostendorf B (2007) Environmental decision support systems: current issues, methods and tools. *Environmental Modelling and Software* **22**, 123–127.

Norris R, Liston P, Mugodo J, Nichols S, Quinn G, Cottingham P, Metzeling L, Perriss S, Robinson D, Tiller D and Wilson G (2005) Multiple lines and levels of evidence for detecting ecological responses to management intervention. In *Proceedings of the 4th Australian Stream Management Conference: Linking Rivers to Landscapes.* (Eds ID Rutherfurd, I Wiszniewski, MJ Askey-Doran and R Glazik) pp. 456–463. Tasmanian Dept of Primary Industries, Water and Environment, Hobart.

NRM South (2009a) *Tasmanian River Condition Index Reference Manual.* NRM South, Hobart.

NRM South (2009b) *Aquatic Life Field Manual. Tasmanian River Condition Index. Book 1.* NRM South, Hobart.

Schofield NSJ and Davies PE (1996) Measuring the health of our rivers. *Water* **23**, 39–43.

Susser M (1977) Judgment and causal inference: criteria in epidemiologic studies. *American Journal of Epidemiology* **105**, 1–15.

Susser M (1991) What is a cause and how do we know one – a grammar for pragmatic epidemiology. *American Journal of Epidemiology* **133**, 635–648.

Improving the utility and sensitivity of estuarine monitoring

Jeff Ross, Christine Crawford, John Gibson, Barry Gallagher, Jason Beard and Stephen McGowan

SUMMARY

This chapter describes the development and key outputs of a study that examined the links between land use, land management and the health of estuarine ecosystems in Tasmania. The project originally involved four major tasks: retrospectively examining the relationships between land use, water quality and quantity, and parameters of estuarine health; assessing change in the condition of selected estuaries in relation to land use change over the period; evaluating indicators of estuarine resource condition and trigger levels of change; and reconstructing a history of estuarine condition back to European settlement. Differences in location, time of year and methods of past sample collection and analysis lead to high levels of variation in the data, which limited our ability to clearly identify land use impacts. This and partner feedback highlighted the need for a standardised framework for the design and interpretation of estuarine monitoring. In response, the contemporary survey in task 2 was modified to identify the key sources of variation so that future monitoring programs could be designed to detect anthropogenic-driven change, with data collection focused on characterising the physical, chemical and biological conditions of the water column and sediments. The findings from historic data and contemporary surveys were used to develop a decision tree and conceptual diagrams to support management of Tasmanian estuaries based on three major drivers of estuarine condition – flushing time, seasonality of critical events and sensitive locations characteristic of different estuary types. This was possible because the study found that multi-estuary characterisation is sufficiently consistent to provide a valid model.

THE BEGINNINGS

A critical initial step following the successful funding of the Landscape Logic project was the research prioritisation workshop held in December 2006 with key stakeholder agencies. Three core questions were identified in consultation with the regional natural resource management (NRM) organisations for the Tasmanian retrospective study.

1   Is the end-of-catchment a good indicator of the whole system in terms of water quality and quantity? What is the impact of pollutants on estuarine systems?
2   How do land use and land management affect water quality and quantities? How do best management practices for various land uses affect water yield and water quality?
3   Are national water quality standards relevant to Tasmania? (Are they relevant at regional and subregional scales? What should the trigger values be here? What is the point of monitoring if it is not relevant to our management actions? How can back-ground or natural levels of pollutants be distinguished from human-induced change?)

These core questions reflect priority issues for the NRM organisations in protecting and/or improving the condition of our natural resources, especially through supporting practice change and on-ground activities. Thus their focus is more towards current issues and manage-ment decisions than the retrospective aspect of Landscape Logic. The estuarine study was designed to address questions 1 and 3 – focusing on the extent to which estuarine ecosystem condition depends on changes in water quality and quantity resulting from upstream catch-ment land use and management.

THE PLAN

To link land use, and resulting changes in water quality and quantity, to the health of estuarine ecosystems, four core activities were planned:
1   compiling existing estuarine datasets and examining the relationships between land use, water quality and quantity, and parameters of estuarine health;
2   re-surveying estuarine condition in selected estuaries to assess change in relation to land use change over the intervening period;
3   evaluating indicators of estuarine resource condition and trigger levels of change;
4   applying a paleo-ecological approach to reconstruct a history of estuarine condition back to European settlement.

INITIAL HURDLES

Differences in location, time of year and methods of sample collection and analysis led to high levels of variation in the data previously collected in Tasmanian estuaries. This limited our ability to use these data to clearly identify relationships between land use and estuarine health in either time or space. These difficulties in detecting relationships between land use and estuarine health highlighted the need for a standardised framework for the design and interpretation of estuarine monitoring, a need confirmed via consultation with our resource management partners, the three Tasmanian NRM organisations. In response to these early discussions, our contemporary survey was modified to identify the key sources of variation so that future monitoring programs could be designed to detect anthropogenic-driven change.

USE OF EXISTING ESTUARINE DATASETS

Prior to the Landscape Logic project, physical, chemical and biological data on Tasmanian estuaries had been collected sporadically (mainly in response to ad hoc programs and isolated public concern), with little effort made by state government to collate the physical and chemical data from Tasmanian estuaries compared with freshwater databases. No long-term monitoring of estuaries had occurred, except for the Tasmanian Shellfish Quality Assurance program for a limited number of variables related to food safety. Furthermore, the available data were largely scattered in unpublished reports and computer files. Consequently, there were very limited data underpinning our general understanding of how Tasmanian estuaries function and the processes that occur within them. Although the distribution of benthic organisms in Tasmanian estuaries had been analysed, no attempt had been made to synthesise the chemical data, particularly nutrients, and primary productivity for individual estuaries or on a regional or state-wide basis. Little, if any, effort had been made to integrate data on river chemistry (in many cases the major source of the nutrients) into an understanding of estuarine behaviour.

While our team's first task of collating all available information on Tasmanian estuaries was essential to the research task, there was significant broader public value in bringing all the data (and corresponding metadata) together into a single location. We also selected data from these sources and compiled them into a single database that allows rapid extraction of data and plotting of relationships between environmental variables for single estuaries, a range of estuaries, or all estuaries in the database. This database for 31 estuaries around Tasmania (Fig. 4.1) is currently being developed so that it can be interrogated by managers through a web-based interface, available at http://www.landscapelogic.org.au/.

Analysis of the pre-existing data highlighted a number of previously unrecognised features of Tasmanian estuaries. In particular, it became clear that nutrients in the estuaries of the northern Tasmanian coastline were higher and responded conservatively with salinity, i.e. the main biogeochemical process occurring within the estuary was merely mixing of the nutrient-rich river water with nutrient-poor marine water and rapid flushing as a result of the high tidal range and generally shallow bathymetry in these estuaries.

In contrast, the larger estuaries of the east coast had consistently lower nutrient concentrations, but in most cases there were scattered records of extremely high chlorophyll a concentrations (>50 μg L^{-1}). This was in contrast to the nutrient-rich northern estuaries, in which chlorophyll a concentrations rarely exceeded 5–10 μg L^{-1}.

A further feature of southern Tasmanian estuaries was the importance of oceanic waters in supplying nutrients to the estuaries, especially in winter. This is due partly to the low rates of nutrient production from the landscape, and to the periodic influx of nutrient-rich subantarctic water into coastal waters.

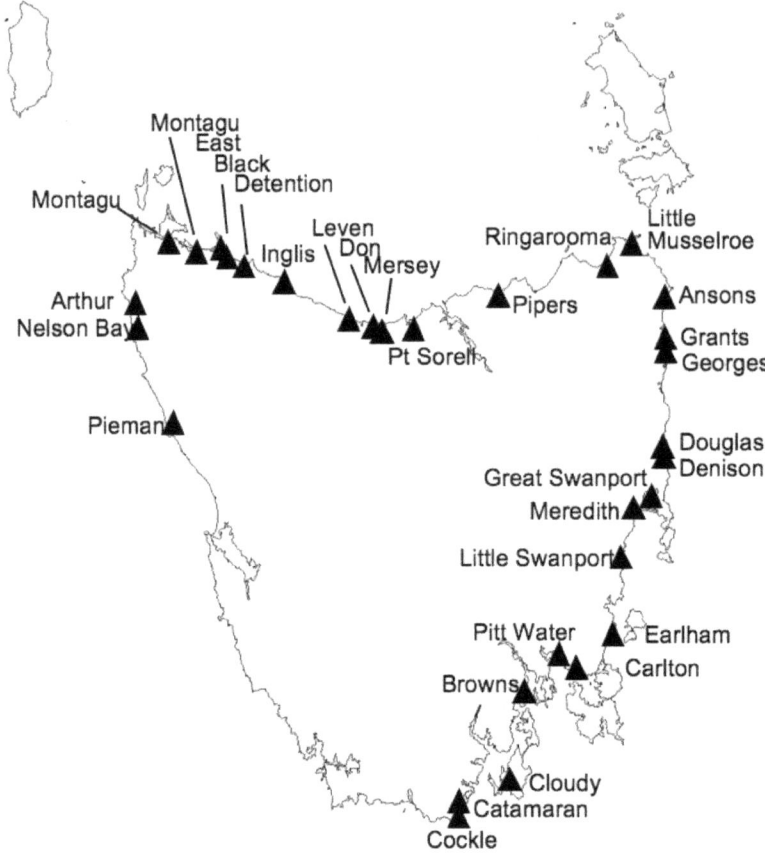

Figure 4.1: Data sources for Landscape Logic Tasmanian estuary research – contributing estuaries and their catchments.

RE-SURVEYING ESTUARINE CONDITION IN SELECTED ESTUARIES

Eleven Tasmanian estuaries representing a range of geomorphologic types and environmental conditions were chosen for re-sampling in detail in order to obtain more information about their physics, chemistry and biology, and to allow comparisons with earlier studies. This re-sampling program resulted in a better understanding of the connection between the physical and chemical behaviour in the estuary, in particular confirming the conservative mixing seen for many solutes in the northern Tasmanian estuaries. However, some solutes, such as soluble reactive phosphorus, did not follow this behaviour, which we took to indicate interactions with the sediments. Other observations included the role of the strong tidal cycle in sediment transport and water column mixing.

Comparison of data from our sampling and earlier studies was equivocal: the natural variability of the estuaries, resulting to a large extent from variations in nutrient loads entering the estuaries from terrestrial sources, made identification of long-term change difficult. Apparent changes in nutrient concentrations in the northern estuaries reflected transient observations, such as a nutrient peak measured on a particular sampling trip possibly reflecting a short-term pulse in nutrient input from the river rather than any longer-term process. A similar situation occurred for the eastern estuaries, though in these cases nutrient levels were low. Analysis of the timing of the occasional high chlorophyll *a* concentrations indicated that they were related to stochastic flood events, and therefore not amenable to analysis of change.

Instead, we compared estuaries around Tasmania using all the environmental data available and assessed the condition of estuaries in relation to their hydrology and geomorphology and human activities in the catchment. Using this information, we were able to classify estuaries according to their vulnerability to anthropogenic activities and to quantify the differences between affected and relatively pristine estuaries.

Sampling was also undertaken to provide a clearer understanding of the parameters that structured the ecosystems – in particular benthic organisms, algae and zooplankton of the estuaries. While some information on the benthic communities was available, the algal and zooplankton data we collected generated genuinely new information. It was clear that there were many drivers influencing the species make-up of these communities. Geography clearly played an important role, Tasmania's east coast being strongly influenced by the East Australian Current while the north coast is subject to westerly currents in Bass Strait. The chemical and physical sediment characteristics were also important, especially salinity gradients and sediment particle size.

ESTUARINE DECISION TREE

From an examination of environmental data collected in estuaries around Tasmania over the last two decades and from more detailed studies of 31 estuaries as part of Landscape Logic, we developed an estuarine decision tree (EDT; Fig. 4.2) based on the vulnerability of Tasmanian estuaries to human-induced change. The EDT aims to support management of Tasmanian estuaries by standardising and simplifying condition assessments and their interpretation.

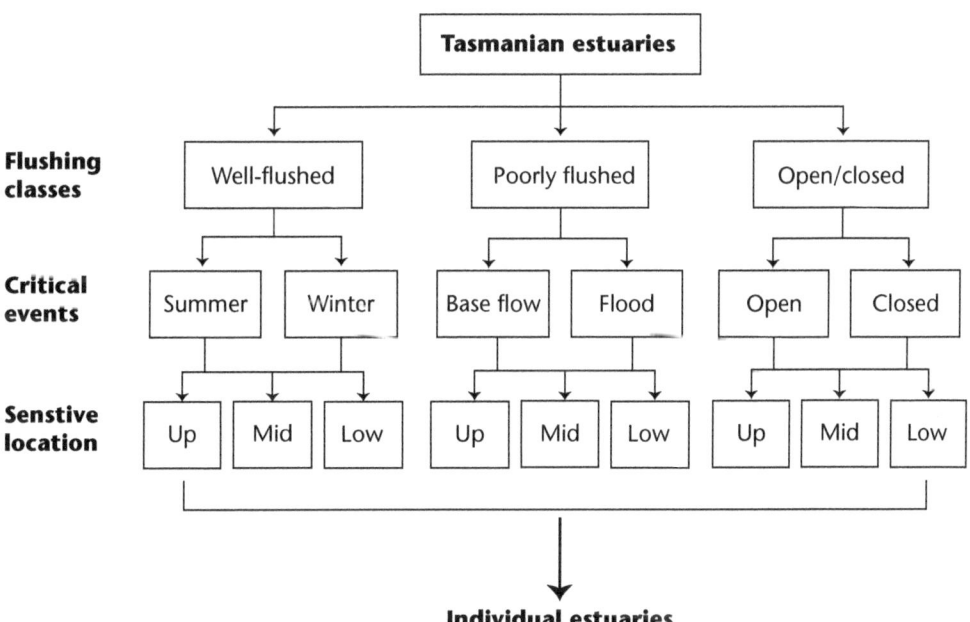

Figure 4.2: Estuarine decision tree.

We concentrated on the sensitivity of estuaries to eutrophication due to increased concentrations of nutrients, especially nitrogen and phosphorus, and organic matter. Derived from diffuse sources due to agricultural activities and human settlements in the catchment, these stressors are significant in many estuaries. The EDT is also applicable to other diffuse sources such as agricultural chemicals. Although sedimentation due to tree clearing and agriculture in the catchment can have a significant impact on estuarine health, we did not investigate this stressor in detail.

Flushing class

Using the known fact (NRC 2000) that estuaries most susceptible to degradation are those that have a poor ability to dilute or flush out pollutants, we used the key hydrological (river flow and tidal range) and physical (volume and geomorphology of the entrance) determinants of flushing time as the primary classification of estuaries for their vulnerability to human impacts. Estuaries were classified as either well flushed and therefore the least susceptible to degradation, poorly flushed and moderately vulnerable, or intermittently open and closed and highly susceptible. We found the distribution of these groups around Tasmania to be relatively consistent.

Well flushed estuaries are mostly located on the north coast, whereas estuaries of the east coast are predominately poorly flushed or intermittently open/closed. Tasmania's west coast estuaries have low tidal range but high river input due to heavy rainfall in the catchment. Sand movement along the coast is an important factor controlling the entrances to these estuaries; those with small catchment areas and/or long riverine estuaries tend to be poorly flushed or open/closed estuaries.

Critical events

The second tier in the EDT identifies the critical times or events when the different flushing classes are most sensitive to anthropogenic impact. Well flushed estuaries generally show strong seasonal patterns, with the largest freshwater flows in winter and spring and low flows in summer. These estuaries are most vulnerable during summer low flows when pollutants take longer to be flushed from the system, temperatures are higher and there are more hours of daylight, resulting in greater biological activity. Although poorly flushed estuaries receive less freshwater flow, flooding can be extensive and at any time of year. Thus the greatest contrast in the condition of poorly flushed estuaries is between base flows and flood events. After floods, nutrients brought into these estuaries during the flood are retained for extended periods and can result in eutrophication, especially in summer. Conversely, a reduction in base flows, such as during drought and through water extraction, can have the opposite effect of reducing nutrient input and hence productivity of an estuary. Intermittently open and closed estuaries are clearly most vulnerable when they are closed, as any nutrient or pollutant entering the estuary cannot be flushed out by tidal activity.

Sensitive locations

The response to anthropogenic inputs varies depending on location within an estuary. In well flushed estuaries the greatest impact has been observed in the upper estuary, where dilution of pollutants by oceanic waters is lowest. Similarly, in poorly flushed estuaries vulnerability increases with distance upstream due to decreasing dilution, whereas in open/closed estuaries human influences are generally evident across the entire lagoon.

Estuary-specific information

Although we have classified estuaries into broad groupings, the key drivers of flushing time and ecological characteristics are clearly going to vary somewhat from estuary to estuary, thus estuary-specific information should be used when available. Data on estuarine condition have been compiled and summarised in fact-sheets for a number of Tasmanian estuaries (www. landscapelogicproducts.org.au) and should be used when available. The full dataset will be available via a web-based interface that allows rapid extraction of data and plotting relationships between environmental variables for single estuaries, a range of estuaries, or all estuaries in the database. For estuaries where no data have been previously collected, users should refer to the generic information provided for the relevant estuary class and/or fact-sheets and data for similar estuaries.

INDICATORS OF ESTUARINE CONDITION

Indicators for monitoring Tasmanian estuarine, coastal and marine environments were recommended in the Tasmanian Indicator Compendium (Mount *et al.* 2006) and summarised by Crawford (2006), including monitoring methods. The specific indicators to be monitored are determined by the objectives of each monitoring program. The Tasmanian indicators are a subset of the national indicator set and are those considered a high priority for monitoring in Tasmania. A range of indicators is required and we recommend that stressors such as nutrients or turbidity, as well as ecological response indicators such as chlorophyll *a*, are monitored (Table 4.1). If a response is detected, the causes can be traced back via the stressors to the pressures.

These indicators include some that are readily measured by the community because they are on site to record unusual events, such as algal blooms or mass mortalities, and minimal training is required, whereas others require considerable experience, funding and specialist interpretation. For smaller estuaries, photographs taken at low tide from identical locations

Table 4.1: Indicators recommended for monitoring Tasmanian estuarine and coastal waters

Indicators recommended for monitoring Tasmanian estuarine and coastal waters
Temperature
Salinity
Dissolved oxygen (esp. bottom waters)
Turbidity
Chlorophyll *a*
Habitat extent
Animal and plant species abundance
Shoreline position
Nutrients in the water column
Toxicants
Pathogens
Algal blooms
Mass mortalities
Litter
Invasive species

over a period of time provide important information on habitat change, such as an increase in sediment or nuisance macroalgae.

To interpret the condition of an estuary using the indicators listed, we recommend other contextual data be collected to aid our understanding of their vulnerability to human-induced change. These include temperature, salinity profiles, basic hydrodynamic properties (tidal range and river flow), bathymetry and entrance state. Temperature, for example, is not generally an indicator of condition, but it does provide contextual information relating to the rate of processes in the estuary. These contextual data are required for the primary classification of flushing class in the EDT.

Our EDT helps identify the sensitivity of an estuary to anthropogenic impact by classifying estuaries according to their rate of flushing. It further aids in prioritisation of where and when to monitor by identifying, for each flushing class, the time and location that the estuaries are most vulnerable, e.g. well flushed estuaries during low river flows over summer in the upper estuary. Generally a biological response, such as increased algal production, is more likely in summer when hours of daylight are longer and water temperatures are higher.

Monitoring, in particular for compliance purposes, requires a sufficient number of samples be collected to be able to statistically show that a change in condition has occurred and that it can be attributed to human activities in the catchment (i.e. is not explained by natural changes in condition). There are numerous publications on sampling design, including ANZECC (2000). However, in practice sufficient funds are often not available to sample at the required intensity, especially in smaller estuaries, and a trade-off between statistical validation and available budget often occurs. In this case signs of deteriorating condition, such as large increases in chlorophyll *a* or turbidity levels, should trigger additional sampling to determine the cause of the changed condition.

THRESHOLDS OF POTENTIAL CONCERN

The principle of monitoring natural resource status is that condition data can be routinely assessed for changes; this allows an early response to minimise or reverse a negative impact. Making management response decisions involves identifying and agreeing on standard values which, once exceeded, will trigger further assessment of the changed conditions (e.g. to ascertain whether a deteriorating condition is temporary or sustained, and to begin to attribute the likely cause). Estuaries naturally have widely fluctuating conditions and it can be difficult to determine whether a change is due to natural events or human activities.

Threshold values, often called trigger values, are values that signal that the environmental health is deteriorating to an unacceptable level, as determined from available data and previous knowledge. We have chosen to call these condition decision points 'thresholds of potential concern' (TPCs), after Biggs and Rogers (2003), to distinguish them from 'trigger values' which have connotations of compliance points as part of national water quality ANZECC (2000) guidelines. The TPCs are our best estimate, from the data available, that environmental conditions are deteriorating and further assessment is required. They should be assessed and improved as more data become available. We have developed TPCs for the indicators of estuaries classified as well flushed and as poorly flushed (Table 4.2). Thresholds for intermittently open/closed estuaries are more difficult to quantify because of the major fluctuations in their physical characteristics and the lack of data available for these systems.

Methods used elsewhere to determine trigger values include expert opinion and using the 80th and 20th percentiles in reference estuaries (ANZECC 2000). Where there are few reference estuaries, using the 25th percentile of all data from a range of estuaries has been recommended (US EPA 2000). In Tasmania, there is an insufficient number of estuaries in each flushing class in near-pristine condition to be used as reference sites and few estuaries have been routinely monitored for 24 months as recommended by ANZECC guidelines. However, most estuaries are not as degraded as those in the USA, as shown in the results of Murphy *et al.* (2003), and the 25th percentile provided values that were too low. We have used the median value (50th percentile) to set TPCs. Expert opinion and other assessments have then been used to further evaluate the suitability of these threshold values.

The ANZECC Guidelines for Fresh and Marine Water Quality (2000) for estuaries in south-eastern Australia have previously been used in Tasmania as default trigger values (i.e. a single value for all Tasmanian estuaries for each indicator). Our data show that these ANZECC trigger values are not suitable for all Tasmanian estuaries and coastal waters, especially well flushed systems (Table 4.2). For example, in well flushed estuaries in winter when river flow is highest, median nitrate plus nitrite (NOx) value of 260 µg/L in the middle to upper estuary corresponds with a chlorophyll *a* median value of 1 µg/L, which contrasts with ANZECC guideline triggers for south-eastern Australian estuaries of 15 µg/L and 4 µg/L, respectively. These results also emphasise the importance of using both pressure/state indicators and ecological/value indicators in a monitoring program. If only nutrients had been measured the system would have been considered to be seriously affected, when in fact high nutrients occurred in the estuary during low light and temperature conditions that were unsuitable for primary production and were rapidly flushed from the estuary before significant uptake by macro- and microalgae could occur.

Importantly, our data show that the trigger value and subsequent interpretation also vary depending on where and when the sample is collected. Therefore, we have developed thresholds that differentiate between locations and times that are likely to respond differently.

Estuarine data collected in several monitoring programs were combined and estuaries were classified according to the EDT. Location in well flushed estuaries was determined from position in the estuary and salinity of <25 or >25; upper and middle estuary sites were combined as they demonstrate similar responses (Table 4.2). Poorly flushed estuaries were classified

Table 4.2: Thresholds of potential concern for Tasmanian estuaries showing trigger values for south-eastern Australian estuaries listed in the ANZECC guidelines (2000)

	Well flushed 50th percentile				Poorly flushed 50th percentile			ANZECC
	Summer		Winter				Flood	
Salinity	<25	>25	<25	>25	Base (>10)		(<10)	All Tas.
Location	Up–mid	Lower	Up–mid	Lower	Upper	Mid–low	Up–mid	estuaries
Chlorophyll a µg/L	1.8	0.8	1.0	0.5	2.6	2.0	0.5	4
Turbidity nsu	4	3	8	3	2	2	5	0.5–10
Diss. oxygen %	94	96	93	97	91	99	95	80
TN µg/L	550	350	955	290	450	360	953	300
NOx µg/L	70	14	243	60	6	2	27	15
Ammonia µg/L	26	15	27	12	16	12	15	15
TP µg/L	64	35	108	29	31	30	25	30
Phosphate µg/L	5	8	6	11	4	7	3	5

according to whether freshwater inflow was base flow or a flood event, determined by salinity >10 or <10, respectively (Table 4.2). They were also grouped according to location; upper and middle site data were combined during flood events because they demonstrate similar responses as floodwaters penetrate well down the estuary, but floods were rarely sufficiently large to result in salinities <10 in the lower estuary. During base flows, however, the middle and lower estuary showed similar results whereas the upper estuary had the least flushing and the greatest signs of deteriorating conditions. TPCs were evaluated for the different classes.

TPCs for other indicators, including habitat extent and animal and plant species abundance, are currently under development. Habitat extent is determined by mapping the habitats (SeaMapTasmania http://www.utas.edu.au/tafi/seamap/). Animal and plant species abundance in estuaries has mostly been assessed using infaunal macroinvertebrate assemblages; they are a useful integrator of estuarine condition because they are generally long-lived and immobile. Toxicants and pathogens are usually monitored only for specific purposes and assessed according to ANZECC guidelines.

The TPCs for physical and chemical parameters of poorly flushed estuaries during base flow (salinity >10) and flood events (salinity <10) and for well flushed estuaries during summer and winter are shown in Table 4.2. The right column shows the trigger values for south-eastern Australian estuaries that are listed in the ANZECC Guidelines (2000).

These TPCs provide a means for managers to interpret estuarine condition data that have been collected as part of estuarine health assessments. They provide a warning that the estuary may be undergoing significant negative impact and that further assessments are required to determine whether this change in condition is a long-term trend and, if so, what is causing the change. Preferably managers will have already identified, as part of the establishment of the monitoring, evaluation and reporting process, the management actions that will be imposed to reverse the decline in condition.

PALEO-ECOLOGICAL RECONSTRUCTION

A historical study was undertaken in Little Swanport estuary (Fig. 4.1) as a test case of applying paleo-ecological approaches to understanding long-term changes in estuarine ecology. Our purpose was to attempt to reconstruct a history of estuarine condition back to European settlement. We collected four cores from Little Swanport estuary that covered sediment formation over the last 138 years. The effects of major floods and marine incursions could be identified within the cores, as well as long-term (multi-decadal) cycles in seagrass and macroalgae abundance. Useful information was obtained by placing short-term environmental data within the longer-term context of estuarine dynamics.

LESSONS LEARNED

The lessons from this process have been categorised into: scale of consideration, the role of indicators of estuarine health, the value of collaboration across disciplines and the value that stakeholders add, particularly in ensuring uptake of research by helping guide the development of products. These lessons are discussed below.

FOCUS VS MULTIPLE CATCHMENTS

At the commencement of the project, one of our NRM organisational partners strongly recommended that research efforts should not be focused on a small number of 'representative' catchments because this approach did not produce a broad understanding of catchments across a region. They advocated for research to be spread across a broad range of catchments. Our estuarine studies have identified pros and cons of both approaches. The detailed paleoecological study described above in the Little Swanport estuary, for example, provided a better understanding of long-term cycles and changes in an estuarine system. However, the complexities of sediment formation in estuaries (deposition and erosion, spatial variability etc.) means that such studies are difficult, time-consuming and expensive.

The multiple estuary approach, of comparing environmental data from 31 estuaries and sampling 11 estuaries in more detail, has greatly improved our knowledge of the function of estuaries around Tasmania. This information has enabled us to develop the EDT which can be used to standardise and simplify condition assessments and their interpretation.

INDICATORS OF ESTUARINE HEALTH

Measures of health of Australian rivers and estuaries have historically focused on water quality, determined from concentrations of nutrients and turbidity. Our study clearly identified that these indicators alone are not sufficient to determine the health of an estuarine system. High nutrient concentrations do not necessarily imply that an estuary is in poor condition as these nutrients may be rapidly flushed from the estuary or may occur when other components essential for excess productivity, such as long hours of daylight and warm temperatures, do not occur. The Duck and Montagu estuaries in north-western Tasmania are good examples – their nutrient concentrations in the mid–upper estuary are markedly higher than ANZECC guideline values, especially in winter, but chlorophyll *a* values are much lower. The residence time of nutrients in these estuaries is short due to the large tidal range and high river flow in winter.

MULTI-COLLABORATIVE PROJECTS

Collaboration with scientists across disciplines, e.g. ecology and sociology, and even across the landscape within a discipline, such as terrestrial, freshwater and estuarine systems, has provided a much wider view of issues and research approaches. This broader knowledge is important for better management of our natural resources and developing integrated catchment management. Some of the lessons from collaborating with researchers from different disciplines, including freshwater ecology and agronomy, involved realising the importance of developing a general understanding and a common language for ecological systems and indicators. Regular half-yearly meetings between researchers involved in the Tasmanian studies and annual meetings of all Landscape Logic participants, with a structured presentation of research updates and informal discussions, has been important in developing these collaborations.

STAKEHOLDER ENGAGEMENT AND PRODUCTS

From the commencement of Landscape Logic there has been an emphasis on involving end users in the development of research products, to encourage uptake of results. As a consequence, an improved mutual understanding and appreciation of the requirements of stakeholders and researchers has developed. This has primarily occurred through face-to-face meetings and discussions at the Landscape Logic annual and biannual workshops and conferences. An important legacy of the Landscape Logic project will be ongoing improved collaborations between researchers, government and end users, in particular in relation to management of estuaries.

Because of this collaboration with end users, the estuarine team recognised early that the NRM organisations required more than the development of a model such as a Bayesian Decision Network (BDN), which had been the original focus of the research. Estuaries are complex systems and a simple, quick and reliable means of characterising the vulnerability of estuaries to human impacts was required, to facilitate better planning and management by the NRM regions. The regions also requested the development of appropriate trigger values for monitoring the condition of estuaries around Tasmania. These were developed from local datasets and knowledge.

These collaborations allowed us to recognise that, for findings to remain relevant to end users, detailed scientific information had to be translated into user-friendly forms that could be readily adapted to suit planning and management responsibilities, without compromising rigour.

Developing indicators and thresholds was somewhat difficult because two of the key parameters that influence estuaries, water quality and quantity, are currently managed under two separate legislative frameworks, the *Water Management Act 1999* and the *Environmental Management and Pollution Control Act 1994*, respectively, even though they are integrally related. Further, these Acts are implemented by two separate divisions within the responsible environmental agency.

This work has been summarised in fact-sheets on the EDT, thresholds of potential concern, conceptual models of Tasmanian estuaries, and a number of estuaries around Tasmania, available from the authors.

REFERENCES

ANZECC (2000) Australian and New Zealand Guidelines for fresh and marine water quality. National Water Quality Management Strategy Paper no. 4, Australian and New Zealand Environment and Conservation Council.

Biggs HC and Rogers KH (2003) An adaptive system to link science, monitoring and management in practice. In *The Kruger Experience: Ecology and Management of Savannah Heterogeneity.* (Eds JT du Toit, KH Rogers and HC Biggs) pp. 59–79. Island Press, Washington.

Crawford C (2006) 'Indicators for monitoring the condition of estuaries and coastal waters'. Internal report. Tasmanian Aquaculture and Fisheries Institute, Hobart. <http://eprints.utas.edu.au/view/authors/Crawford,_C,.html>.

Mount RE, Carr E, Dowson G, Gales R, Morris A, Middleton N, Crawford C, Butler E, Thompson PA, Shields D, Hunter J and Eriksen R (2006) *Tasmanian NRM Estuarine, Coastal and Marine Resource Condition Indicator Compendium.* National Land and Water Resource Audit, Canberra.

Murphy R, Crawford C and Barmuta L (2003) 'Estuarine health in Tasmania, status and indicators: water quality'. Technical Report no. 16. Tasmanian Aquaculture and Fisheries Institute, Hobart.

NRC (2000) *Clean Coastal Waters: Understanding and Reducing the Effects of Nutrient Pollution.* National Academy Press, Washington DC. <www.nap.edu/catalog/9812.html>.

US EPA (2000) *Estuarine and Coastal Marine Waters: Bioassessment and Biocriteria Technical Guidance.* Office of Water. US Environmental Protection Agency, Washington.

5

Understanding the effectiveness of vegetated streamside management zones for protecting water quality

Philip Smethurst, Kevin Petrone and Daniel Neary

SUMMARY

We set out to improve understanding of the effectiveness of streamside management zones (SMZs) for protecting water quality in landscapes dominated by agriculture. We conducted a paired-catchment experiment that included water quality monitoring before and after the establishment of a forest plantation as an SMZ on cleared farmland that was used for extensive grazing. In a second study, we monitored water quality during the harvesting of a 20-year-old plantation in an SMZ. We found concentrations of bacteria, sediment and phosphate were lower in the buffered paired catchment, but that lower nitrogen concentrations could not be attributed to the intervention. Harvesting caused no appreciable increase in sediment delivery to the stream and we found it to be a minor source compared to other disturbances (road drainage and cattle disturbance).

Simulation of hillslope processes and stream flow illustrated that uptake by SMZ vegetation was a more important nitrate-mitigating process than denitrification. Because this model thoroughly integrates climate and within-soil processes, its usefulness should also be tested for other nutrients and chemicals. For example, the observed decrease in phosphate concentrations in stream water due to the SMZ treatment was probably predictable. However, additional modelling approaches are needed to simulate SMZ effects on sediment and bacteria.

Key factors contributing to the success of the studies were the cooperation of land owners who understood the research aims and were willing to collaborate with on-ground actions essential to the study, convergence of the interests and resources of multiple partners, the proximity of research staff and laboratory to the field site, early peer review of aims and methods, and experimental designs that controlled many potentially confounding factors (e.g. farm management, roads, catchment geomorphology). Some of our results and interpretations were compromised by limited pre-treatment data during a period of high inter-annual variability in rainfall and flow. Longer-term datasets would have improved assignment of measured water quality changes to the watershed management interventions. Apart from climate variability, these changes were also driven by high variability in other natural characteristics (e.g. size, soils and topography of each catchment), multiple interacting drivers of change, long time-lags between intervention and response, and non-linear processes driving undefined thresholds of change. Our key results and limitations are presented as a guide to planning future research.

BACKGROUND

Water quality in many agricultural landscapes worldwide is poorer than desired, presenting a risk to stream biota, people and livestock. Water quality comparisons across broad land use categories show Tasmania is no exception. For example, Chapter 2 (this volume) and Broad and Corkrey (2011), summarising nitrogen and phosphorus contamination in a wide range of Tasmanian catchments, found that streams with highest nutrient concentrations and loads also have the highest proportion of intensive agriculture. In contrast, streams draining mainly native or plantation forestry have relatively low concentrations of inorganic nutrients. Chapter 3 (this volume) and Magierowski et al. (2010) found a similar trend with aquatic habitat quality related to land use, stream water quality and riparian shading. Livestock access to streams is a primary concern for land managers, because it provides a point source of contamination by nutrients, sediment and bacteria. Water temperature increases are also a concern where trees have been removed from streamsides since they affect physical, chemical and biological processes of stream biota (Neary 2002).

Such concerns have motivated some natural resource managers, including farmers, to consider SMZs for buffering water quality from the potentially adverse affects of up-slope agricultural activities. The term 'SMZ' is preferred here to synonymous terms such as 'riparian buffers', because it is not limited to a particular landscape position, hydrological function or management system (see Neary et al. 2010 for a discussion of terminology). The water quality benefits of SMZs at plot and hillslope scales have been amply demonstrated (Collier et al. 2001; Mayer et al. 2007; Zhang et al. 2010), along with improved aesthetics, property values, native habitat, stock safety, wood production and carbon sequestration. However, water quality benefits at a catchment scale have not been clearly established because of temporal changes in up-slope land use, insufficient proportion of streams with SMZs, a preference for buffering low-risk high-order large rivers rather than higher-risk small headwater streams, and the associated disturbances related to roads and drains (Cotching, pers. comm.; Sutton et al. 2010).

Overall, the adoption of SMZs in Tasmania has been very low for a variety of reasons. For example, among 146 landholders who responded to a survey of riparian management in six Tasmanian catchments, only 75 500 trees or shrubs had been planted (Curtis et al. 2009; Chapter 6, this volume). Assuming 2000 trees or shrubs were planted per ha, it implies 38 ha had been planted in total. This planted area was probably less than 0.01% of the total cleared area in these catchments, compared to 5–10% coverage expected if all streams had been buffered. A common deterrent to farmer adoption is that the SMZ area is perceived to be lost from agricultural production. This deterrent led Robins (2002) to describe several potential income-generating activities for this zone, one of which was commercial wood production. Commercial forestry in this zone is accepted and used widely in the US and Europe, but not in Australia (Smethurst 2008). A possible deterrent to SMZ plantation use in some parts of Australia are state-based codes of forest practice that discourage commercial forestry next to streams. While these codes can be revised periodically to account for new information and changes to forestry practices (Smethurst 2008), the investment risk is generally borne by the landholder and involves dependence on changing policies or regulations.

While the effect of removing streamside vegetation has been documented in several studies internationally (e.g. Coweeta Hydrologic Laboratory; Swift and Messer 1971), we found only one study where the effect of tree establishment had been quantified in an agricultural catchment (Newbold *et al.* 2010). We found no studies that reported the use of fast-growing *Eucalyptus* or *Pinus* plantations for SMZs in Australia. This highlighted a need for water quality research at a small catchment scale that documented the effects of commercial plantation forestry in SMZs. Our interest was shared by a number of collaborators – a forest practices regulator (Forest Practices Board, Tasmania), the forest industry (Cooperative Research Centre for Forestry and its partners), a state government body charged with encouraging forestry on private land (Private Forests Tasmania), natural resource management organisations (partners in Landscape Logic) and CSIRO's Water for a Health Country Flagship and its Sustainable Agriculture Flagship.

PAIRED-CATCHMENT EXPERIMENT

DESCRIPTION

We established a paired-catchment study in a single paddock in a mixed grazing and native forest landscape in southern Tasmania (Figs 5.1, 5.2). Fertilisers were not used on pastures or native forest for several years prior to and during the study. Water quality measurements for one year prior to SMZ establishment in 2008 provided pre-treatment data. Changes as a result of the SMZ were determined by comparison with these pre-SMZ measurements and with the adjacent reference catchment that was not fenced and did not have trees planted. We monitored stream flow, temperature, salinity and turbidity every 15 minutes, and several other water quality parameters (e.g. oxygen, bacteria, nitrogen and phosphorus) about three-weekly or less frequently using grab samples. Several storms were automatically sampled every few hours for several days for measurements of various water quality parameters. Further details of methods are provided in Smethurst *et al.* (2011).

RESULTS

Our data showed that cattle exclusion using the fenced SMZ reduced the delivery of bacteria, phosphate and sediment to the headwater stream (Figs 5.3, 5.4, 5.5). The buffering effect on turbidity (cloudiness) became quite noticeable within a year of SMZ establishment, and was most apparent during storms that coincided with a grazing event during a relatively wet winter–spring period. Nitrogen may have been attenuated by tree establishment, but pre-SMZ variability in nitrate concentration indicated that some of this effect might not have been SMZ-related.

Figure 5.1: The paired-catchment experiment showing catchments with (1) and without (2) streamside management zones of eucalypts and acacias planted in 2008. Additional SMZs shown adjacent (left) and below these catchments are one year older. Photo: P. Smethurst.

Figure 5.2: A streamside management zone showing a plantation of *Acacia melanoxylon* (blackwood), *Eucalyptus globulus* (blue gum) and *Eucalyptus nitens* (shining gum) in the paired-catchment experiment. Seedlings were planted on spot-cultivated mounds next to pits that intercepted overland flow. Photo: P. Smethurst.

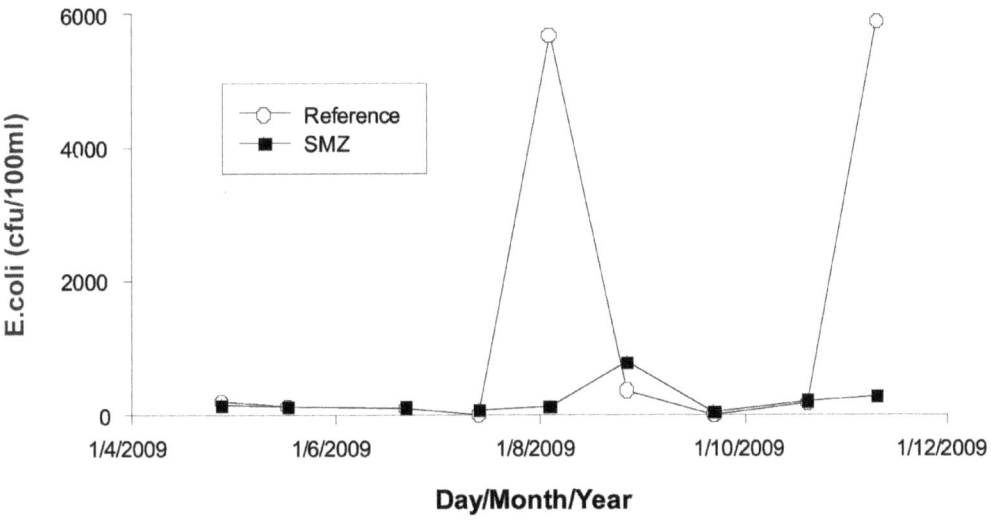

Figure 5.3: Bacterial counts (*E. coli*) in paired catchments with (SMZ) and without (Reference) streamside management zones showing higher values in the reference catchment on two occasions coinciding with severe cattle disturbance of the reference riparian zone during a very wet high flow period.

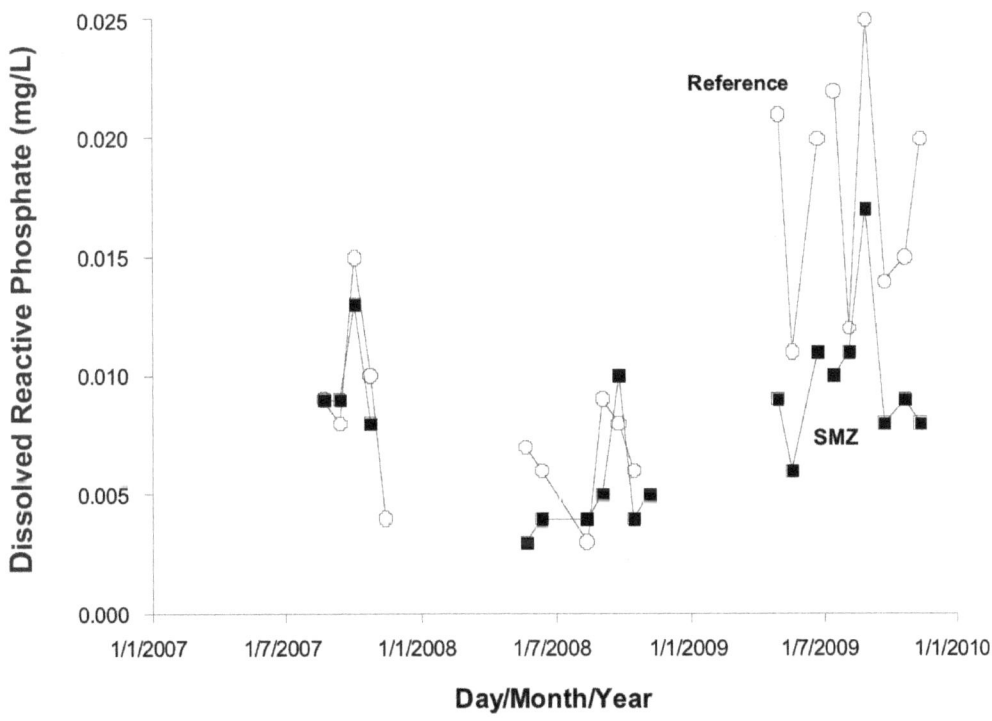

Figure 5.4: Phosphate concentrations in paired catchments with (SMZ) and without (Reference) streamside management zones showing similar concentrations for two years prior to and during SMZ establishment in mid-2008, but higher in the reference catchment during the wet winter of 2009.

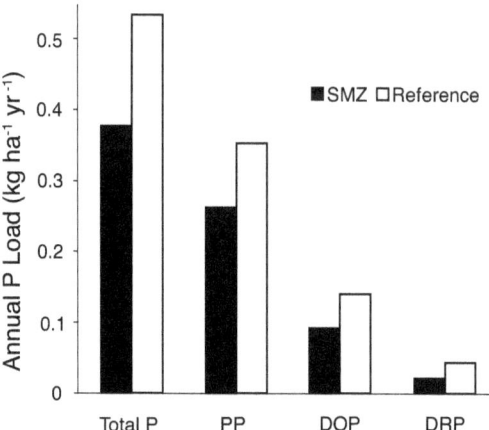

Figure 5.5: Annual loads (export in stream water per ha of catchment) of various forms of phosphorus in the paired-catchment experiment showing lower values in the treated (with SMZ) catchment during the wet winter of 2009. PP = particulate P, DOP = dissolved organic P, DRP = dissolved reactive phosphate.

The benefit of the SMZ for reducing turbidity was most evident on one high flow occasion when grazing was associated with off-scale turbidity readings in the reference catchment (>300 nephelometric turbidity units, NTU). During the same storm, values in the SMZ catchment were 50–140 NTU. During the dry-weather period preceding the winter storms, turbidity was approximately 16 NTU without and 5 NTU with an SMZ. The only clear negative effect of SMZ establishment on water quality parameters was a small transient increase in turbidity during one storm soon after cultivation, attributed to sediment originating from cultivated mounds used for tree planting. Turbidity at that time peaked around 29 NTU and turbidity without the SMZ was 11 NTU. Cultivation pits (Fig. 5.2) might have exacerbated tunnel erosion in the SMZ catchment, but this was not assessed.

Reduced phosphate concentrations in SMZ stream water were probably due to diversion of overland flow through the soil via the cultivation pits and other surface roughness features. A portion of phosphates in water percolating through soil would have been sorbed to the soil particles.

Sediment trapping by pits created during cultivation for trees likely contributed to reduced delivery of these contaminants to the stream, along with the presence of tall ungrazed remnant pasture. In addition, below-ground processes were likely important for attenuating the delivery of dissolved nitrogen and phosphorus (Smethurst *et al.* 2011). Observations and simulation modelling predicted that nitrate removal via denitrification occurred throughout the catchment where a water table existed, which was up to about halfway up the hillslope. However, nitrogen uptake by pasture and trees is likely to become a far more important nitrogen removal process than denitrification in this catchment. Since trees tend to reduce soil moisture and groundwater levels, denitrification is not expected to increase with SMZ establishment (Smethurst *et al.* 2011). Because this model thoroughly integrates climate and within-soil processes, its usefulness for simulating other nutrients and chemicals should be tested. For example, the observed decrease in phosphate concentrations in stream water due to the SMZ treatment was probably predictable using the model. However, additional overland flow modelling approaches are needed to simulate SMZ effects on sediment and bacteria.

HARVESTING AN SMZ PLANTATION

By good fortune, our team's work coincided with the harvesting of an SMZ plantation. Details of the opportunity and the water quality impacts we measured are outlined below.

DESCRIPTION

We identified the opportunity in 2009 to conduct water monitoring as part of a commercial plantation harvest in an SMZ on farmland in north-western Tasmania (Fig. 5.6). The plantation was located on a tributary of the Pet River, near Burnie, Tasmania. The predominantly north-flowing tributary included an upstream SMZ plantation that was not to be harvested and could therefore serve as a reference. Pre-harvest monitoring using grab samples started in December 2008. The larger eastern side of the SMZ was harvested in March–April 2009, after which the grab sampling program was augmented by continuous monitoring above and below the harvested reach every 15 minutes. Rainfall at Burnie for 2009 was 11% below average, but included several large storms that caused many trees in the remaining western side of the SMZ to be blown over. Salvage harvesting of this area was conducted November 2009 to January 2010. Harvesting was conducted as a routine commercial operation in accordance with the Tasmanian Forest Practices Code using tracked tree-felling and log-forwarding machinery. Machinery was required to stay out of a 5 m machinery exclusion zone next to the tributary stream. The plantation was well-stocked and had grown vigorously, with some trees in excess of 60 cm diameter over bark at 1.3 m height.

RESULTS

Our data showed total harvesting of a 20-year-old pulpwood plantation of *Eucalyptus nitens* in an SMZ was conducted without adversely affecting turbidity during the operation or subsequently during a wet winter period that included several storms (Figs 5.7, 5.8, 5.9; Neary *et al.* 2010). Turbidity is a measure of the cloudiness of water due to suspended sediment. Overall, the erosion risk at this site was considered low because the soil is a red-brown, well-structured, basalt-derived Ferrosol soil on low-gradient slopes. Upstream of the harvested area, which

Figure 5.6: A view upstream of the SMZ harvesting experiment showing (1) harvesting adjacent to the tributary, (2) the dam sampled below the harvested reach (dam 13 in Fig. 5.7) and (3) the upstream unharvested SMZ plantation. Photo: P. Smethurst.

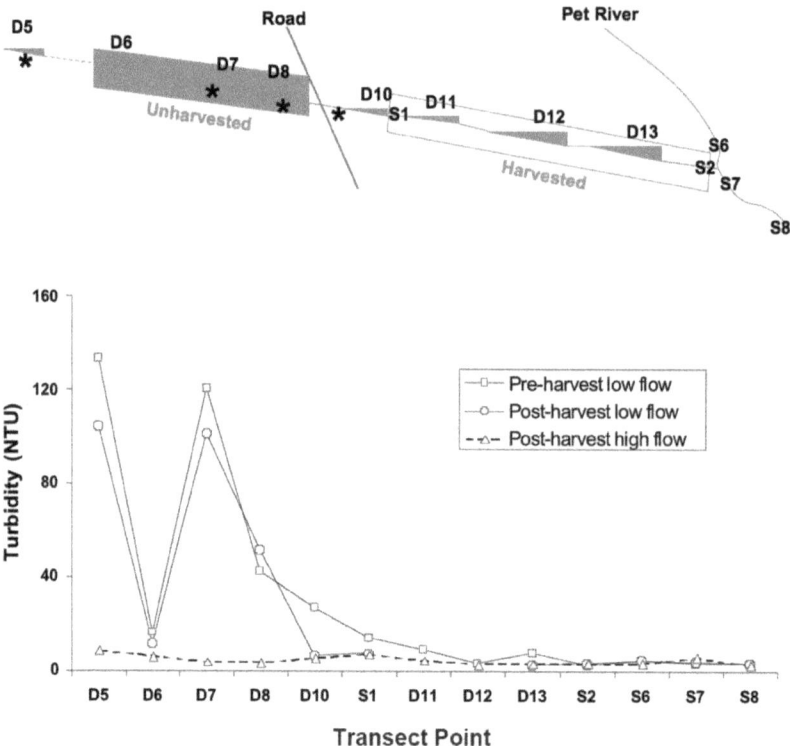

Figure 5.7: Schematic of the harvested catchment (top) showing the relative layout of dams, roads, streams, plantations and regular cattle access points (*). During the flow season, water flowed from dam 5 (D5) to stream sampling point 8 (S8) and beyond. Plantation SMZs are shown unharvested (shaded) and harvested (unshaded). Turbidity measurements along the catchment are shown for three occasions (bottom).

included an SMZ that was not harvested, turbidity was seriously affected by cattle access and road runoff. Most of the sediment generated by cattle disturbance and the road was captured in a dam above the reach of the harvested SMZ. Hence, in-stream dams throughout the catchment acted as filter traps.

Continuous monitoring in the dam just above the harvested SMZ (D10, Fig. 5.7) started soon after harvesting during low flows. Stream water there usually had turbidity values an order of magnitude higher than those observed below the harvested SMZ (D13, Figs 5.7, 5.8). Turbidity at the upper dam was more responsive to storms than the dam at the bottom of the SMZ. However, on occasions during high flows, connectivity increased between the dams and similar values were recorded for several days when turbidity in the lower dam reached a maximum of 33 NTU. Turbidity in the dam just above the harvested SMZ (D10, Figs 5.7, 5.8) was probably a combination of newly delivered sediment and resuspended sediment that had previously accumulated, but the relative contributions were not determined.

SUCCESS FACTORS

Several technical, organisational, collaborative and coincidental factors contributed to the success of our team's research.

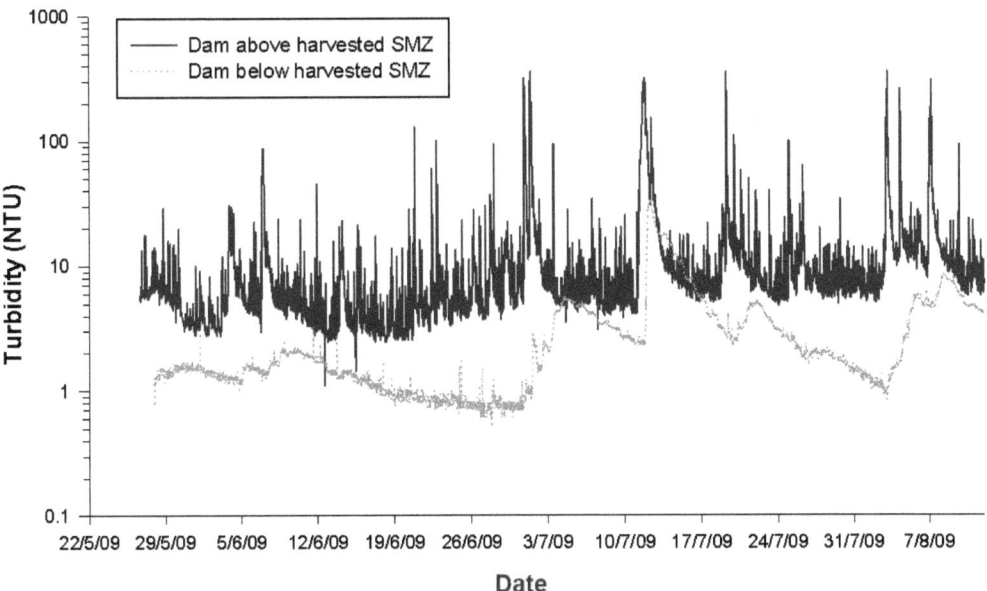

Figure 5.8: Turbidity trends before and during the winter of 2009. High values in the dam above the harvested reach were associated with sediment generated upstream by the road or cattle disturbance. For short periods, similar turbidity values were measured in the dam below the harvested reach, which coincided with storms and dam overflows that probably resulted in flow- and water-quality connectivity between the two dams.

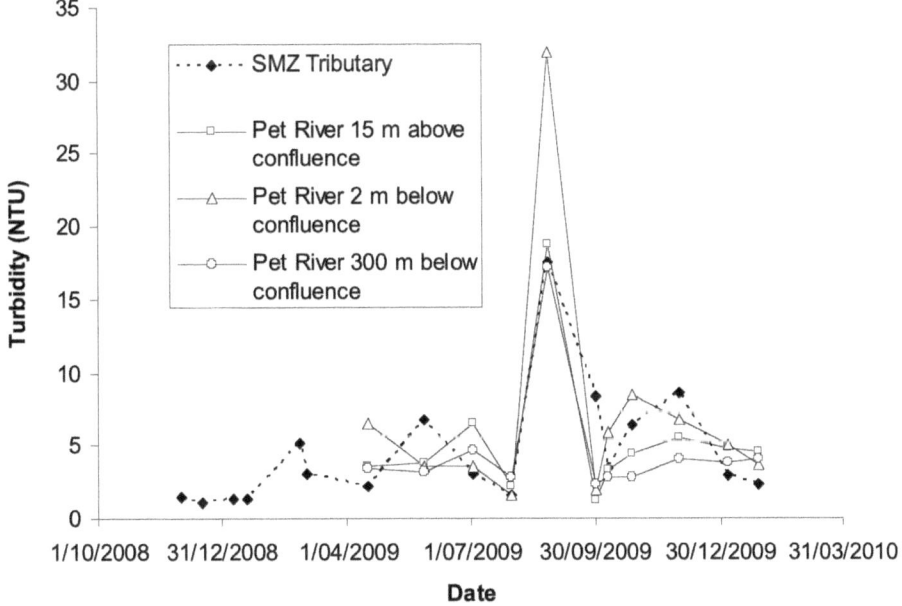

Figure 5.9: Patterns of turbidity around the confluence of the Pet River (which had little upstream SMZ harvesting) with the tributary containing the SMZ. Harvesting periods were March–May 2009 and November 2009–January 2010. Changes in turbidity in the tributary appear unrelated to harvesting, and values were generally similar to those above and below the confluence in the Pet River. Observed variations in turbidity are probably due to an interaction of rainfall patterns, landscape management unrelated to the SMZ and stream turbulence.

TECHNICAL FACTORS

There were six factors to which we attribute the technical success of our team's research. These are experimental design, peer review of the experimental objectives and methods, catchment options and selection, measurement systems, staff and other resources, and proximity to research sites. These factors are discussed below.

Experimental design

Paired catchments are a strong experimental design for catchment-scale studies, especially if the catchments are identical or similar in size and shape and an adequate pre-treatment measurement period is included when catchment replication is low (Brown *et al.* 2005). The paired-catchment method is schematically explained for flow in Figure 5.10. Two approximately identical catchments are identified. Flow is monitored across a range of rainfall conditions for several years to build a relationship between the two catchments for flow, i.e. a

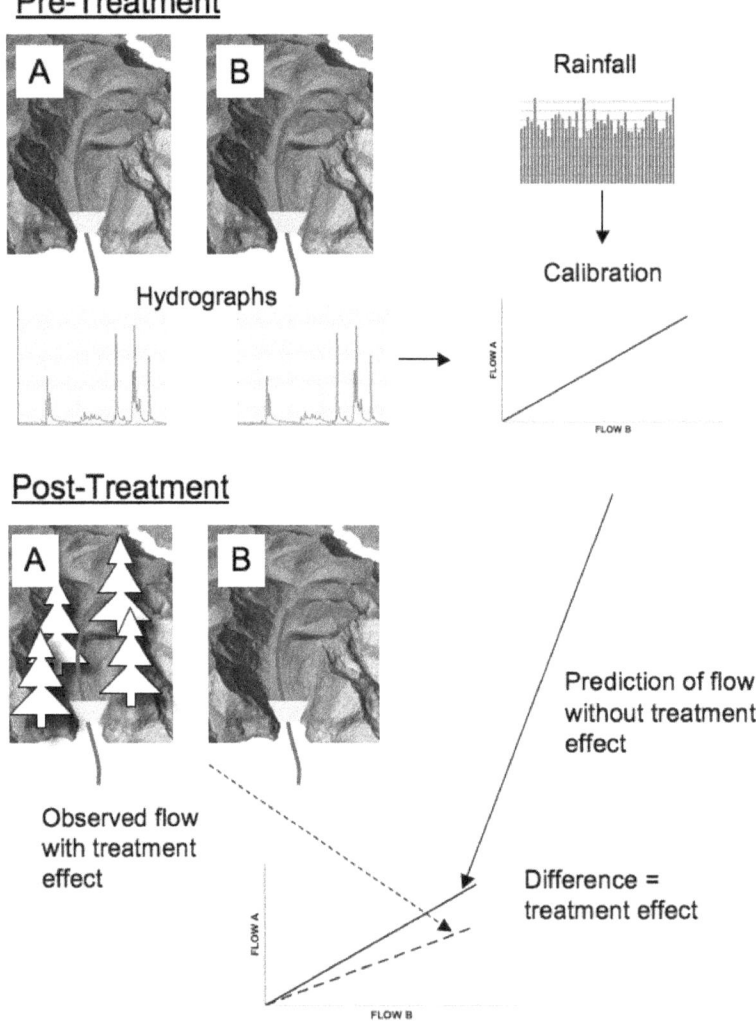

Figure 5.10: Schematic of the paired-catchment method.

pre-treatment calibration of flows. One catchment is then treated with a vegetation change, e.g. plantation establishment, fire or another management treatment, and flow monitoring continues for a few more years. Observed flow in the treated catchment is plotted against flow in the untreated reference catchment, along with that expected from the original calibrations; the difference is due to a treatment effect. These analyses are usually done with annual data, but monthly, seasonal, high flow or low flow examples are also available. Water quality comparisons are also possible.

In the current study, we attributed pre- and post-treatment changes in the flow ratio between treated and reference catchments to a treatment effect, e.g. vegetation change. If data are available, more sophisticated empirical regression techniques could be used to relate the difference in flows to season and rainfall. For example, Bren and Hopmans (2007) presented outputs for the conversion of eucalypt native forest or pasture to *Pinus radiata* plantation. An overview of empirical relationships for vegetation changes in South African studies is provided by Scott *et al.* (2000).

Our establishment experiment included replication of reference catchments for some measurements and one year of pre-treatment measurements for the paired treated and reference catchments. For our harvesting study, a paired-catchment approach was not possible, but a nested longitudinal design enabled some degree of temporal and spatial controls for water quality.

Peer review of experimental objectives and methods

We sought peer review at an early stage by coordinating an on-site inspection and discussion with an expert panel of several catchment hydrologists familiar with forestry and agriculture. We employed standard methods of measurement, which facilitated the set-up and maintenance phases.

Catchment options and selection

Initially, we selected only one tree establishment catchment for study in a pre-treatment versus post-treatment comparison. Expert input encouraged division of this larger catchment into several smaller subcatchments and selection of two for a paired-catchment design. This change of design enabled us to avoid potentially confounding hydrological effects of roads and farm tracks and to work with smaller catchments, which in turn provided logistical, technical and budgetary efficiencies.

We spent considerable time searching Australia (by telephone and email) for a suitable harvesting catchment, since we needed an SMZ plantation that had reached a harvesting age and was scheduled for harvesting within a year. This constraint required an SMZ plantation that had been established 15–20 years earlier, when the concept of streamside tree establishment was not typically considered. Fortunately, the best of three potential sites located in three different states was in Tasmania, which was close enough to the CSIRO research base in Hobart to facilitate its use.

Measurement systems

The experimental design was matched by adequate measurements provided by automatic and manual instruments and technical support, that in turn enabled data acquisition.

Staff and other resources

Our team consisted of scientists with complementary skills in hydrology (K. Petrone), agriculture, forestry and soil science (P. Smethurst), and soil science and hydrology (D. Neary). We were well supported by several technicians with complementary skills. Apart from their agricultural experience, the farmers already had an interest in plantation forestry, and Private Forests Tasmania provided advice and funding to establish the SMZ plantation. The research budget primarily came from federal sources (CRC and CERF programs) and was boosted by industry funds via the CRC for Forestry. Modern telecommunications and air travel facilitated key staff contributing from distant locations, networking generally, and awareness and access to relevant literature.

Proximity

The establishment research site was reasonably close (40 minutes drive) to the CSIRO research base in Hobart, which allowed frequent visits for set-up, servicing, maintenance and measurements. By comparison, the harvesting research site was four hours drive away and this limited the effort that could be put into this site. Fortunately, one of the team (B. Cotching) was based within 15 minutes of the site, which greatly assisted site measurements and visits.

COLLABORATIVE FACTORS

There was a very strong convergence of organisational interests that included setting up a demonstration site of an SMZ plantation for farmers, the general public and other stakeholders, improving the level of confidence of natural resource managers (including regulators and farm foresters) in SMZ plantations as a potential land use, and publishing scientific papers and other communication materials that document knowledge advances in this field. Of primary importance for selection of research sites was the involvement of Private Forests Tasmania staff and the farmers who were contacted for possible collaboration.

The success of our team was greatly assisted by the fact that everyone involved shared the research vision and worked very constructively towards its implementation. The productive working relationships between the farmers and researchers were also of major importance.

COINCIDENCE

It was fortunate that the above-mentioned factors coincided in space and time to ensure a successful outcome. An example is that the harvesting site was probably the only highly suitable site in Australia, and it was a coincidence that it became available when a satisfactory monitoring site was being sought.

LIMITING FACTORS

All projects operate within a number of constraints which significantly affect the nature of their work. The constraints that our team was aware of were imperfections implicit in paired-catchment designs, time, distance and budget.

IMPERFECTIONS OF PAIRED-CATCHMENT DESIGNS

Knowledge of process gained from catchment-scale studies can provide considerable insight that is not gleaned from modelling studies alone. Used carefully, the paired-catchment design is the best option available for detecting treatment effects at a catchment scale and it performs best when its limitations are acknowledged. Considerations we made when employing this research design were:

- paired catchments can lead to substantial errors when applied to other catchment conditions because derived relationships are empirical (Bren and Hopmans 2007);
- climate variability that coincides with a potential treatment effect can reduce the usefulness of the pre-treatment calibration. This factor increases the necessity for modelling approaches (Moore and Scott 2006);
- similar problems arise when rainfall patterns do not match for the treatment–reference comparisons (Liu *et al.* 2004);
- extrapolation of small-scale paired-catchment results to much larger catchments can be problematic if vegetation changes and rainfall patterns across the larger catchment are not well defined (Best *et al.* 2003; Brown *et al.* 2005);
- very few paired catchments are identical in geology, soils and topography.

These differences invariably affect comparisons to some degree.

Long pre-treatment calibration periods and replication through space and time help address many of these limitations of the paired-catchment approach. Although the paired catchments we used in the establishment study were adjacent and selected by an expert panel, they were not identical. They did not have the same proportions of rock types (syenite and mudstone; discovered after catchment selection) or vegetation (native forest and pasture) and did not have identical size, slope or shape. These factors could variously affect the interpretation of results for several water parameters and need to be considered when interpreting results. For example, the SMZ catchment consistently had much higher electrical conductivity (EC, salinity) values than the non-SMZ catchment. This observation could indicate that syenite rock (formed under highly saline conditions) weathers to yield highly saline water.

The SMZ catchment was also larger than the non-SMZ catchment, yet the catchments had similar stream length. It could therefore be expected that the catchments differ to some degree in wetness and biogeochemical behaviour. For example, the SMZ catchment most likely has a higher incidence of tunnel erosion. A dramatic example of a difference in catchment behaviour was observed when tunnel erosion was first detected in the SMZ catchment during an early winter storm when a sudden drop in EC coincided with an increase in turbidity (Fig. 5.11). Several days later, after most storm rainfall, a similar rapid increase in turbidity was observed in the reference catchment while there was only a minor decrease in EC. This is a demonstration that turbidity and EC patterns in response to rainfall were substantially different between the two catchments, and this could not be attributed to an SMZ effect. Thus, the differences in these non-linear responses and undefined threshold relationships between catchments had implications for the interpretation of results.

TIME

For most flow and water quality parameters, our study was restricted to 1.5 years pre-treatment and 1.5 years post-treatment monitoring. The net effect was that the limitations and associated risks of the paired-catchment design were substantial since the full range of rainfall, soil water and management conditions were not provided pre- and post-treatment. Time limitations also put pressure on skill development, because some team members were unfamiliar with hydrological methods. Some aspects of the study (e.g. the detailed nitrogen component, Smethurst *et al.* 2011) had their time-frames significantly reduced because of delays in finalising grant contracts. Fortunately, other aspects of the research had commenced two years earlier.

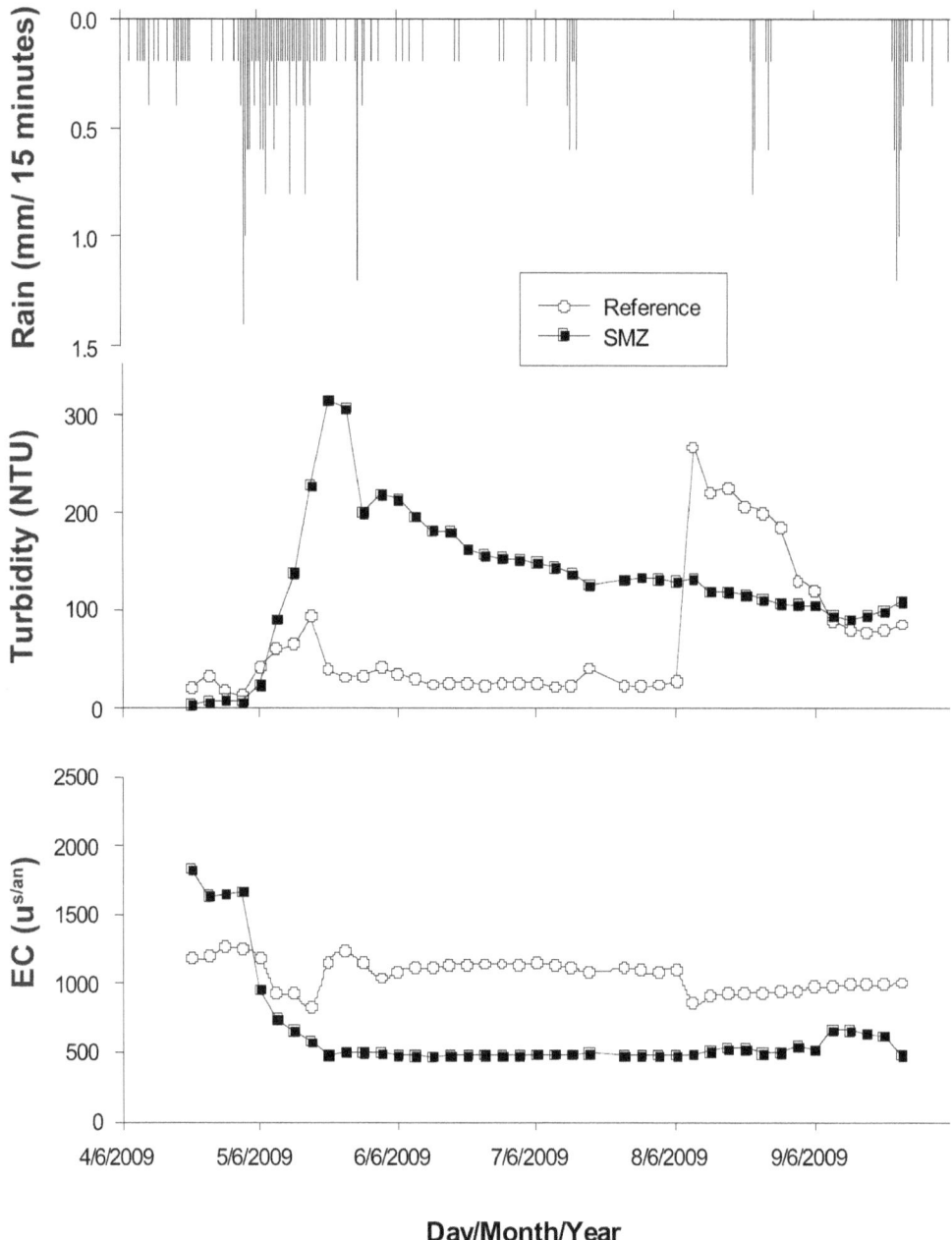

Figure 5.11: Rain, turbidity and EC patterns in the paired-catchment experiment during a storm early in the flow season during 2009 when tunnel erosion was identified as influencing flow and water quality patterns differentially in each catchment.

DISTANCE

The main hydrologists involved in the study were located elsewhere in Australia or overseas, which limited interactions and in-person participation. While within-team communication was very good and locally based staff were very adaptable and quick-learning, co-location of all staff in Hobart would have reduced communication overheads and may have improved outcomes.

BUDGET

As with most research, budget limitations contributed to the other issues listed above, e.g. low replication, short pre- and post-treatment periods and limitations on staff time, instrumentation and sampling. This necessitated frequent appraisal of the effectiveness of expenditure in terms of the value of data likely to be gained.

IMPLICATIONS OF THE RESEARCH

Here we distinguish between two key audiences for our research: land managers and other researchers.

IMPLICATIONS FOR LAND MANAGERS

Exclusion of livestock from streams can substantially improve water quality. Marked water quality benefits can be expected only where SMZs are established over a large proportion of headwater stream length, and even narrow buffers can deliver substantial benefits. In this study, SMZ width ranged from 10 m to 30 m and the area was fenced to exclude livestock access. Even with SMZs, major sources of poor water quality will remain where livestock continue to have access to the stream and drainage from roads and tracks is not diverted across pasture or other vegetation before it reaches the stream (Grayson *et al.*1993; Croke and Mockler 2001; Lane and Sheridan 2002). These are factors that contribute to the commonly observed overall negative effect of agriculture on water quality (Chs 2 and 3, this volume).

Logistical and financial considerations are important for establishing SMZs. Establishment of an SMZ requires a definition of situation-specific objectives such as aesthetics, property values, stock safety, commercial options, water quality, habitat enhancement and the use of native or introduced plant species. Second, potential financial and non-financial costs and benefits need to be weighed for different options. Third, detailed plans are needed for establishment and long-term maintenance of the SMZ. Fencing and off-stream watering points for livestock and stream crossings can be expensive and therefore a major consideration.

Commercial tree-farming for wood production or carbon credits is an option compatible with many commercial and environmental objectives. Dry-weather crash-grazing might be an option for utilising grass within the SMZ once trees have reached a suitable size, but the effects on water quality have not been evaluated. Potential practitioners of these management techniques are encouraged to seek advice from agroforestry regulators and advisors.

Our results show that, at least at some sites, forest plantations in an SMZ can be harvested to the water's edge without adversely affecting turbidity. While these results should encourage broader adoption of SMZs on farmland, care with harvesting and use of best management practices will be needed to ensure that it does not lead to unacceptable sedimentation. Harvesting methods may need to be adapted to suit different conditions of erosion risk (e.g. soil type and slope).

Prior to the initiation of the research described here, farm forestry advisors of Private Forests Tasmania had worked with the farmers at our research sites to plan or establish the

SMZ plantations. This partnership ensured excellent transfer of lessons about practices and outcomes, which are now being shared with other farmers in similar situations.

IMPLICATIONS FOR RESEARCHERS

While SMZ plantation practices are common in some countries, very few examples exist in Australia where the water quality impact of harvesting techniques can be evaluated. We believe additional Australian examples would improve the confidence in adopting these SMZ practices. We recommend that future research in this field draw on our experience to strengthen research planning and maximise the return on investment.

While many researchers, land managers and regulators are now aware of the research described here, there is no indication yet that the research has had an impact on policy. However, regulators were involved with the research from its inception and results are expected to help improve codes of forest practice during the next decade as they are reviewed in each Australian state and territory.

REFERENCES

Best A, Zhang L, McMahon T, Western A and Vertessy R (2003) 'A critical review of paired catchment studies with reference to seasonal flows and climatic variability'. CSIRO Land and Water Technical Report 25/03. CRC for Catchment Hydrology Technical Report 03/4. Murray-Darling Basin Commission, Canberra, Publication 11/03. <http://www.clw.csiro.au/publications/technical2003/tr25-03.pdf>.

Bren L and Hopmans P (2007) Paired catchment observations on the water yield of mature eucalypt and immature radiata pine plantations in Victoria, Australia. *Journal of Hydrology* **336**, 416–429.

Broad ST and Corkrey R (2011) Estimating annual generation rates of total P and total N for different land uses in Tasmania, Australia. *Journal of Environmental Management* **92**, 1609–1617.

Brown AE, Zhang L, McMahon TA, Western AW and Vertessy RA (2005) A review of paired catchment studies for determining changes in water yield resulting from alterations in vegetation. *Journal of Hydrology* **310**, 28–61.

Collier KJ, Rutherford JC, Quinn JM and Davies-Colley RJ (2001) Forecasting rehabilitation outcomes for degraded New Zealand pastoral streams. *Water Science and Technology* **43**, 175–184.

Croke J and Mockler S (2001) Gully initiation and road-to-stream linkage in a forested catchment, southeastern Australia. *Earth Surface Processes and Landforms* **26**(2), 205–217.

Curtis A, Race D and Sample R (2009) 'Management of riparian zones in Tasmanian agricultural catchments'. Technical Report no. 11. Landscape Logic, Hobart. <http://www.landscapelogic.org.au/publications/Technical_Reports/No_11_Tas_riparian.pdf>.

Grayson RB, Haydon SR, Jayasuriya MDA and Finlayson BL (1993) Water-quality in mountain ash forests: separating the impacts of roads from those of logging operations. *Journal of Hydrology* **150**(2–4), 459–480.

Lane PN and Sheridan GJ (2002) Impact of an unsealed forest road stream crossing: water quality and sediment sources. *Hydrological Processes* **16**, 2599–2612.

Liu S, Leslie LM, Speer M, Bunker R and Mo X (2004) The effects of bushfires on hydrological processes using a paired-catchment analysis. *Meteorology and Atmospheric Physics* **86**, 31–44.

Magierowski RH, Davies PE and Read SM (2010) 'The Tasmanian River Condition Bayesian Network'. Technical Report no. 26. Landscape Logic, Hobart.

Mayer PM, Reynolds SK, McCutchen MD and Canfield TJ (2007) Meta-analysis of nitrogen removal in riparian buffers. *Journal of Environmental Quality* **36**, 1172–1180.

Moore RD and Scott DF (2006) Response to comments by PF Doyle on 'Camp Creek revisited: streamflow changes following savage harvesting in a medium-sized, snowmelt-dominated catchment'. *Canadian Water Resources Journal* **31**, 135–138.

Neary DG (2002) Environmental sustainability of forest energy production. In *Bioenergy from Sustainable Forestry: Guiding Principles and Practices*. (Eds J Richardson, T Smith and P Hakkila) pp. 37–67. Elsevier, Amsterdam.

Neary DG, Smethurst PJ, Baillie BR, Petrone KC, Cotching WE and Baillie CC (2010) Does tree harvesting in streamside management zones adversely affect stream turbidity? Preliminary observations from an Australian case study. *Journal of Soils and Sediments* **10**, 652–670.

Newbold JD, Herbert S, Sweeney BW, Kiry P and Alberts SJ (2010) Water quality functions of a 15-year-old riparian forest buffer system. *Journal of the American Water Resources Association* **46**, 299–310.

Robins L (2002) *Managing Riparian Land for Multiple Uses*. RIRDC Publications, Canberra.

Scott DF, Prinsloo FW, Moses G, Mehlomakulu M and Simmers ADA (2000) 'A re-analysis of the South African catchment afforestation experimental data'. Report 810/1/00, CSIR Report ENV-S-C 99088. Water Resources Commission, South Africa.

Smethurst P (2008) 'Summary of Australian codes of forest practice as they pertain to managing commercial plantations in stream-side buffers on cleared agricultural land'. CRC Forestry Technical Report no. 178. <http://www.crcforestry.com.au/publications/downloads/TR178-Smethurst-FORPRINT.pdf>.

Smethurst PJ, Petrone KC, Baillie CC, Worledge D and Langergraber G (2011) 'Streamside management zones for buffering streams on farms: observations and nitrate modelling'. Technical Report no. 28. Landscape Logic, Hobart. <http://www.landscapelogicproducts.org.au/site/products/57-technical-report-no-28-streamside-management-zones-for-buffering-streams-on-farms-observations-and-nitrate-modelling>.

Sutton AJ, Fisher TR and Gustafson AB (2010) Effects of restored stream buffers on water quality in non-tidal streams in the Choptank River Basin. *Water Air Soil Pollution* **208**, 101–118.

Swift LW and Messer JB (1971) Forest cuttings raise stream temperatures of small streams in the southern Appalachians. *Journal of Soil Water Conservation* **26**, 111–116.

Zhang X, Liu X, Zhang M, Dahlgren RA and Eitzel M (2010) A review of vegetated buffers and a meta-analysis of their mitigation efficacy in reducing nonpoint source pollution. *Journal of Environmental Quality* **39**, 76–84.

6

Management of Tasmania's riparian zones by rural landholders

Allan Curtis and Digby Race

SUMMARY

Riparian areas play important roles in protecting water quality and providing habitat. In this chapter we describe social research that explored the assumed relationships between recommended property management practices and these ecological functions in six Tasmanian catchments, carried out through a survey mailed to riparian landholders and interviews with landholders and other stakeholders. Non-farmers (compared to those who identified as farmers by occupation) comprised the majority of riparian landholders and, while they had strong conservation values, few had been engaged in natural resource management (NRM) programs. Most survey respondents had implemented some practices expected to enhance riparian condition but our findings suggest there is considerable scope to engage landholders in additional work as most respondents expressed a strong stewardship ethic, being prepared to put the long-term health of their land ahead of short-term economic gain. Our study also found that while riparian areas are highly valued by landholders, they are perceived to be difficult to manage and any activity is heavily scrutinised. Many landholders are concerned that recommended practices such as fencing and stock exclusion add to the complexity of property management, particularly management of floods, weeds, feral animals and fire. As a result of these concerns, about half of those surveyed who have removed willows, a dominant invasive exotic species, have not replaced them with endemic vegetation. Given that the river health research (Chapter 3, this volume) found that shading is a critical factor in restoring ecological condition, public investment in willow removal may be having a perverse ecological outcome.

INTRODUCTION

Riparian areas have important ecological functions and are assumed to have an important influence on water quality. Within the Landscape Logic project, biophysical scientists explored the assumed relationships between recommended property management practices and these ecological functions and water quality outcomes (see Chs 2 and 3, this volume). These practices include controlling stock access to waterways and replanting native vegetation along stream banks. These actions are expected to provide habitat, filter nutrients from overland flow into streams and provide shade that lowers water temperature and light and reduces the likelihood of algal blooms.

Working with the biophysical scientists, the Landscape Logic social research team examined the social acceptability of recommended practices. As part of this research, we examined the extent to which recommended practices were being implemented and the factors influencing implementation. Data were collected through a survey mailed to riparian landholders in six Tasmanian catchments and a series of interviews with riparian landholders, industry representatives and government staff working in those catchments. Preliminary findings were refined through workshops with NRM organisation staff and industry representatives and through peer review of a project technical report (Curtis *et al.* 2009). These findings also provided advice to NRM organisation staff about how to more effectively engage landholders in riparian management.

In the following sections we explain what we did and why, identify our key findings and discuss their implications for NRM. We conclude with a discussion of our experience of integrated science within this project and our wider experience as social researchers in the Landscape Logic project.

RESEARCH APPROACH

Our Landscape Logic colleague, Dr Bill Cotching (Chapter 2, this volume), identified six Tasmanian catchments (Coal, Jordan, Quamby Brook, Macquarie, Pet and Inglis-Flowerdale) where substantial public investment had been made to support landholder implementation of practices expected to improve the condition of riparian areas (Fig. 6.1). Data collection was undertaken using two methods.

- A survey was mailed to a random selection of rural landholders in each of the six catchments owning more than 1 ha of riparian land within 50 m of a recognised waterway. The mailing list of 310 landholders was generated from cadastral and other spatial data by the Spatial Analysis project (Chapter 13, this volume) and a 65% survey response rate achieved.
- Semi-structured interviews were held with 43 rural landholders and 14 industry representatives and government staff working in the six catchments, the list of interviewees generated through consultation with the Landscape Logic knowledge broking team (Chapter 19, this volume) and staff from Tasmania's three regional NRM organisations. Interviews were conducted between March and May 2009.

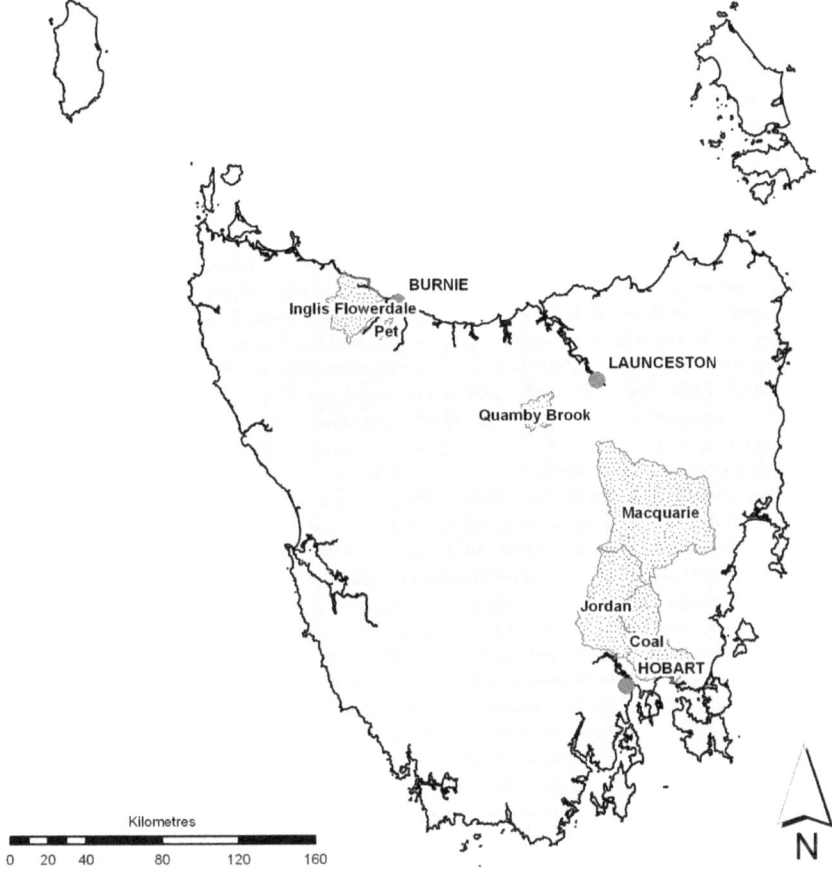

Figure 6.1: Location of the six selected catchments in Tasmania.

Engaging landholders in practice change is difficult because there is a potentially large set of factors influencing decisions which vary with each technology and landholder context. Even the concept of implementation is problematic. For example, when does a trial of practice change represent implementation? We also need to acknowledge that there are factors beyond the control of NRM practitioners that affect landholder implementation and that need to be considered when engaging individual landholders. For example, values (either held values or values that landholders attach to particular assets on their property or the district) are powerful influences on behaviour (Curtis and Mendham 2011) and remain relatively stable. Pannell *et al.* (2006) proposed a framework for exploring implementation, that addresses four broad sets of factors:

1   the nature of the practice or technology;
2   the personal characteristics of the landholder and their immediate family;
3   the wider social, economic and environmental context of the landholder;
4   the nature of any intervention or learning process.

Drawing on the literature on landholders' adoption of new practices (Vanclay 1992; Barr and Cary 2000; Cary *et al.* 2002), our research experience (e.g. Curtis and Robertson 2003; Curtis and Mendham 2011) and the ideas and experience of Landscape Logic researchers and end user partners, we identified a limited number of factors for inclusion in the mail survey which were likely to explain differences in the level of implementation of recommended NRM practices. These were:

- values attached to the waterway and adjoining wetlands;
- knowledge of NRM;
- attitudes about roles and responsibilities of key stakeholders;
- assessment of issues relevant to NRM;
- confidence in recommended practices;
- involvement in a short course, government NRM programs and property management planning;
- constraints to better management;
- long-term plans for the property;
- other social and farming variables including property size, occupation, landcare membership, absentee or resident owner, on- and off-property work, on- and off-property income, enterprise mix, age and gender.

Current recommended practices (CRPs) for riparian management that are expected to lead to improved water quality outcomes were also included in the survey. Practices identified through discussions with Landscape Logic researchers and partners include the replacement of willows by native vegetation, fencing to manage stock access, buffering streams from cultivation and installing off-stream watering points for livestock. Property management action beyond riparian areas also has an impact on water quality and was therefore within the scope of Landscape Logic research in Tasmania. For example, as tillage practices and fertiliser application rates influence soil erosion and nutrient runoff, which in turn may affect water quality, survey questions also explored implementation of these practices.

The landholder and stakeholder interviews occurred after the survey and were intended to allow the research team to explore key issues identified from analysis of the survey data. Interviewees were purposefully selected to provide a cross-section of landholders, including large- and small-scale landholders, long-time and new residents, commercial and hobby farmers, dairy farmers and wool producers, men and women, younger and older landholders and participants and non-participants in NRM programs. Using these criteria, potential interviewees were identified from suggestions by staff at the regional NRM organisations, local government, agribusiness and 'stock and station' agents. All landholders interviewed had properties with riparian frontage. Most interviews were of one to two hours duration and were digitally recorded with the interviewee's permission. Interviewees were informed of the research objectives and provided with an overview of some key findings from the survey. The subsequent conversations explored the following topics:

- management of the riparian area and their wider property;
- extent to which CRPs for improving water quality were desirable and feasible;
- involvement in NRM programs that focus on improving water quality.

In addition to landholders, the research team interviewed 14 other stakeholders from regional NRM organisations, local government, company staff and representative organisations. Stakeholder interviews explored the importance of water quality to their industry or organisation and their perceptions of the characteristics of successful NRM programs.

FINDINGS

SUBSTANTIAL ON-GROUND WORK IMPLEMENTED

The survey explored implementation of 20 practices that had been identified in consultation with Landscape Logic partners. These questions distinguished between three periods of the landholders' management of their property – during the period of their management, during the last five years, and during 2008. Over the longer management period, most survey respondents had erected fencing to enhance regeneration of native vegetation, erected fencing to manage stock access to waterways or adjoining wetlands and established off-stream watering points for stock. During 2008, around two-thirds of all respondents had spent time controlling pest animals and non-crop weeds in their frontage or adjoining wetlands. Survey data (Table 6.1) suggests that, for these practices, most respondents were undertaking substantial amounts of work (as opposed to symbolic work) that could be expected to improve resource conditions. For example, more than half of all respondents had fenced parts of their property to encourage regeneration of native vegetation. In most instances, a substantial proportion (median of 15 ha) had been fenced (median property size of 81 ha) (Table 6.1).

Although more than a third of all survey respondents had planted trees and shrubs along the waterway or in adjoining wetlands during their management period, much of that work was on a small scale and appears to have been largely symbolic. A small proportion of respondents said they had removed willows and only half indicated they had replaced willows with native vegetation (Table 6.1). Most respondents to this question had said it was

Table 6.1: Implementation of selected current recommended practices by Tasmanian riparian landholders, n=146

Recommended practices	n	Implemented (%)	Work Implemented to Oct. 2008 (median)
Practices undertaken during the period of property management			
Number of native trees and shrubs planted along the waterway or in adjoining wetlands (including direct seeding)	99	38	325 trees/shrubs
Area of land fenced for natural regeneration of native vegetation	114	57	15 ha
Length of fencing erected to manage stock* access to the waterway or adjoining wetlands (km)	84	61	2 km
Number of off-river/stream watering points established for stock*	81	65	6
Removed willows	112	26	No data**
Removed willows and replaced them with native vegetation	110	13	No data**
Practices undertaken this year (2008)			
How much time did you or others spend controlling pest animals and non-crop weeds in your river/creek frontage or adjoining wetlands (days worked)?	115	68	6 days

* Only respondents identifying as livestock enterprises are included for practices focused on livestock management.

** Survey items exploring willow removal only asked respondents to indicate if they had/had not undertaken an action or it was not applicable to their situation.

Source: Curtis *et al.* (2009).

'Not applicable', suggesting these respondents had no willows and/or thought willows should not be removed.

SCOPE FOR IMPROVED ENGAGEMENT AND IMPLEMENTATION

Positive values

Most survey respondents demonstrated a strong stewardship ethic, with almost two-thirds (64%) agreeing with the statement, 'Reduced production in the short term is justified where there are long-term benefits to the environment'. This item has been used in surveys in other catchments, including a 2008 survey of river frontage owners in the Goulburn Broken Catchment in Victoria, where a smaller proportion (35%) agreed with the statement (Curtis *et al.* 2008). This finding suggests there is considerable scope to engage Tasmanian landholders in additional work.

The top five values that survey respondents attributed to their river and creek frontages and adjoining wetlands (habitat for native fauna, stabilising stream banks and preventing erosion, aesthetics, stock water, adding to the market value of the property) included environmental, social and economic values, reinforcing the importance of appealing to a range of different values. A small number of significant relationships were identified between statements in the survey exploring values attached to the property and implementation of recommended practices. While these findings suggest values are not a powerful driver of landholder behaviour in this study, there were examples where values did predict behaviour. For example, respondents giving a higher rating to the statement, 'The river or creek frontage or adjoining wetland was important because it provided access to stock for water', were significantly less likely to have limited stock access to those areas for no more than a week at a time in 2008. Those who valued these areas because they provide important shade and shelter for stock were significantly more likely to have erected fencing to manage stock access to the waterway or adjoining wetlands, installed off-stream watering points for stock and removed willows.

On-property income and government funding

Many interviewees reported that the difficult business environment, coupled with continuing drought, had left Tasmanian farmers with little financial reserves for investing in on-ground NRM. Several interviewees said that they had deferred their NRM activity until conditions (business and climatic) improved. Only 37% of survey respondents said they made an on-property net profit in the 2006/07 financial year. For those reporting a profit, the median profit level was $15 000. The median off-property income before tax for the respondent and their partner of $45 000 was three times the median on-property profit of $15 000, suggesting that off-property incomes are higher than on-property incomes for most respondents. However, research, including in this study, suggests that landholders are more likely to spend on-property income rather than off-property income to implement NRM. There was a significant positive relationship between on-property profitability and two practices implemented during the period of property management: land fenced for natural regeneration of native vegetation and establishing off-stream watering points for stock. The low level of on-property profitability and income suggests that landholders in the six catchments, particularly farmers, have limited capacity to invest in recommended practices, especially those with high public benefits.

All landholders who undertook NRM practices in their riparian area reported that it had been undertaken with some financial and/or logistical support from a locally coordinated project (e.g. Landcare, local government or non-government organisation). Overall, 23% of respondents said they had received support through federal or state NRM programs and regional NRM organisations in the past five years. This finding is similar to that in the Goulburn Broken study of river frontage owners in Victoria in 2007 (Curtis *et al.* 2008). In the Tasmanian study, support through federal or state NRM programs and regional NRM organisations in the past five years was significantly linked to higher implementation of two recommended practices: trees and shrubs planted along the waterway or wetlands during the period of management and land fenced for natural regeneration of native vegetation.

Occupational identity

Farmers were the largest single occupational grouping in this study but comprised only 39% of all river frontage survey respondents (Table 6.2). Although farmers in this study owned larger properties than non-farmers, with the exception of the Macquarie catchment the aggregate length of river or creek frontage owned by non-farmers was greater than that owned by farmers.

It seems that NRM organisations have focused on farmers and have not effectively engaged the larger cohort of non-farmers. For example, 42% of farmer respondents compared to 11% of non-farmers said that in the last five years federal or state government programs or the regional NRM organisation had supported work on their river/creek frontage or adjoining wetlands. There were similar differences between farmers and non-farmers in the proportions of each cohort participating in Landcare, and farmers were significantly more likely to be involved in property management planning.

Those identifying as farmers were very different from non-farmers in terms of their concerns about property issues, values attached to river frontages, attitudes about NRM roles and responsibilities and self-assessed knowledge of NRM. Where a difference in concerns existed between farmers and non-farmers, farmers were significantly more concerned about each of the seven property-related issues. Farmers were also more knowledgeable, less likely to accept a duty of care for biodiversity or a stronger role for government in ensuring compliance with the *Environment Management and Pollution Control Act 1994* and more likely to give a higher rating to the economic values of their river or creek frontage and adjoining wetlands.

Table 6.2: Comparing respondents who are farmers and non-farmers by occupation, Tasmania riparian landholder survey 2008, n=146

Topic	Median farmer n=54	Median non-farmer n=84
% all respondents	39	61
Age	52 years	56 years
Area of the property	400 ha	42 ha
Hours of on-property work last year	56 hr	8 hr
Days of paid off-property work last year	0 days	195 days
Pre-tax on-property profit 2006/07	$15 000	$5000
Total off-property pre-tax income for respondent and partner 2006/07	$25 000	$55 000
Landcare member/participant	41%	11%
Member or involved with a local commodity group	33%	2%
Government- or regional NRM organisation-supported work on river/creek frontage or adjoining wetlands in past 5 years	42%	11%

Farmers were significantly more likely to report that they had implemented four recommended practices – fenced land for natural regeneration of native vegetation, erected fencing to manage stock access to the waterway or adjoining wetlands, tested soils for nutrient status in paddocks where they had applied fertiliser and tested the quality of the main water source for stock or irrigation purposes. Apart from the practices related to fertiliser use, these results appear counterintuitive given that non-farmers gave a higher rating to environmental values. Part of the explanation is that farmers are more knowledgeable about NRM practices, spend more time on their property and are more connected to the local community and the networks that facilitate the exchange of information and resources, as well as the establishment of local norms about what a 'good farm looks like in this district'.

Investing in human and social capital is effective

The 'capitals' concept is a widely accepted and useful framework (Australian Bureau of Statistics 2002) that can aid our understanding of the complex web of factors that affect landholder capacity (ability) to implement recommended practices (Pannell *et al.* 2006). The skills, abilities and wellbeing of the population form our human capital (Castle 2002); social capital refers to the social relations, networks, trust and norms that arise between people when they interact and that can lead to further benefits (Sobels *et al.* 2001).

Our previous research provided considerable evidence of the positive impact of investment in capacity-building. These activities include short courses, including those related to property management, and the activities of Landcare and commodity groups which focus on raising awareness, improving knowledge and building confidence in recommended practices (Curtis and De Lacy 1996; Curtis and Mendham 2011).

The findings in this study (Table 6.3) affirm the value of many existing NRM policy instruments that engage and build human and social capital and suggest ways of enhancing NRM outcomes in future. For example, only 21% of survey respondents said they had completed a relevant short course in the past five years and participation varied significantly across the six catchments. Only 22% of respondents said they were a member of or involved with a local Landcare group; again, participation varied significantly across the six catchments. By comparison, around 40% of rural properties across Victoria have a Landcare participant involved in their management.

Table 6.3: Levers that work: the influence of investment in human and social capital on implementation of riparian management, Tasmania riparian landholder survey 2008, n=146

Levers: investing in human and social capital	Significant positive relationships* with implementation of recommended practices
Concern about issues	14 practice items
Higher knowledge of NRM	12 practice items
Confidence in CRP	4 practice items
Property planning	8 practice items
Local action planning	6 practice items
Landcare membership	4 practice items
Short course	3 practice items
Government support	2 practice items

* All relationships reported in Table 6.3 are statistically significant at the 95% confidence level using a variety of pair-wise tests, including the Kruskal-Wallis Rank Sum Test; Spearman rank order correlations and proportions test.

Source: Curtis *et al.* (2009).

Existing recommended practices seen as problematic

While survey respondents and interviewees placed a high value on their riparian areas, these areas were also seen as problematic as they are difficult to manage, landholder actions are heavily scrutinised by others and many landholders, particularly those who are farmers by occupation, are concerned that recommended practices that involve fencing and the exclusion of stock add to the complexity of property management. The suite of practices recommended for riparian areas is also inconsistent with many landholders' views about 'best practice' management. Around 40% of respondents were concerned that fencing to manage stock access would lead to increased pest plants, make it difficult or more costly to water stock, create habitat or cover for pest animals, and result in more damage and expense from flood events. These concerns suggest it will be difficult to engage the 39% of respondents who said they had not fenced to manage stock access to the waterway or adjoining wetlands during their management period.

Nevertheless, many landholders had implemented practices intended to improve efficiency and profitability that also had NRM benefits, including strategies to optimise on-farm nutrient management (to maximise returns from nutrient inputs), fencing steep river banks to manage stock access (to minimise injury to livestock), removing willows (to reduce the risk of flooding and improve access to manage pest plants and animals) and installing off-stream watering points for livestock (to improve livestock health/management). Landholder acceptance of these practices suggests that more effort is needed, particularly through research with landholders, to identify and test practical and cost-effective options likely to lead to long-term commitment to active management of riparian areas, that will lead to improved ecological conditions.

CONCLUSIONS

IMPLICATIONS FOR POLICY AND MANAGEMENT

While most survey respondents had implemented some work expected to enhance riparian condition, our findings suggest there is considerable scope to engage landholders in additional work, provided engagement is informed by improved understanding of the social context. Non-farmers comprised the majority of riparian landholders and managed most of the riparian lands. Although they had strong conservation values, non-farmers were less engaged in NRM programs than farmers. It seems that NRM practitioners need to do something different if they wish to engage the non-farmer cohort. Survey findings about the issues of concern to landholders, the values they attach to river frontages and adjoining wetlands, their attitudes to their NRM roles and responsibilities and their knowledge of NRM practices are likely to provide NRM practitioners with considerable insight into how to achieve higher rates of adoption.

Most survey respondents demonstrated a strong stewardship ethic. However, with very low on-property incomes, especially in the case of farmers, much of the activity was dependent on public investment. Given the positive link between on-property income (but not off-property income) and implementation of accepted practices, there appears to be a persuasive argument for increased levels of cost-sharing for work with high public benefit, particularly during times of reduced on-property income, including extended droughts.

Attending property management planning courses and belonging to a Landcare group were good predictors of the adoption of recommended practices, but only a small proportion of landholders were engaged through these activities. Again, these data suggest there are opportunities to improve implementation of recommended practices through investment in platforms that engage and build human and social capital.

Although riparian areas are highly valued by most landholders, they are also perceived as problematic. These concerns seem to have limited the effectiveness of past investments in willow removal, in that half of the survey respondents who had removed willows had not replaced them with native vegetation. As shown in Chapter 3 (this volume), research in these catchments found that shading is important to rehabilitation of river ecological condition, suggesting that if willows are removed but not replaced by endemic vegetation then restoration of water quality is unlikely. This suggests a potential waste of public resources, as most willow removal in Tasmania has occurred as a result of public investment. A strong case can be made that willow removal should not be approved without a binding agreement to replace willows with endemic vegetation. Figure 6.2 provides a visual comparison of two different riparian management options after willow removal.

There needs to be a determined effort to engage landholders in processes that identify and test the efficacy of approaches to riparian management that are acceptable to landholders, particularly farmers. An important step might be to abandon the idea of pre-1788 resource condition targets in favour of negotiated 'improved' resource condition targets and steps to achieve those outcomes (Curtis and Lefroy 2010). For example, rather than fencing riparian corridors and excluding stock, a practical alternative might involve establishing riparian paddocks, providing off-stream watering points and rock areas where stock can periodically cross the waterway. Negotiations with landholders would include discussions about the size of the paddock and the timing and extent of grazing. Monitoring of resource condition trends would be part of a comprehensive evaluation of the strategies adopted and would contribute to further refinement of both the goal-posts and the strategies employed.

LESSONS ABOUT INTEGRATED RESEARCH

Multi-disciplinary teams

Working as part of a multi-disciplinary research program had many benefits, including providing access to landholder databases held by the spatial scientists in the Landscape Logic project, access to key informants, particularly with the support of the knowledge broking team in the Landscape Logic project, and regular interactions with scientists to explore the assumed links between property management and environmental outcomes. These interactions improved the efficiency of the research process and the quality of research outcomes. They included input into the development of the survey instrument, peer review of the draft technical report and stakeholder comments at workshops where preliminary findings were explored.

Bayesian Decision Networks

Bayesian Decision Networks (BDNs) were the primary integration tool used in the Landscape Logic project. Exposure of the social research team to BDNs had important benefits and suggested some lessons for their further application to social research. Influence diagrams, the first step in developing a BDN, proved to be a useful first step in integration in that this shared process helped to clarify concepts and understanding of causal relationships, identify states and thresholds for key variables and capture the contributions from different disciplines and practitioners. Exposure to BDNs also had the benefit of improving the research team's ability to communicate and explore the survey data with landholders and fellow researchers.

(a)

(b)

Figure 6.2: Two different approaches to riparian management after willow (*Salix fragilis*) removal. (a) Revegetated and fenced stream banks. (b) Uncontrolled stock access and little fringing vegetation.

However, efforts to apply BDN analysis to the Tasmanian riparian data were limited by the number of survey respondents, which was below the 200 cases considered critical for BDN analysis. In another case (Ticehurst *et al.* 2011), building and populating a BDN using survey data was more effective as there was a large dataset (n=503) and the social researchers and modellers developed a shared understanding of BDN analysis, the social dataset, protocols for accessing the database and the release and authorship of research products.

The role of social science in landscape research

The social research team was engaged after the Landscape Logic project had commenced, replacing the team contracted originally. This meant we were not involved in the formative processes of setting the research agenda and building relationships across the discipline groups. The prevailing view at the time we commenced was that the social research team would provide a 'service' role to the biophysical projects by providing data on the social context of specific areas of interest, to aid interpretation of data and uptake of results. By the conclusion of the project, however, the social research team had altered the perceptions of Landscape Logic researchers and partners on the role of carefully designed social research in understanding the processes that underpin landholder adoption, such as the role of social norms (Chapter 16, this volume), hence its role in scoping research issues and defining research questions. Our experience in Landscape Logic helped us identify needs for future integrated social research, including studies at catchment scale to model the extent to which predicted levels of adoption would be likely to result in achieving environmental condition targets.

REFERENCES

Australian Bureau of Statistics (2002) 'Measures of Australia's progress 2002 (Catalogue no. 1370.9)'. ABS, Canberra.

Barr N and Cary J (2000) 'Influencing improved natural resource management on farms'. Bureau of Rural Sciences, Canberra.

Cary J, Webb T and Barr N (2002) 'Understanding landholders' capacity to change to sustainable practices: insights about practice adoption and social capacity for change'. Bureau of Rural Sciences, Canberra.

Castle EN (2002) Social capital: an interdisciplinary concept. *Rural Sociology* **67**, 331–349.

Curtis A and De Lacy T (1996) Landcare in Australia: does it make a difference? *Journal of Environmental Management* **46**, 119–137.

Curtis A and Lefroy T (2010) Beyond threat and asset-based approaches to natural resource management in Australia. *Australasian Journal of Environmental Management* **17**, 6–13.

Curtis A and Mendham M (2011) Bridging the gap between policy and management of natural resources. In *Changing Land Management: Adoption of New Practices by Rural Landholders*. (Eds D Pannell and F Vanclay) pp. 377–397. CSIRO Publishing, Melbourne.

Curtis A and Robertson A (2003) Understanding landholder management of river frontages: the Goulburn Broken. *Ecological Management and Restoration* **4**, 45–54.

Curtis A, Race D, Sample R and McDonald S (2008) 'Management of water ways and adjoining land in the Mid-Goulburn River: landholder and other stakeholder actions and perspectives'. Report to the Goulburn Broken Catchment Management Authority. Institute for Land, Water and Society, Report no. 40. Charles Sturt University, Albury, NSW.

Curtis A, Race D and Sample R (2009) 'Management of riparian zones in Tasmanian agricultural catchments'. Technical Report no. 11. Landcape Logic, Hobart. <http://www.landscapelogicproducts.org.au/site/products/33-management-of-riparian-zones-in-tasmanian-agricultural-catchments>.

Pannell DJ, Marshall GR, Barr N, Curtis A, Vanclay F and Wilkinson R (2006) Understanding and promoting adoption of conservation technologies by rural landholders. *Australasian Journal of Experimental Agriculture* **46**, 1407–1424.

Sobels JA, Curtis A and Lockie S (2001) The role of Landcare group networks in rural Australia: exploring the contribution of social capital. *Journal of Rural Studies* **17**, 265–276.

Ticehurst JL, Curtis A and Merritt WS (2011) Using Bayesian Networks to complement conventional analyses to explore landholder management of native vegetation. *Environmental Modelling and Software* **26**, 52–65.

Vanclay F (1992) The social context of farmers' adoption of environmentally-sound farming practices. In *Agriculture, Environment and Society: Contemporary Issues for Australia.* (Eds G Lawrence, F Vanclay and B Furze) pp. 94–121. Macmillan, Melbourne.

Spatial diagnosis of catchment water quality: using multiple lines of evidence

Kirsten Verburg, Hamish Cresswell, Ulrike Bende-Michl, John Gibson and Peter Hairsine

SUMMARY

Targeting actions to improve water quality and designing monitoring to evaluate the effectiveness of those actions are challenges faced by catchment managers the world over. Both tasks require an understanding of how a catchment 'works' in terms of sediment and nutrient delivery into the stream network. This can be translated into five key questions.

1   Which nutrients or sediments contribute to water quality problems?
2   What is their origin?
3   Where in the catchment are their source areas?
4   Along which hydrological pathways are the materials transported to the waterways?
5   When are these processes likely to occur?

This chapter describes how multiple lines of evidence can be used to focus on these questions. A literature review and case study in the Duck River catchment in north-western Tasmania identified a range of different and often complementary types of information that could be used. These include water quality monitoring over space and time, tracer studies and various types of modelling. It was found that these approaches had different strengths as applied to each of the five key questions and that, to form a diagnosis, a combination of different lines of evidence was necessary. In many research studies, this process is applied informally. To make this more transparent we developed the *Guide to Spatial Diagnosis of Catchment Water Quality* as a framework for organising available evidence and integrating that knowledge into a spatial diagnosis. The case study from the Duck River catchment is used to demonstrate how the framework evolved, illustrate its use in organising evidence to support decision-making and design monitoring programs, and reflect on the value of the study for landholders and managers in that catchment. While this chapter focuses on sediment and nutrients, the approach and many of the methods are transferrable to other constituents.

INTRODUCTION

The challenge of targeting actions to address water quality issues within catchments and designing monitoring to evaluate the effectiveness of those actions was expressed by our natural resource management (NRM) partners at the inception of the Landscape Logic project as the following questions:

1. where in the landscape to target specific water quality-related NRM interventions;
2. how to decide on the appropriate water quality management options to use in an identified location;
3. how to predict the likely impact of interventions and assess their actual impact after implementation.

To answer these questions we required an understanding of how a catchment 'works' with respect to sediment and nutrient delivery into the stream network. It was therefore decided that our team would focus on the development of spatial conceptual catchment 'models'. The form of these conceptualisations was left open, but it was acknowledged that spatial conceptual catchment models are powerful in providing a focus for integrating different forms of knowledge. This was to facilitate communication with local stakeholders and provide a sound basis for local decision-making. The conceptual models can also highlight relevant processes that are not yet well understood.

Implicit in the development of spatial conceptual catchment models are a number of research challenges:

- how to identify and incorporate primary functional processes operating within a catchment;
- how to adequately represent sources, sinks and connectivity;
- how to maximise the use of existing data;
- what extra high-information content data could provide maximum value;
- the best approaches for field verification through nutrient monitoring.

As a special focus, our team explored the value of continuous high-frequency nutrient monitoring to help conceptualise pathways, sources and sinks. This activity was in response to international studies that had analysed high-frequency water quality monitoring data (daily or hourly) and suggested that what previously appeared to be random noise from monthly or less frequent monitoring may be attributable to processes (Harris and Heathwaite 2005; Jordan *et al.* 2005). This suggested that if high-frequency monitoring could indeed capture this process information, it might provide new insights into how catchments function (Kirchner *et al.* 2004) and identify the causes of water quality problems. We deployed high-frequency monitoring equipment capable of analysing samples *in situ* within the stream and on-site using a bank-side analyser.

At the start of the project we envisaged that the identification of primary functional processes and representation of sources, sinks and connectivity would make use of the 'response units' concept. Response units are landscape units forming spatial patterns that relate to dominant (or unique) hydrochemical process dynamics (Bende 1997). During the course of the project this approach was found useful for setting up a dynamic process model (the JAMS/J2000-S model; Bende-Michl *et al.* 2007). However, incorporating this approach into a spatial conceptual catchment model to disaggregate the catchment into zones differentiating expected nutrient sources and transport mechanisms proved more difficult. Alternative methods were developed for this purpose ('initial spatial conceptual modelling', and 'spatial source area likelihood estimation'). The response unit approach and related methods of tracking nutrient delivery from individual functional hydrochemical response units to a catchment outlet were therefore not pursued to the extent initially envisaged.

To provide a framework for the conceptualisation process, five key questions were identified.

1   Which forms of nitrogen, phosphorus or sediments contribute to water quality problems?
2   What is their origin?
3   Where in the catchment are their source areas?
4   Along which hydrological pathways are the materials transported to the waterways?
5   When are these processes likely to occur?

Answering these questions effectively constitutes the development of a spatial catchment diagnosis which, similarly to a medical diagnosis, combines lines of evidence and is used to inform management and to determine the best strategies for monitoring the effectiveness of interventions.

A literature review of methods to determine sources, sinks and pathways of nitrogen, phosphorus and sediments within catchments identified a myriad of methods. These methods were found to have different strengths in relation to the five questions above. In addition, the roles of different methods varied. Some methods helped to develop hypotheses, whereas other methods were more suited to hypothesis testing. It became clear that multiple lines of evidence were often needed to build an objective diagnosis. This led to the development of a multiple lines of evidence framework for spatial catchment diagnosis (Table 7.1; Verburg *et al.* 2009). Discussions with NRM partners in Landscape Logic showed that there was a need for clarity about which methods were most useful, and when and where. This led us to view our project as focusing on the development and evaluation of a number of methods for spatial catchment diagnosis and resulted in the development of *A Guide to Spatial Diagnosis of Catchment Water Quality* (Verburg *et al.* 2010; 'the Guide' from here on). The multiple lines of evidence framework is central to the Guide and can assist catchment managers in the cost-effective choice of methods.

Most of the method development and evaluation concentrated on a case study catchment in north-western Tasmania. The Duck River catchment was chosen following a thorough selection process (Cresswell and Lefroy 2007) which also influenced the focus of additional work by other projects in Landscape Logic (Chs 2, 3, and 4, this volume). The Guide includes a

Table 7.1: Multiple lines of evidence framework for spatial catchment diagnosis

Method	Constituent/ form	Origin	Source areas	Pathways	Timing
Initial spatial conceptual modelling	*	*	*	*	*
Spatial source area likelihood estimation			**	*	
Spatial snapshot surveys	*/**	*	**		*
Event monitoring	*/**			*/**	**
High-frequency monitoring	*/**	*		*/**	**
Isotope and tracer analyses		**	*	*/**	
Modelling	*	*	*	*	*

* Some information, or allowing hypotheses to be formed.
** Detailed information, area of strength.

worked example of its application in the Duck River catchment. Additional evaluations of a subset of methods were performed in the Jordan River catchment in south-eastern Tasmania and the Meander River catchment in northern Tasmania.

BUILDING A SPATIAL CATCHMENT DIAGNOSIS USING MULTIPLE LINES OF EVIDENCE

Detailed descriptions of the various methods and the lines of evidence they provide are included in the Guide (Verburg *et al.* 2010). Here we present a few examples that reflect on the method development and illustrate lessons from their application in the case study catchments.

INITIAL SPATIAL CONCEPTUAL MODELLING

Research and NRM projects usually start with gathering information from existing reports and maps. The difficulty with this process is that the different pieces of information have commonly been compiled with different intentions for different audiences. Combining the information into a spatial conceptual model that links what happens on the land with responses in-stream is therefore not straightforward. Hence the review of existing reports and maps seldom goes beyond summaries of the available information. Our work in this area focused on describing the methodology that an experienced soil and landscape scientist would use to synthesise the available information into an initial spatial conceptual model. We refer to this as an *initial* model, because it should be the starting point for further evaluation, indicating gaps in knowledge where other lines of evidence are needed. Because of this role, the spatial conceptual model does not need to be expressed as a complex diagram. A simple but structured narrative accompanied by maps, diagrams or photos will suffice.

Initial spatial conceptual models developed as part of the project followed a similar approach. First, existing reports and maps were studied with a view to disaggregating the catchment into zones that reflect similar geomorphology and soils or land use. For each zone, this information was interpreted in terms of likely hydrology and possible source areas and transport processes. Particularly useful were the reconnaissance soil maps and associated reports recently updated by the Tasmanian Department of Primary Industries, Parks, Water and the Environment (DPIPWE; map resources available at www.dpiw.tas.gov.au/inter.nsf/WebPages/EGIL-53E2YX?open). A detailed soil survey by Hubble and Bastick (1995) covered only a small part of the Duck River catchment, but helped significantly in the interpretation of the plains zones of the catchment. Also valuable were the land capability surveys, environmental flow assessments and groundwater conceptual model reports produced by DPIPWE. The desktop integration of existing information was followed by field visits with a local expert (where possible) for field verification of the model. The visual observations of these visits proved valuable, especially in relation to establishing the relative discharges of different streams and to assessing land use intensity and its impacts.

In the Duck River catchment, the disaggregation resulted in seven different zones (Fig. 7.1). While the hydrological interpretations of these zones were not entirely new, presenting the information in a spatially organised way and backed with evidence from reports and field observations did improve understanding of key source areas and transport processes within the catchment (Cresswell and Cotching 2011). It predicted that the central plains zones would be the most likely sources of nitrogen and phosphorus to the overall catchment loads, with saturated surface runoff and subsurface flow to drains being responsible for the high connectivity in this area. The initial spatial conceptual model also made it immediately obvious that any management decisions would need to take into account the relative position within the catchment (hills vs plains).

In the Jordan River catchment the initial spatial conceptual modelling provided fewer new insights. Many of its findings were already included in existing reports, including the Jordan Rivercare plan (Sprod *et al.* 2002) and the environmental flow assessment report (Davies *et al.* 2005). In this catchment, sources such as gullies and bank erosion dominate sediment supply and are relatively easily identified. This experience pointed out that the initial conceptual modelling approach, as outlined above, has more to offer in catchments where water quality

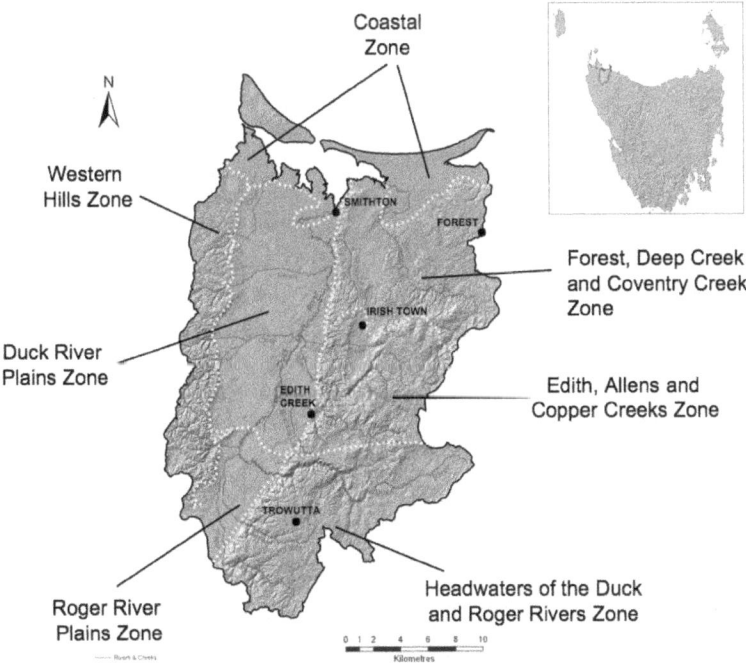

Figure 7.1: Location and broad catchment disaggregation of the Duck River catchment.

issues are related to diffuse sources. There it can provide a spatial assessment of likely sources and an understanding of the transport processes involved.

SPATIAL SOURCE AREA LIKELIHOOD ESTIMATION

The need for development of a methodology to identify source areas within the catchment was recognised early and defined as one of the milestones for our team. It was envisaged that this would be a rule-based approach, but the exact approach was less clearly defined. During the course of the project we explored various options. We first looked at using the response units concept, but found it difficult to come up with a meaningful disaggregation without first modelling the whole catchment with a process-based distributed model. As most catchment management organisations would not be in a position to model the catchment in detail, this did not seem a suitable approach. We then tried to set up some rules from first principles (e.g. road and farm tracks, stock access, runoff from wet areas near stream), but defining rules for their mapping was not straightforward.

A literature review carried out to aid our reflection on these approaches identified an index approach that was widely used in the US for mapping the risk of phosphorus loss on farms (Sharpley *et al.* 2003; Buczko and Kuchenbuch 2007). A number of factors that are known to indicate the availability of nutrient sources or the likelihood of transport to a receiving water body are rated. The factors are then combined using a weighted additive, multiplicative or multiplicative-additive approach. Within Australia, a similar approach was developed for nitrogen and phosphorus loss in Victoria by Melland *et al.* (2007) and converted to a spatial approach (still at farm scale) by Hill (2008). Spatial applications of similar approaches at the catchment scale were reported in two conference papers (Newham *et al.* 2002; Drewry *et al.* 2007; see Drewry *et al.* 2011 for an update). This approach seemed promising and we pursued the development of a methodology that would enhance the various factors with information from terrain analysis. Availability of a hydrologically sound digital elevation model proved to be a challenge, but this approach was abandoned mainly because the step of combining factors seemed too subjective. Many different versions of index approaches have been published and all use slightly different weights to combine not only the factors but also the impacts of different pathways. Justifications for the different weights are seldom given and, even when provided, it would be difficult to customise a model for a particular catchment.

So in the final year of the project we switched direction again and developed what is presented in the Guide as the spatial source area likelihood estimation (SSALE) method (Section 3.2, Verburg *et al.* 2010). The SSALE method uses a series of decision trees that identify areas with a high likelihood for source availability or a high likelihood for transport of the source material along different pathways. The decision trees are based on soil, land use, rainfall and terrain characteristics that are known to influence the availability, mobilisation and transport of nutrients and sediments. Many match factors used in the index approaches. The decision trees and the decision thresholds contained in them are, however, fully transparent and can be adjusted to reflect conditions specific to particular catchments (Fig. 7.2). They can be used in a non-technical context for natural resource or catchment management group discussions or be implemented in a GIS environment like ArcGIS. For this reason we developed an ArcGIS toolbox implementation of the method.

Our experience with the SSALE method in the Duck River catchment was that it helped narrow down the parts of the plains that were at most risk of nutrient losses (Fig. 7.3). Defining meaningful thresholds is, however, an iterative process that is quite closely linked with the

Transport – surface runoff

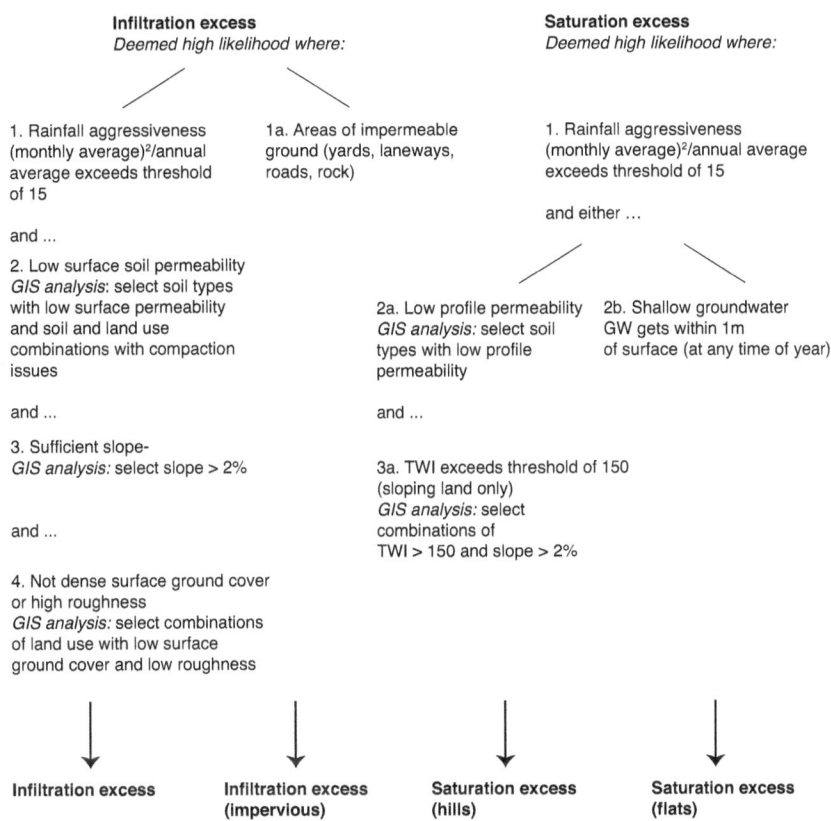

Figure 7.2: Spatial source area likelihood estimation (SSALE) decision tree for surface runoff with thresholds as set for the Duck River catchment. (TWI = Topographic Wetness Index)

initial conceptual model of a catchment. Therefore, the two methods cannot be regarded as two independent lines of evidence.

Being developed so late in the project, the SSALE methodology is still new and experience in setting thresholds needs to be built through its application in a range of different catchments. For now, the method outcomes should probably be considered more of a hypothesis than firm evidence (i.e. as marked * in Table 7.1).

SPATIAL SNAPSHOT SURVEYS

Spatial snapshot surveys, also referred to as longitudinal or synoptic sampling, are used to obtain information about the spatial location of point and non-point sources within the catchment. DPIPWE carried out a number of such surveys as part of its State of Rivers reports in the late 1990s and early 2000s. We quickly realised the value of these surveys, using the report for the Duck River catchment (DPIWE 2003) to prepare for our first reconnaissance trip in the catchment. Over the course of the case study, four snapshot surveys were carried out in the Duck River catchment with an expanding set of sampling sites (up to 50). These surveys were

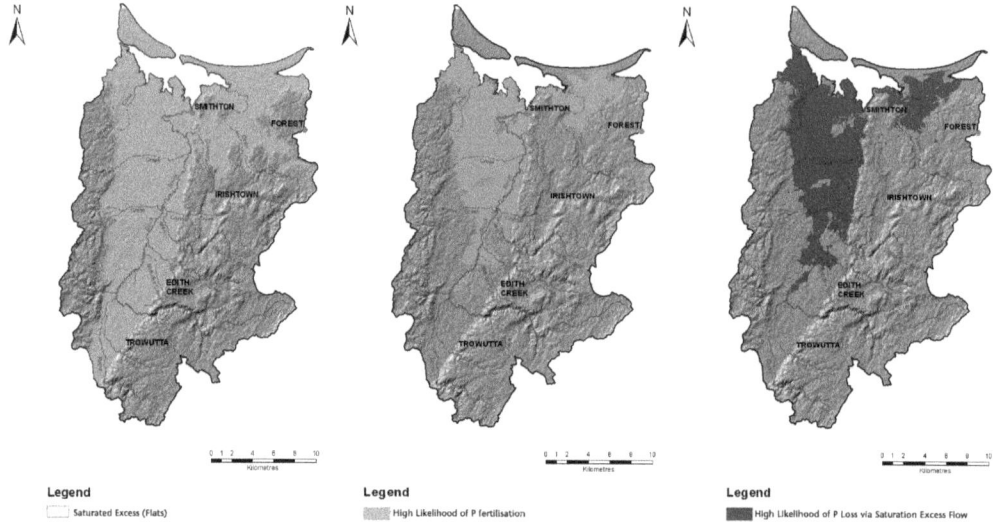

Figure 7.3: Derivation of critical source areas for phosphorus transport via saturation excess runoff. a) Areas of high likelihood of saturation excess runoff. b) Source availability likelihood. c) Combined critical source areas.

performed during stable base flow in summer (low flow conditions) and winter (high flow conditions). Samples were analysed for different forms of nitrogen and phosphorus, electrical conductivity, total suspended sediments, turbidity and pH, with dissolved organic carbon determined on a subset of samples.

Maps of the base flow concentrations of total phosphorus and nitrate (Fig. 7.4) confirmed the expectations from the initial conceptual modelling and source area analysis that transport and source likelihood differ between the plains and the hills in this catchment. High total phosphorus concentrations were predominantly found in the lower part of the Duck River plains; there were only a few 'hotspots' for total phosphorus in the hills and the location of these varied in the repeat snapshots. The snapshot surveys showed clear differences between summer and winter nitrate concentrations in most locations, except for a few sites in some streams just below the escarpment (Duck River and Spinks, Birthday, Copper and White Water creeks) that had high year-round nitrate concentrations.

Longitudinal analysis showed a marked increase in nitrate concentrations at the escarpment, accompanied by an increase in electrical conductivity (EC) and a decrease in dissolved organic carbon (DOC) (Fig. 7.5). Tributaries flowing into the Duck River above the escarpment could not explain the increase in EC. An influx of nitrate-rich, high-EC but low-DOC groundwater was suspected, which was consistent with the finding that the changes were more marked in summer. This was analysed using isotope and tracer analysis (see following section). It is interesting that in this case the sampling process itself contributed to our understanding, as visual observations of the changing water characteristics (clarity and colour) between sites during the survey prompted us to check what was happening at the escarpment.

As we did not collect discharge data in conjunction with the snapshot sampling, we could not do a load balance for every section of the river (Grayson *et al.* 1997), limiting our analysis of relative contributions to catchment loads. For assessment of these we had to rely on field

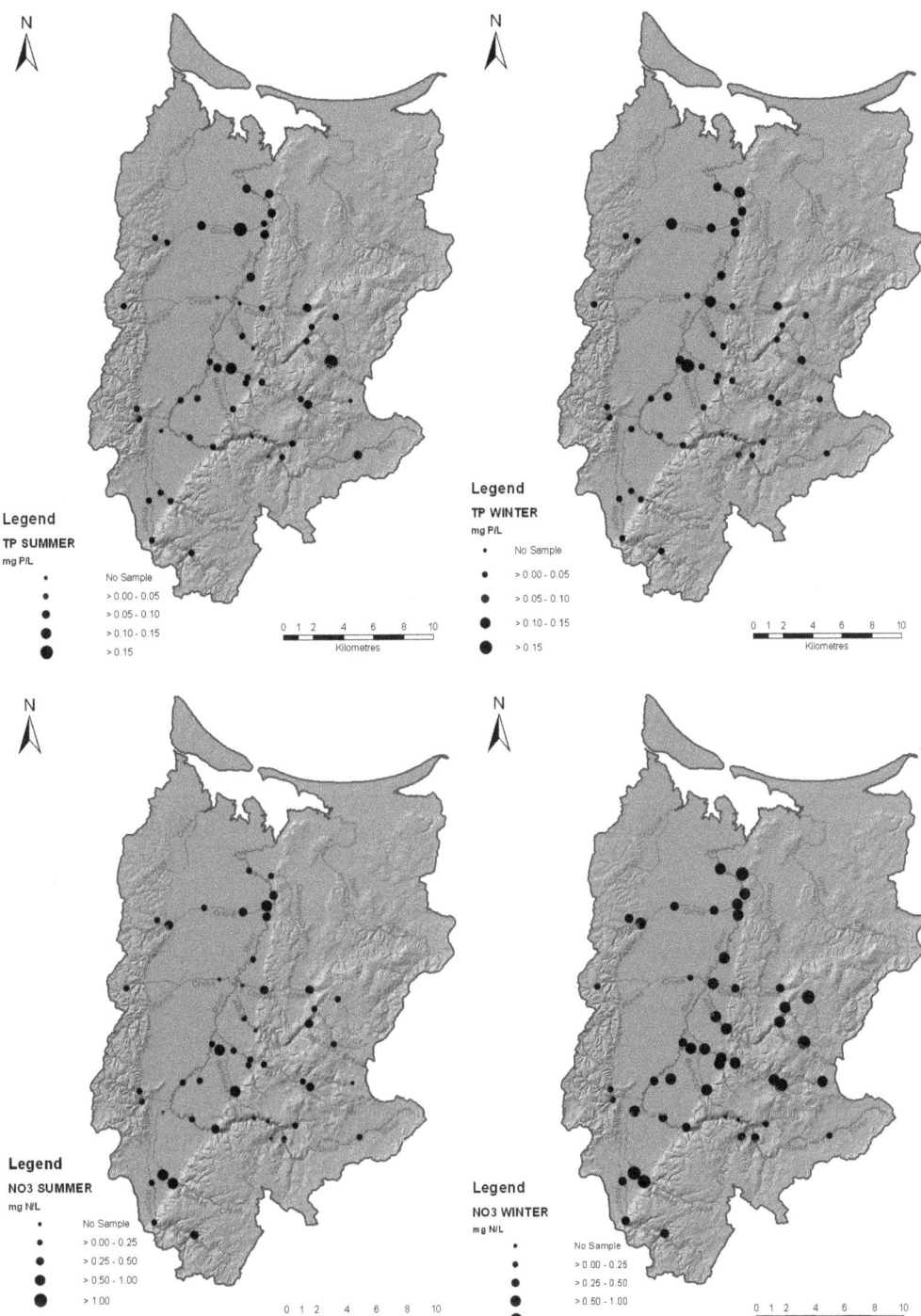

Figure 7.4: Total phosphorus (top) and nitrate (bottom) concentration observed during base flow in summer (March–April 2009) and winter (June 2009).

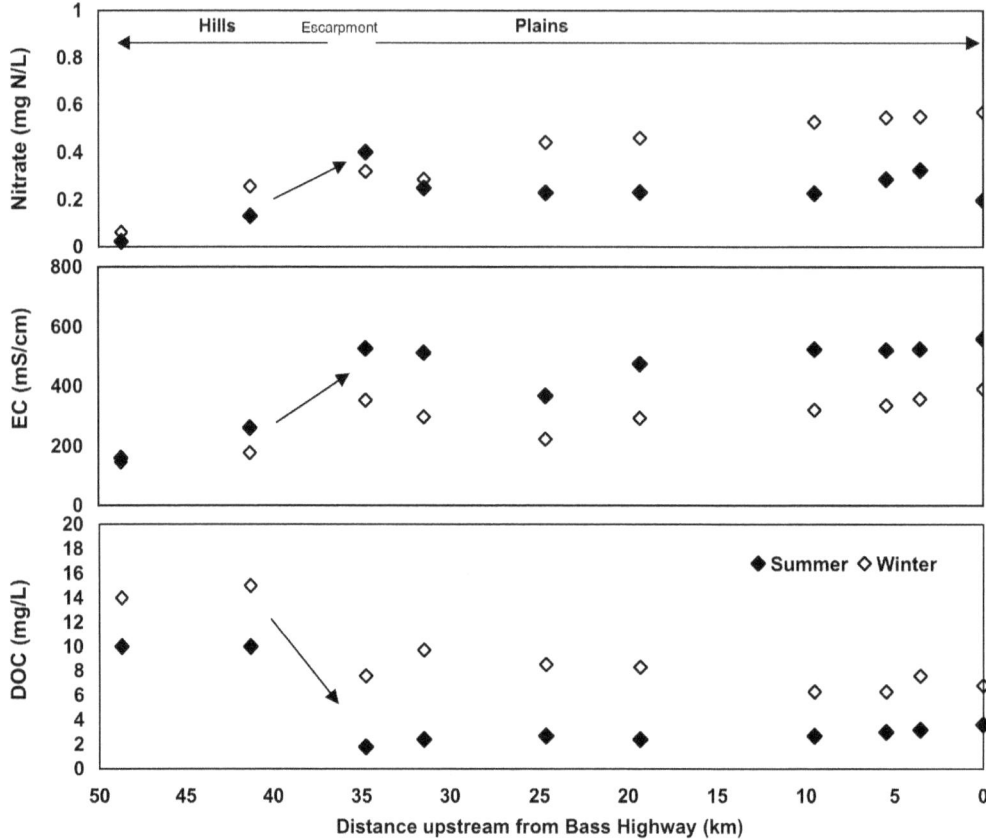

Figure 7.5: Duck River transects of nitrate, electrical conductivity (EC) and dissolved organic carbon (DOC) during the summer (March–April 2009) and winter (June 2009) snapshot surveys.

observations made for the initial conceptual model and a few targeted measurements made during the final snapshot survey. The fact that the method is only suitable for stable (base) flow conditions, and most export from the catchment occurs in response to rainfall events, to some extent justified the lack of more precise load estimations. The benefits of the surveys were mainly derived from the identification of source areas on the basis of sudden increases in concentration and of general spatial patterns. However, as relative event flow contributions may be different from relative base flow contributions, it may be advantageous to complement the spatial snapshot surveys with event monitoring in select locations. Time and budget did not allow this in our case study, except for observations made at the catchment outlet using the high-frequency nutrient monitoring equipment.

The snapshot samples were also analysed for the different forms of nitrogen and phosphorus. This proved useful when we analysed the data for clusters with similar chemical compositions, which often reflect the combined impact of subcatchment geology, climatic conditions and land use (Close and Davies-Colley 1990). This analysis highlighted the different nature of one of the tributaries (Geales Creek) joining the Duck River in the lower plains zone. As with other sampling sites in this zone, the water was characterised by elevated total phosphorus levels. The dissolved reactive phosphorus levels in Geales Creek were, however, uncharacteristically low. Turbidity was found to be extremely high at most sampling times. Field observations (red iron staining along the creek) and further geochemical analyses suggested that acid sulfate soils were likely to affect water quality in the Geales Creek subcatchment. The low dissolved reactive phosphorus levels were assumed to be linked to the formation of iron oxyhydroxide-phosphate complexes, which in turn contributed to the high turbidity levels.

ISOTOPE AND TRACER ANALYSES

When we discussed with local landholders the possibility of the occurrence of groundwater springs along the escarpment, they pointed to a spring on Birthday Creek. DPIPWE confirmed the presence of a large spring on Duck River and anecdotal evidence on the internet suggested the presence of springs in the Copper Creek area.

Isotopes and tracers can sometimes be used to identify the origins of nutrients and sediments and/or the source of the water that carries them. These methods work because the constituents that act like tracers (e.g. chloride, sodium, radon) and the isotopes of water (δ^2H, $\delta^{18}O$) or its constituents (e.g. $\delta^{15}N$) are conservative or behave in a predictable way.

In our case study, radon analysis of selected samples confirmed the inflow of groundwater into Duck River and Birthday Creek at the escarpment where radon activity readings were clearly elevated. Analysis of geochemical data from the same samples using ternary diagrams showed that the samples clustered into two groups (Fig. 7.6); those influenced by springs at the escarpment, and others. This provided convincing support for the hypothesis that nitrate-rich groundwater was flowing into Duck River and a number of creeks at the escarpment. Data for Copper, Whitewater and Coventry creeks from the State of Rivers report (DPIWE 2003) confirmed these creeks were also spring-fed with groundwater from the same or similar origin.

Knowing that groundwater discharging as springs was the source of high nitrate concentrations at the escarpment provided information about the pathway this nitrate had taken (Table 7.1), but did not tell the origin of the nitrate. The hills above the escarpment have been farmed for at least 80 years and the initial conceptual modelling showed that the soils are very permeable and rainfall is high (~1600 mm/yr). Nitrate leaching over an extended period could have reached deeper groundwater layers under these hills. There are also areas of karst which may have allowed faster transport to the groundwater body. However, the nitrate may have a natural origin, as there is evidence of high nitrate groundwater in other, pristine karst areas in Tasmania (Ian Houshold and Don Rockliff, DPIPWE, *pers. comm.*). Data from the nitrate isotope pilots did not enable resolution of this question. Most samples fell into the range of $\delta^{15}N$ where fertiliser and soil mineralisation origins cannot be distinguished (Fig. 7.7). Samples from sites close to the springs and from the springs themselves did not appear to have a different nitrate isotope signature. Other tracers or markers may need to be applied to explore the origin of the nitrate.

The nitrate isotope analyses did suggest that three samples (taken in winter) were likely affected by animal waste. This included samples from a drain in the plains near Birthday Creek under dairy land use and from the creek itself downstream of that point. The analysis demonstrated that, while nitrate concentration levels in Birthday Creek were already high due to its groundwater spring origin, additional nitrate was contributed in the plains downstream of drains, most likely from a dairy-related origin.

HIGH-FREQUENCY MONITORING

The project also explored the value of continuous high-frequency nutrient monitoring to help conceptualise pathways, sources and sinks. Various high-frequency nutrient analysers are available (Bende-Michl and Hairsine 2010; Bende-Michl *et al.* 2010), although few have been

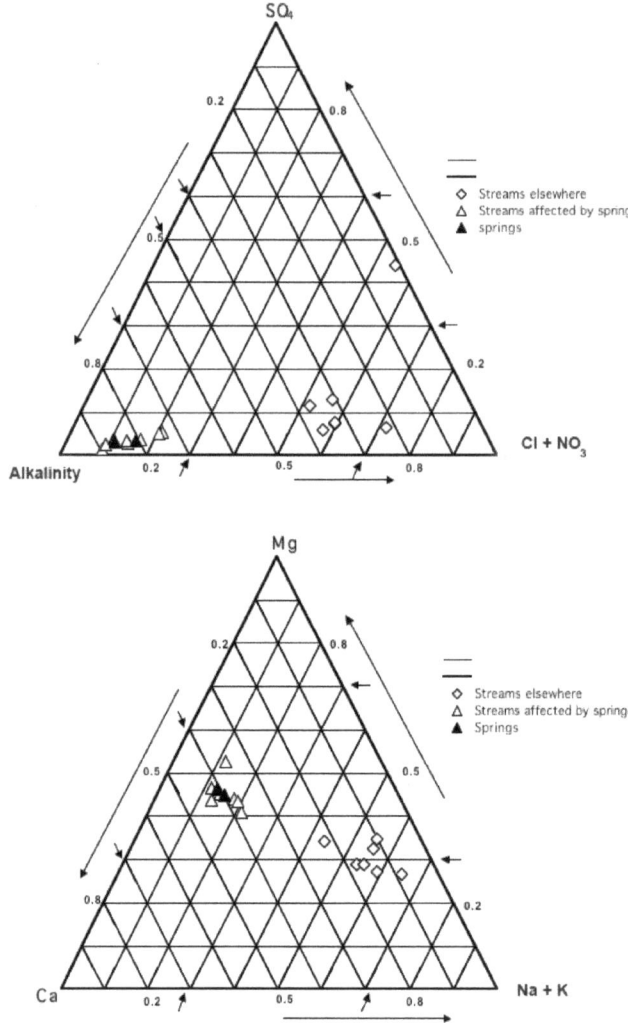

Figure 7.6: Ternary diagrams of geochemistry: data from March 2010 spatial snapshot survey combined with State of Rivers data (DPIWE 2003) for Copper, Coventry and Whitewater creeks (all three affected by springs).

tested or used for unattended operation in environmental settings, such as along the bank of a river. Two sites were selected. One was on Duck River near the Scotchtown weir (the flow-monitoring point of the Tasmanian DPIPWE), located upstream of tidal influences. The second site was on Geales Creek, a tributary flowing into Duck River just upstream of Sotch-town weir. Both sites were instrumented with a UV/Vis spectrophotometer, an optical *in situ* instrument that measures the absorbance of light in the UV and visible ranges. This instrument provided nitrate concentrations at a five-minute time-scale. A second instrument, a wet chemistry analyser, was installed in a shed at the Duck River site. Samples were pumped to the instrument, passed through a 20 μm filter then analysed for total nitrogen and phosphorus, reactive phosphorus, nitrate and ammonia at an (approximately) hourly time-scale. Operation of the wet chemistry analyser proved time-consuming and a high degree of technical skill and

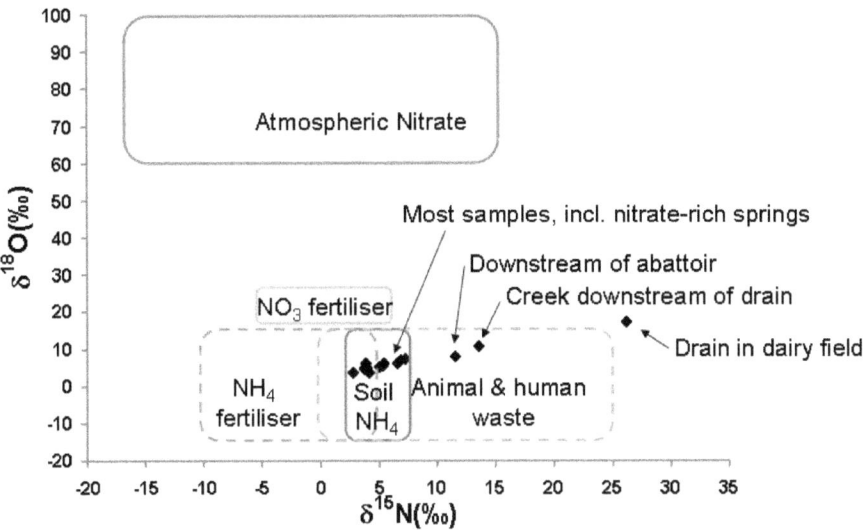

Figure 7.7: Dual $\delta^{15}N$-$\delta^{18}O$ nitrate isotope data from the first pilot (August 2009).

resourcefulness was required to fine-tune the instruments to their operating environment and to diagnose instrument failures.

Despite these operational issues, the study collected a detailed two-year dataset that proved useful in refining the conceptual model. In particular, it provided a better understanding of the timing, duration and magnitude of nutrient exports from the catchment and allowed these to be linked to different hydrologic conditions (Ulrike Bende-Michl *et al.*, CSIRO, unpub. data).

High-frequency measurements of discharge and phosphorus concentration exhibited changing patterns during the winter high flow seasons. In 2008 the early winter events were relatively small but nevertheless presented the highest phosphorus concentrations (Fig. 7.8). The first five of 20 events during that winter season contributed almost half of the winter phosphorus export. This suggests that during the preceding summer there was a build-up of nutrients in the landscape, readily available for transport with the first few events. Later in the season transport

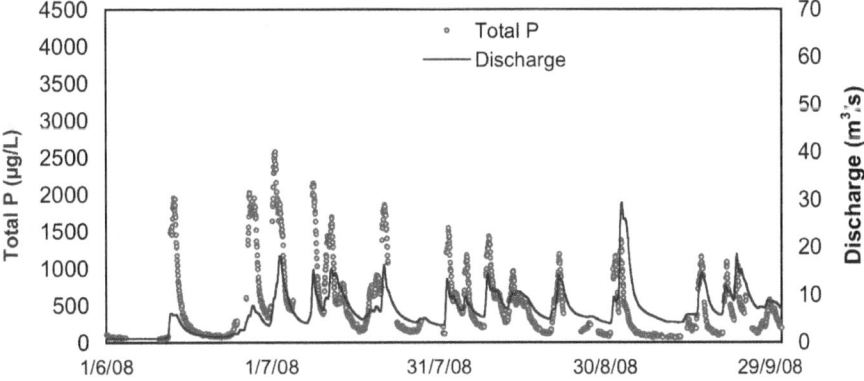

Figure 7.8: Discharge and phosphorus concentration during the high flow period at Scotchtown station, June–September 2008.

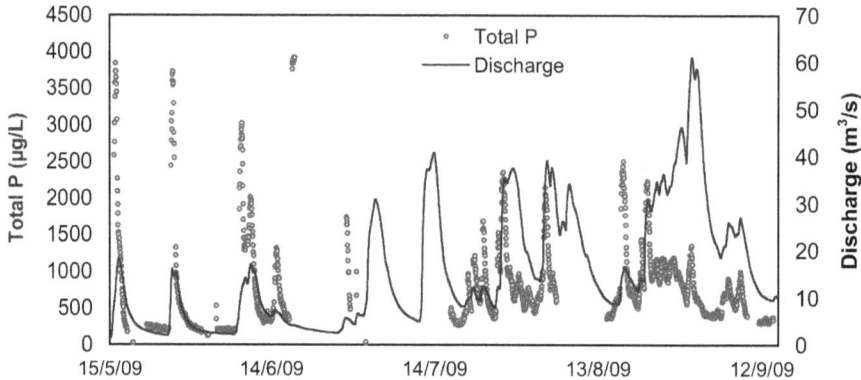

Figure 7.9: Discharge and phosphorus concentration during the high flow period at Scotchtown station, May–September 2009.

became supply-limited, with similarly sized discharge events transporting less phosphorus. Similar decreasing trends in peak concentrations were observed in runoff events from a small 12 ha dairy catchment in the neighbouring Montagu River catchment (Holz 2010).

The second year of the study was significantly wetter and significantly increased the amount of phosphorus export. The first events of 2009 again had the highest phosphorus concentrations for the year (Fig. 7.9). In July–August 2009 consistently wet weather is presumed to have caused more parts of the catchment to become connected, increasing phosphorus supply and maintaining a high level of phosphorus export. As in 2008, a very large event late in the season clearly became supply-limited early, although in this case the very high flows later in the event may have expanded the contributing catchment area even further, causing another peak of phosphorus export. The Duck River catchment is clearly a system in which winter phosphorus export dynamics are governed by a build-up of phosphorus supply during dry weather periods and rapid export during events, with the contributing catchment area determined by the sequence and magnitude of events.

Nitrate during winter was equally responsive to flow and its dynamics were event-driven. Nitrate concentrations typically increased with discharge, which is different from the dilution effect observed in the small 12 ha catchment (Holz 2010). This may be related to the combination of different hydrological pathways in the larger catchment. The possibility that this relates to the size of the catchment is also being explored. Some supply limitation over time was observed, but this was not as distinct as for phosphorus.

Ammonia concentrations during winter were still event-driven to some extent but were not systematically related to discharge magnitude. Nor did they exhibit a pattern like that observed for phosphorus and nitrate. The magnitudes of the concentration peaks were more random and were presumably governed by incidental supply (Granger *et al.* 2010). The ammonia peaks were shorter in duration than those of phosphorus or nitrate.

During the low flow period, the high-frequency monitoring identified diurnal variations in nutrient concentrations. These variations were not regular and seemed to reflect a mixture of riverine nutrient transformations (possibly including the growth of algae within the river during daylight hours) and transfers from point sources.

The improved understanding gained from high-frequency monitoring was that fluctuations in nutrient concentrations during winter are much larger than previously observed and that the magnitude of the nutrient releases is strongly affected by hydrological conditions.

Understanding nutrient supply and the contributing catchment area will be critical for design of improved management. In relation to monitoring design, the findings emphasise the importance of capturing the smaller early winter flow events that currently may not trigger flood sampling due to their lower discharge but that feature high phosphorus concentrations.

MODELLING

The complementary role of three catchment water quality models used in Landscape Logic projects were explored by Bende-Michl *et al.* (2009). The benefits of a lumped, semi-distributed conceptual catchment model are rapid estimations of sediment and nutrient loads at subcatchment and catchment scales, while those of a fully distributed process model are aimed at detailed process understanding of source areas and nutrient delivery. Some models can contribute to a spatial catchment diagnosis, either in developing hypotheses or testing them. For example, a distributed model set up for a subcatchment could explore the stream responses to nutrient transport via different pathways, and compare these with the conclusions drawn from high-frequency monitoring.

Our experience in the Duck River case study has highlighted, however, that it is important to use multiple lines of evidence when model predictions and observations do not match. The acid sulfate soil issues in the Geales Creek subcatchment emerged only after a number of independent observations were combined. Previous simulation with a lumped, semi-distributed conceptual model had identified the need to increase model coefficients describing export from this subcatchment (Broad and Cotching 2009). It is important not to jump to conclusions on the basis of such simulations alone and it is critical that the reasons for the increased export coefficients are explored. The high nitrate inputs from groundwater springs are another issue that would likely have been missed if we relied on modelling alone. A model is an excellent way to integrate existing knowledge and understanding, but each model application is based on an underlying spatial conceptual model. Ideally, this conceptual model should be based on one or more of the other lines of evidence in Table 7.1.

SYNTHESIS: SPATIAL CATCHMENT DIAGNOSIS

It is difficult to be prescriptive about how to assemble a spatial catchment diagnosis. Usually the knowledge gained from the application of different methods and consideration of how this adds to the evidence (Table 7.1) will build a picture of what is happening in the catchment. Documenting the evidence within the multiple lines of evidence framework can be a useful way to formalise this process (e.g. Table 4.2 of the Guide) or the findings can be presented in a flow diagram to illustrate how the 'story' evolved (e.g. Figure 4.26 of the Guide).

The question is often raised of what to do with contradictory lines of evidence. A literature review highlighted that, in other fields where multiple lines of evidence are used, some have taken a (semi-) quantitative approach with relative weights assigned to the different lines of evidence on the basis of study design, number of reference or control sites, for example, or study quality (Norris *et al.* 2005). Most studies, however, use informal applications with 'best professional judgement' (Burton *et al.* 2002; Chapman *et al.* 2002) providing the synthesis. The spatial diagnosis of catchment water quality lends itself best to informal application, especially as it covers multiple aspects (the five key questions) and spatially distributed processes. Most often we are not trying to prove or disprove one hypothesis, but are building up a consistent picture or conceptual model.

Nevertheless, it can be useful to reflect on the accumulated evidence, pose some hypotheses and test them. A possible approach is shown in Table 7.2, adapted from a method used by

Table 7.2: Two hypotheses on the origin of nitrate in Birthday Creek and lines of evidence supporting (+) or not supporting (-) those hypotheses

Background: Birthday Creek flows from the escarpment to Duck River through flat dairy country. High nitrate concentrations were found both upstream just below the escarpment and downstream before the junction with Duck River. The available lines of evidence were checked against two hypotheses on the origin of nitrate in Birthday Creek	Dairy origin	Groundwater origin
Initial spatial conceptual modelling: groundwater influx not considered, surface runoff and subsurface flow via drains from dairy land considered likely pathway and origin	+	–
Spatial snapshot survey: concentrations of nitrate at both sites in Birthday Creek found to be similar, suggesting nitrate levels were already high when creek entered dairy land use area	–*	+*
Spatial snapshot survey: combined analysis of nitrate, EC and DOC provided strong suggestions of influx of high nitrate groundwater, potentially affecting both sites	–*	++
Geochemical analysis: geochemistry of samples at both sites fitted within the cluster of samples from springs and spring-fed creeks	–	++
Radon isotopes: radon analysis confirmed influx of groundwater near upstream site	0	++
Nitrate isotopes: nitrate isotopes confirmed presence of nitrate from animal waste origin mixed in at the downstream site, especially in winter	++ (downstream site)	– (downstream site)
Consistency of evidence		The evidence for a dairy origin of nitrate initially appears inconsistent, but a modified conceptual model that allows for nitrate from a groundwater (spring) origin at the upstream site and an additional source of nitrate from a dairy origin at the downstream site resolves this

* The fact that both sampling sites on Birthday Creek had similarly high nitrate concentrations could, in absence of flow measurement, be interpreted as supporting a common origin at the base of the escarpment. If flow had been measured and a longitudinal analysis of loads carried out, the conclusions may have been different.

the Causal Analysis and Diagnosis Decision Information System (CADDIS; www.epa.gov/ caddis) developed by the US EPA to identify stressors that cause biological impairments in aquatic ecosystems (US EPA 2000). Available evidence can refute or diagnose outright, or support or weaken a case to various degrees. Scores are then evaluated in terms of consistency between lines of evidence. Where lines of evidences appear to contradict, this can relate to uncertainties inherent in the method, sampling procedures or interpretation, but it is also possible that the conceptual model is not complete. This should be an incentive to revisit the conceptual model, rather than reject the evidence.

CONCLUSIONS

The development of the Guide and its application to the Duck River catchment showed that understanding of catchment function improved when available water quality data were interpreted within the context of a spatial conceptual model of the catchment and complemented by additional investigative monitoring. Both the high-nitrate spring water entering several streams at the escarpment in the Duck River catchment and the acid sulfate soil issues in the

Geales Creek subcatchment would not have been identified without application of multiple lines of evidence. Both have implications for future management of the catchment.

That conceptual understanding evolves as new lines of evidence become available was illustrated by the findings of high nitrate in Birthday Creek. The high nitrate was initially attributed to local intensive farming operations, but identifying a year-round high-nitrate spring further upstream made this link less firm. Subsequent analyses with a dual $\delta^{15}N$-$\delta^{18}O$ nitrate isotope technique changed understanding again, with an animal waste origin for the nitrate at the downstream site increasingly likely, at least in winter. This example clearly demonstrates that further investigation of seemingly contradictory lines of evidence is better than rejection or selective use of evidence.

The evolving catchment understanding and the rigorous and transparent process of using multiple lines of evidence made a significant impression on local landholders and industry representatives. When faced with the problem of MCPA (herbicide) detections in the river towards the end of our work, they were motivated to advocate a similarly rigorous approach to identifying source areas, pathways and timing for MCPA delivery to the river. While the Guide focuses on nitrogen, phosphorus and sediments, its approach and many of the methods are transferrable to other constituents.

From an NRM agency perspective, the Guide provides staff with a framework for developing and defending their investment decisions using a transparent and systematic approach. It is also likely to contribute to building the science capability of staff.

Reflecting on our team's work, it is clear that the focus was set by our end user partners at the start. What evolved was our understanding of what their questions really meant and the implications for project outputs likely to be adopted by the NRM organisations. The fact that the NRM organisations were with us on this journey, providing feedback along the way, contributed greatly to the development of our work. Increased clarity of goals also came from presenting many times to various audiences. Explaining our directions and approach was a very effective tool for organising our thoughts and getting to the essence of the task.

The value of the Landscape Logic project was that its integration of science and NRM partners forced us to reflect on how science could most effectively assist managers make decisions on choice of catchment management actions and monitoring. Being involved with the users of our research emphasised the need for clarity, which is often seriously lacking in scientific information typically provided to them. It is hoped that the Guide to spatial diagnosis of catchment water quality and the multiple lines of evidence framework that underpins it will improve the application of scientific information to NRM investment.

ACKNOWLEDGEMENTS

This study was funded by the Commonwealth Environment Research Facilities (CERF) Program, through the Australian Department of the Environment, Water, Heritage and the Arts (DEWHA) and by CSIRO. We would like to thank the following people: Seija Tuomi, Chris Drury, Danny Hunt and Gordon McLachlan (CSIRO) and Rob Wells for their help with sampling, instrument installation and maintenance; James Foley, John Gallant and Trevor Dowling (CSIRO), Lachlan Newham (ANU), Gang-Jun Liu and Grant Dickins (RMIT) for help with GIS and terrain analysis; Ted Lefroy, Greg Pinkard, Regina Magierowski, Greg Holz (University of Tasmania), Aniele Grun (NRM South), Andrew Baldwin and Debbie Searle

(NRM North), Sue Botting (Cradle Coast NRM), Pat Feehan (Goulburn Broken CMA), Bill Cotching (TIAR), Kate Hoyle, Kate Wilson, Donald Rockcliff and Ian Houshold (DPIPWE), Carol Kendall, Megan Young and Steven Silva (USGS) and Brent Henderson (CSIRO) for helpful discussions and/or input into selected methods. Comments by Warren Hicks and Olga Barron on an earlier draft of the chapter were also much appreciated.

REFERENCES

Bende U (1997) Regional hydrochemical modelling by delineation of chemical hydrological response units (CHRUs) within a GIS: an approach of observing man-made impacts in the Broel River catchment (Germany). *Mathematics and Computers in Simulation* **43**, 305–312.

Bende-Michl U and Hairsine PB (2010) A systematic approach to choosing an automated nutrient analyser for river monitoring. *Journal of Environmental Monitoring* **12**, 127–134.

Bende-Michl U, Kemnitz D, Helmschrot K, Krause P, Cresswell H, Kralisch S, Fink M and Flügel WA (2007) Supporting natural resources management in Tasmania through spatially distributed solute modelling with JAMS/J2000-S. In *MODSIM 2007 International Congress on Modelling and Simulation*. December 2007. (Eds L Oxley and D Kulasiri) pp. 2354–2360. Modelling and Simulation Society of Australia and New Zealand. <http://www.mssanz.org.au/MODSIM07/papers/43_s47/SupportingNaturals47_Bende-Michl_.pdf?>.

Bende-Michl U, Broad ST and Newham LTH (2009) Complementary water quality modelling to support natural resource management decision making in Australia. In *18th World IMACS Congress and MODSIM09 International Congress on Modelling and Simulation*. July 2009. (Eds RS Anderssen, RD Braddock and LTH Newham) pp. 2342–2348. Modelling and Simulation Society of Australia and New Zealand and International Association for Mathematics and Computers in Simulation. <http://www.mssanz.org.au/modsim09/F12/bende-michl.pdf>.

Bende-Michl U, Verburg K, Hairsine PB, Tuomi S and Cresswell HP (2010) 'High frequency nutrient monitoring: considerations for instrument selection'. Landscape Logic Fact-sheet.

Broad ST and Cotching WE (2009) Assessing the spatial variation of dairy farm total phosphorus losses in the Duck River, NW Tasmania. In *18th World IMACS Congress and MODSIM09 International Congress on Modelling and Simulation*. July 2009. (Eds RS Anderssen, RD Braddock and LTH Newham) pp. 3471–3477. Modelling and Simulation Society of Australia and New Zealand and International Association for Mathematics and Computers in Simulation. <http://www.mssanz.org.au/modsim09/I8/broad.pdf>.

Buczko U and Kuchenbuch RO (2007) Phosphorus indices as risk-assessment tools in the U.S.A. and Europe: a review. *Journal of Plant Nutrition and Soil Science* **170**, 445–460.

Burton GA, Chapman PM and Smith EP (2002) Weight-of-evidence approaches for assessing ecosystem impairment. *Human and Ecological Risk Assessment* **8**, 1657–1673.

Chapman PM, McDonald BG and Lawrence GS (2002) Weight-of-evidence issues and frameworks for sediment quality (and other) assessments. *Human and Ecological Risk Assessment* **8**, 1489–1515.

Close ME and Davies-Colley RJ (1990) Baseflow water chemistry in New Zealand rivers 2. Influence of environmental factors. *New Zealand Journal of Marine and Freshwater Research* **24**, 343–356.

Cresswell HP and Lefroy EC (2007) 'Selecting catchments in Tasmania for the nutrient and sediment management project'. Technical Report no. 6. Landscape Logic, Hobart.

Cresswell HP and Cotching WE (2011) 'Duck River catchment conceptual model'. Technical Report. Landscape Logic, Hobart.

Davies P, Koehnken L, Barker P and Cook L (2005) 'Jordan River: environmental flow regime assessment. Freshwater systems'. Technical advice on water. North Barker and Associates, Hobart.

DPIWE (2003) 'State of the River Report for the Duck River Catchment. Water Assessment and Planning Branch'. Technical Report no. WAP 03/08. Dept of Primary Industries, Water and the Environment, Hobart.

Drewry JJ, Newham LTH and Greene RSB (2007) An index-based modelling approach to evaluate nutrient loss risk at catchment-scales. In *MODSIM 2007 International Congress on Modelling and Simulation*. December 2007. (Eds L Oxley and D Kulasiri) pp. 2326–2332. Modelling and Simulation Society of Australia and New Zealand. <http://www.mssanz.org.au/MODSIM07/papers/43_s47/AnIndex-Baseds47_Drewry_.pdf>.

Drewry JJ, Newham LTH and Green RSB (2011) Index models to evaluate the risk of phosphorus and nitrogen loss at catchment scales. *Journal of Environmental Management* **92**, 639–649.

Granger SJ, Bol R, Anthony S, Owens PN, White SM and Haygarth PM (2010) Towards a holistic classification of diffuse agricultural water pollution from intensively managed grasslands on heavy soils. *Advances in Agronomy* **105**, 83–115.

Grayson RB, Gippel CJ, Finlayson BL and Hart BT (1997) Catchment-wide impacts on water quality: the use of 'snapshot' sampling during stable flow. *Journal of Hydrology* **199**, 121–134.

Harris G and Heathwaite AL (2005) Inadmissible evidence: knowledge and prediction in land and riverscapes. *Journal of Hydrology* **304**, 3–19.

Hill PA (2008) Managing N losses in rural landscapes. The use of spatial information and risk management. Frameworks in balancing production and conservation: a dairy case study. PhD thesis. Australian National University, Canberra.

Holz GK (2010) Sources and processes of contaminant loss from an intensively grazed catchment inferred from patterns in discharge and concentration of thirteen analytes using high intensity sampling. *Journal of Hydrology* **383**, 194–208.

Hubble GD and Bastick C (1995) 'The soils and land use of Mowbray Swamp and bordering heathland, north-west Tasmania'. Divisional Report no. 122. CSIRO Division of Soils.

Jordan P, Arnscheidt J, McGrogan H and McCormick S (2005) High-resolution phosphorus transfers at the catchment scale: the hidden importance of non-storm transfers. *Hydrology and Earth System Sciences* **9**, 685–691.

Kirchner JW, Feng XH, Neal C and Robson AJ (2004) The fine structure of water-quality dynamics: the high-frequency wave of the future. *Hydrological Processes* **18**, 1353–1359.

Melland R, Smith A and Waller R (2007) *Farm Nutrient Loss Index. An Index for Assessing the Risk of Nitrogen and Phosphorus Loss for the Australian Grazing Industries*. User manual for FNLI Version 1.18. Dept of Primary Industries, Ellinbank, Victoria.

Newham LTH, Jakeman AJ, Letcher RA, Heathwaite AL, Smith CJ and Large D (2002) Integrated water quality modelling: Ben Chifley Dam Catchment, Australia. In *Integrated Assessment and Decision Support. Proceedings of the 1st Biennial Meeting of the International Environmental Modelling and Software Society*, June 2002. pp. 275–280. <http://www.iemss.org/iemss2002/proceedings/pdf/volume%20uno/224_newham.pdf>.

Norris R, Liston P, Mugodo J, Nichols S, Quinn G, Cottingham P, Metzeling L, Perriss S, Robinson D, Tiller D and Wilson G (2005) Multiple lines and levels of evidence for detecting ecological responses to management intervention. In *Proceedings of the 4th Australian Stream Manage-*

ment Conference: Linking Rivers to Landscapes. (Eds ID Rutherford, I Wisznieuwski, MJ Askey-Doran and R Glazik) pp. 456–463. Dept of Primary Industries, Water and the Environment, Hobart.

Sharpley AN, Weld JL, Beegle DB, Kleinman PJA, Gburek WJ, Moore PA Jr and Mullins G (2003) Development of phosphorus indices for nutrient management planning strategies in the United States. *Journal of Soil and Water Conservation* **58**, 137–152.

Sprod D, Paxton M and Hammond R (2002) 'Jordan River rivercare plan'. Southern Midlands Council, Oatlands, Tasmania.

US Environmental Protection Agency (2000) 'Stressor identification guidance document'. EPA-822-B-00-025. Environment Protection Agency, Washington.

Verburg K, Cresswell HP, Bende-Michl U and Hairsine PB (2009) Using multiple lines of evidence to guide the management of catchment water quality. In *18th World IMACS Congress and MODSIM09 International Congress on Modelling and Simulation.* July 2009. (Eds RS Anderssen, RD Braddock and LTH Newham) p. 2464. Modelling and Simulation Society of Australia and New Zealand and International Association for Mathematics and Computers in Simulation. <http://www.mssanz.org.au/modsim09/F12/verburg.pdf>.

Verburg K, Cresswell HP and Bende-Michl U (2010) 'A guide to spatial diagnosis of catchment water quality'. Technical Report no. 20. Landscape Logic, Hobart.

Lessons from integrated bio-economic modelling in the George catchment, Tasmania

Marit Kragt

SUMMARY

Catchment-scale natural resource management (NRM) involves complex decisions that affect a wide variety of values, issues and stakeholders. Designing efficient NRM policies requires assessment of the environmental impacts as well as the costs and benefits of management interventions in an integrated manner. An integrated assessment approach provides useful guidance on bringing together issues and knowledge from multiple disciplines and stakeholders. Despite the need for integrated assessment, there are few comprehensive studies that integrate biophysical models with rigorous economic cost–benefit analyses. Cost–benefit analysis is an economic framework to assess and compare the total social costs and benefits of management interventions. However, the environmental modelling that has underpinned cost–benefit analysis has typically been poor, reducing its credibility and usefulness when evaluating the efficiency of alternative policy outcomes.

This study aims to demonstrate an approach to better integrate scientific data about environmental changes with economic information in a cost–benefit framework. We used a variety of tools to inform the integrated assessment process, such as the CatchMODS hydrological modelling framework, choice experiment non-market valuation surveys, and expert interviews, and developed a Bayesian network (BN) model to evaluate the costs and benefits of different NRM actions. The study highlights the challenges involved with integrating multi-disciplinary knowledge about hydrological, ecological and economic systems. Scientists and economists have different ways of describing a system and typically use different variables to describe similar processes. An important lesson from the model development process was the need for interdisciplinary workshops and experienced facilitators to improve communication and knowledge exchange between different disciplines. The graphical depiction of knowledge in BNs contributed to bringing together scientific and economic information.

INTRODUCTION

Natural resource management typically involves complex decision problems. There may be a variety of environmental, social and economic issues and values at stake, with perspectives diverging between multiple stakeholders. It is widely acknowledged that catchment NRM requires 'whole of systems' approaches to address the multiple impacts of policy decisions (Letcher and Giupponi 2005) and integrate these into an overall management philosophy, process and plan (Dept of Environmental Affairs and Tourism 2000). Integrated catchment management thus requires an analysis of the environmental, social and economic impacts of management changes, in order to work towards the goal of balancing economic development with environmental protection.

The objective of the study described in this chapter was to develop a modelling tool that links biophysical and economic knowledge in one framework. The tool was developed for a case study catchment in Tasmania. It sought to incorporate water quality changes in response to land management and land use changes, impacts on ecosystem assets as a result of those management and water quality changes, and economic changes focusing on management costs and non-use values. An integrated Bayesian network (BN) model was developed that enables an evaluation of the trade-offs between multiple impacts in a cost–benefit analysis.

RESEARCH FRAMEWORK

This study used an integrated assessment approach to assess the changes in environmental and socio-economic systems resulting from catchment NRM changes. Integrated assessment seeks to integrate and share scientific and stakeholder knowledge drawn from multiple disciplinary backgrounds, in order to evaluate a decision problem from different perspectives and provide support for its solution (TIAS 2009). Different tools, methods and procedures are needed to inform the different phases of the assessment process, for example biophysical modelling tools, participatory methods and cost-effectiveness analysis (De Ridder *et al.* 2007). Our research framework broadly followed an iterative integrated assessment approach (Fig. 8.1), which recognises that catchment systems continuously evolve, changing the context and leading to the emergence of new issues and values.

Integrated catchment management calls for targeted investments to achieve the greatest environmental, social and financial net benefits (NWI 2004). If integrated assessment is to support the development of economically efficient NRM, all the marginal social costs and

Research step	Tools and methods

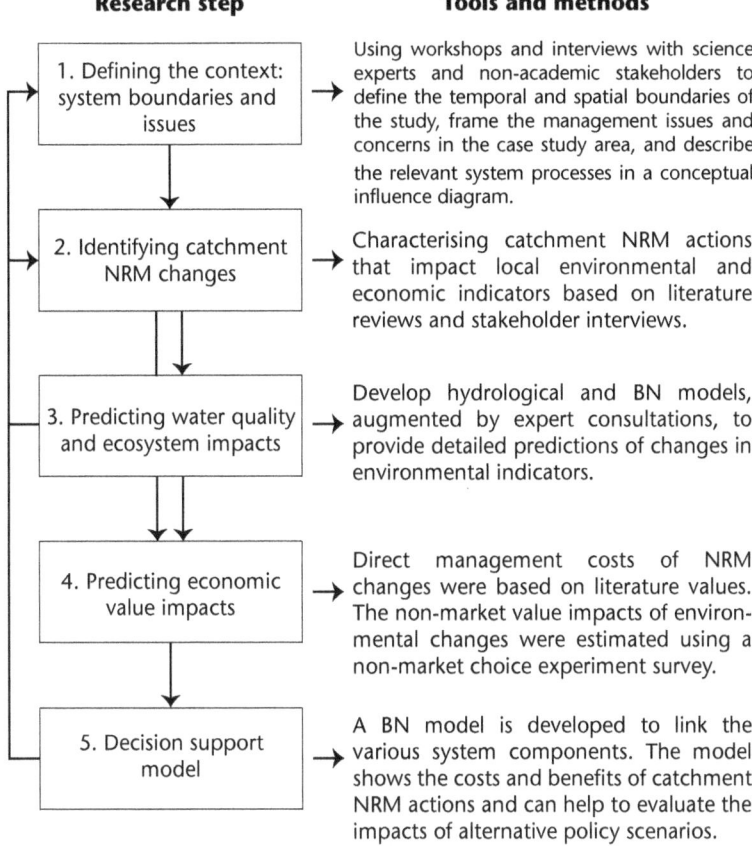

Figure 8.1: Analytical steps in the research process. Based on Kragt *et al.* (2010).

benefits of alternative NRM actions need to be assessed. However, despite the policy interest and identified need for integrated assessment, there are few studies that integrate environmental impacts with economic analysis in a robust framework to guide NRM decisions (Croke *et al.* 2007).

To evaluate the net benefits of alternative policy investments, economists often use a decision framework based on cost–benefit analysis. In a cost–benefit analysis, the recognised marginal social costs and benefits of a proposed project, and the alternative management actions, are systematically assessed and expressed in monetary terms (Hanley and Barbier 2009). This allows a direct comparison of the trade-offs between different environmental impacts, such as on water quality and biodiversity.

Although originating from very different backgrounds, cost–benefit analysis broadly follows the same steps as integrated assessment. First, the scope of the proposed policy action or project is defined, including the geographical scale, the population that will be affected, the time period over which effects are expected, and what alternative policies are available (steps 1 and 2 in Fig. 8.1). Then the physical impacts of the policy changes need to be quantified (step 3) and expressed in monetary terms (step 4). The evaluation phase consists of comparing the net present value of all the relevant cost and benefit impacts that arise over time as a result of the proposed actions. The objective of a cost–benefit analysis is to clearly lay out the trade-offs between the impacts of alternative policy decisions in a consistent and transparent way.

Traditionally, cost–benefit analysis has focused on financial analysis and its scientific underpinning has often been poor (Brouwer *et al.* 2003). The limited integration of biophysical modelling into traditional cost–benefit analysis studies reduces their flexibility to assist in the formulation and assessment of efficient policies. Our study aimed to integrate sound biophysical analyses into a cost–benefit analysis of catchment management changes. Each of the steps shown in Figure 8.1 will be briefly discussed in this chapter.

RESEARCH CONTEXT AND MANAGEMENT CHANGES

Defining the context requires an understanding of the issue under consideration, the system variables that are important and the interrelationships between them (Jakeman *et al.* 2006). Acknowledging the diversity of perspectives about catchment management issues, our study engaged with multiple academics from the Landscape Logic team along with public and other stakeholder representatives to gain a shared understanding about the research context.

The George catchment in north-eastern Tasmania was chosen as a suitable study area because scientific monitoring data were available and its environmental assets were highly valued for their resource (agriculture, forestry, fisheries, tourism) and existence values. Researchers from various disciplines were involved in developing a conceptual influence diagram to describe the management actions, biophysical drivers and processes of the George catchment system. The conceptual model development was based on 11 discussion workshops and 31 structured interviews with a range of natural scientists, policy-makers and community stakeholders (see Kragt 2010 for details). This approach ensured the conceptual model would match the scientific and policy context of the system.

Initially, our assessments were aimed at gaining an understanding of the George catchment system as a whole. This resulted in a very complex conceptual model with several dozen variables and processes (Kragt *et al.* 2011). A complete assessment of all the processes and interactions between variables was clearly outside the scope of the study. Subsequent workshops and consultations with local NRM officers were therefore aimed at identifying a parsimonious model that would represent the interactions between the most relevant management actions in the George catchment and the most important environmental variables that affect human welfare (Kragt and Bennett 2009; Kragt *et al.* 2011). The iterative process of developing a conceptual model took approximately nine months and resulted in a conceptual model for the George catchment that incorporated catchment management actions, water quality changes and three ecosystem indicators: native riparian vegetation, number of rare native species and the area of seagrass in the estuary (Fig. 8.2).

The possible future NRM actions we included in the model needed to match the scientific, political and socio-economic context of the George catchment system and had to be relevant to a variety of stakeholders. Our consultation process with scientists, NRM officers and local community members identified the four most important management actions that would affect the George catchment environment as:

- streambank engineering works;
- riparian zone management through limiting stock access to rivers and establishing buffer zones;

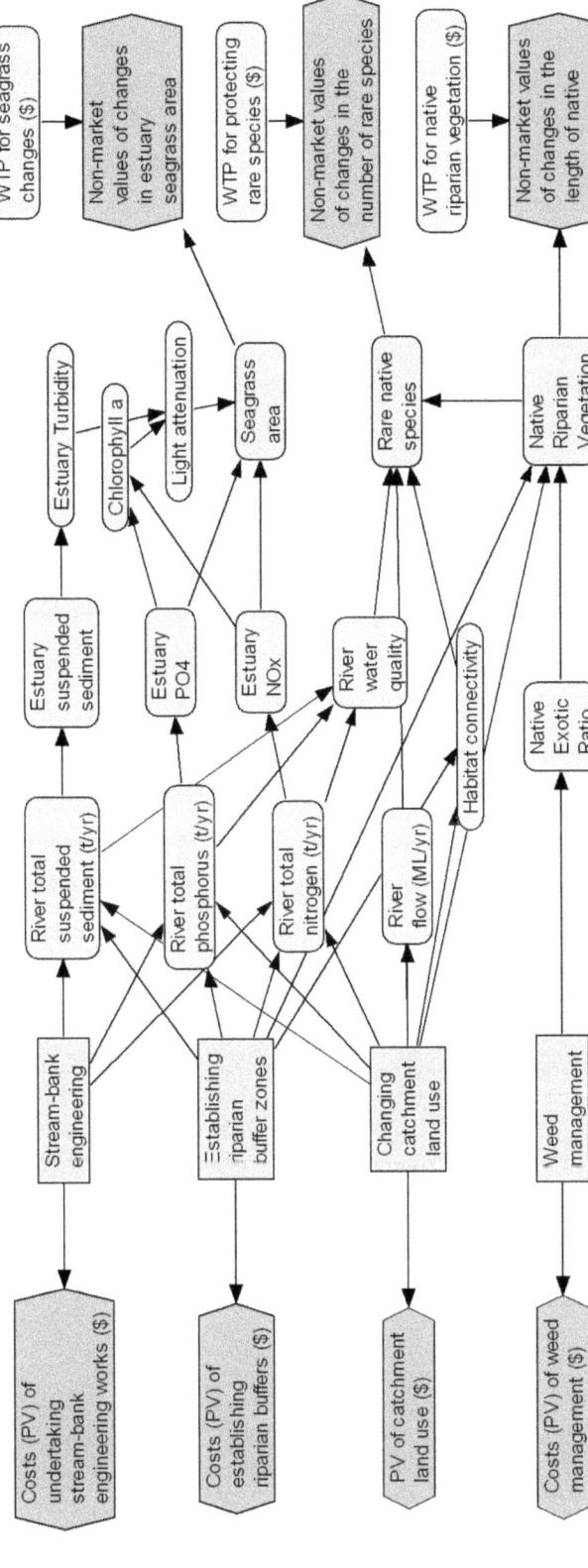

Figure 8.2: Conceptual model for the George catchment, incorporating four management actions (streambank engineering, creating riparian buffer zones, land use changes and weed management) and three environmental attributes (seagrass, rare native species and native riparian vegetation), where WTP is willingness to pay and PV is present value.

- changed catchment land use;
- vegetation management through weed removal.

Some of these actions are already being implemented in the George catchment on a small scale, which increases the validity of management scenarios used in the economic study.

ENVIRONMENTAL IMPACTS

Dominant potential impacts of the policy actions specified in the conceptual model needed to be assessed. Science-based modelling tools were used to predict the changes in the water quality and ecosystem indicators shown in Figure 8.2.

WATER QUALITY MODELLING

We developed a physically based, semi-distributed catchment model for the George catchment to predict water quality changes (Fig. 8.3). The model was based on the Catchment Scale Management of Diffuse Sources framework (CatchMODS; Newham *et al.* 2004). CatchMODS requires a relatively small number of parameters and has been successfully tested in other parts of Australia (Newham *et al.* 2004; Drewry *et al.* 2005; Vigiak *et al.* 2009). The George catchment model is a spatially explicit model that maps the impacts of different management actions on river flows, sediment delivery and nutrient loads, calculated as steady-state averages. Model development was based on GIS data (BRS 2003; Davies *et al.* 2005; DPIW 2005b) and existing monitoring information (Bureau of Meteorology 2007; DPIW 2007) augmented by soil nutrient sampling from different locations within the catchment. More detailed information about the water quality model development is provided in Kragt and Newham (2009). The George-CatchMODS model can serve as a stand-alone decision support tool, to help managers assess the impacts of different NRM actions on water quality variables.

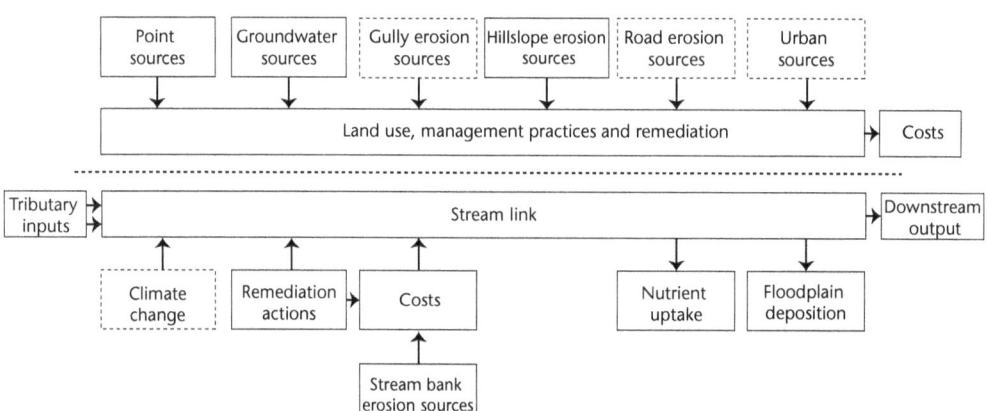

Figure 8.3: Water quality modelling framework for the George catchment (adapted from Newham *et al.* 2004). Dotted boxes depict scenarios that can readily be included in the model, but were not assessed in the present analysis.

ECOLOGICAL MODELLING

Our subsequent modelling was aimed at predicting ecological changes and presented a range of challenges. There were insufficient observational data available on ecosystem changes in the George catchment to develop deterministic models. The ecological modelling therefore used BN modelling frameworks that can accommodate a range of different data sources (monitoring data and expert opinion). Individual BNs were developed to predict changes in the three final assets (seagrass, native riparian vegetation and rare native species). These individual components were based on a wide review of literature, available monitoring data, and interviews with a range of natural science experts (see Kragt 2010 for detail).

Experts found it difficult to predict changes in ecosystem assets without full information about the processes in the George catchment. The individual BNs were therefore targeted at capturing the most likely *range* in levels (rather than precise values) for each of the ecosystem variables given NRM changes. It took several iterations to detail the individual networks and collect enough information for scientists to provide expert opinions in the presence of uncertainty (see Kragt *et al.* 2010 for an illustration of this process). The ecological predictions were used as inputs into the economic modelling.

To adequately populate the conditional probability tables for all variables, we need to know the probability that a certain state is observed at every possible combination of the input variables. Within the time-frame of our study, it was not feasible to collect sufficient ecological information to fully specify the probability relationships between all the variables in the conceptual model and ecosystem assets. Although economic values were estimated for the states of all individual ecosystem variables, only the native vegetation subsection of the conceptual model was developed into a fully functioning BN because of time and data constraints.

ECONOMIC VALUES

Integrated assessment modelling studies often focus on natural systems, with a sparse representation of socio-economic costs and benefits (Ward 2009). Assessing the efficiency of alternative policy scenarios requires an economic valuation of the relevant costs and benefits of NRM actions.

The management costs of NRM in the George catchment are relatively easy to estimate. Direct costs are associated with, for example, material for fencing off riparian buffer zones or riverbank engineering works (Kragt *et al.* 2010). An indirect cost of NRM changes may arise from impacts on production values for industries operating in the catchment (e.g. changes in land use affecting agricultural production, changes in the number of fish caught in the river). Market prices for such 'directly used' goods can provide information about their use values (Fig. 8.4). In this study, the opportunity costs of alternative land uses were included in the analysis. The impacts of environmental changes effected through NRM on production values were difficult to quantify because of the lack of ecological data. Cost impacts on oyster harvests, for example, were therefore not included in this study.

Cost–benefit analysis places emphasis on estimating all relevant benefits and costs, including values of environmental goods and services that are not traded in markets. A complete assessment of the value impacts of catchment NRM needs information about the non-market values affected (Hanley and Barbier 2009). Non-market values include non-consumptive use of

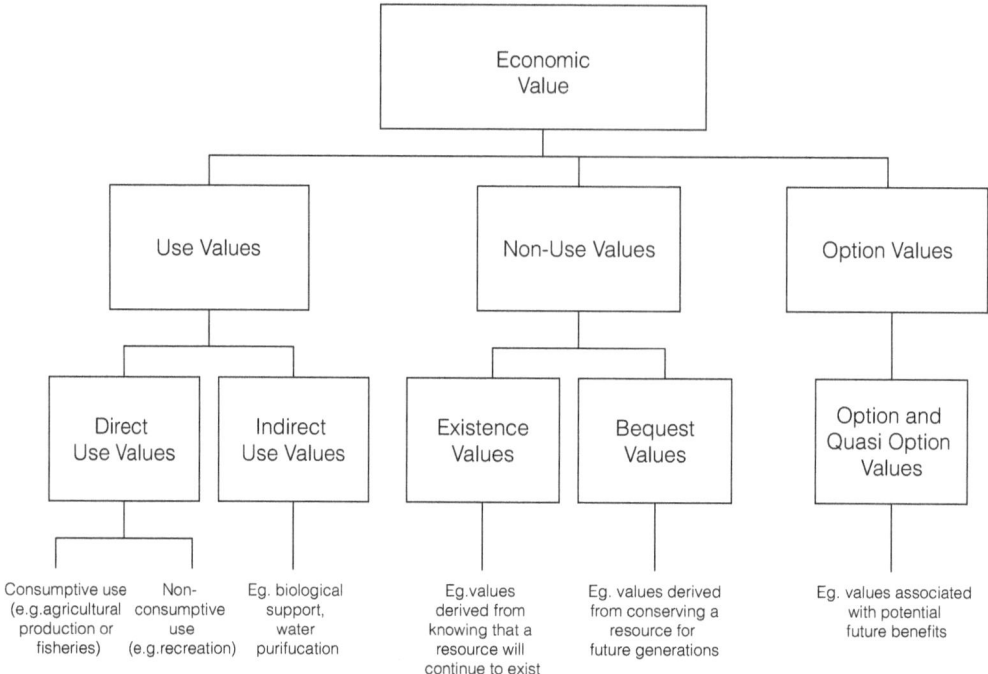

Figure 8.4: Components of economic value (based on Turner *et al.* 2003).

the resource (e.g. recreational activities), indirect use values (supporting services such as reduced soil degradation), non-use values (e.g. existence and bequest values) and option values (value of retaining the option to use a resource in the future) (Fig. 8.4).

Predicting changes in non-market costs and benefits requires the use of non-market valuation techniques. Although there are challenges involved in estimating non-market values (Hanley and Barbier 2009), not accounting for non-market values of environmental impacts may lead to a misallocation of resources and less-efficient decision-making (Bennett 2005).

In this study, we estimated the non-use value impacts of changed catchment NRM using a choice experiment survey (Bennett and Blamey 2001; Alpízar *et al.* 2001). The choice experiment aimed to estimate the aggregate non-use values that Tasmanian communities derive from seagrass, riverside vegetation and rare species. We did not aim to 'unpack' the non-use values into different components, since that would have been largely a theoretical rather than a policy-relevant exercise. The choice experiment survey was developed using a combination of literature review, interviews with science experts and regional natural resource managers, and feedback from focus group discussions (Kragt 2009b). The ecological modelling described above was used to predict 'best case' and 'worst case' scenarios for each of the three indicators of the George catchment environmental conditions. We used the ecological predictions as attribute levels of the choice options in the choice experiment survey (Kragt and Bennett 2011). An example choice question is shown in Figure 8.5.

The survey was administered in the main urban centres of Tasmania (Hobart and Launceston) and in the George catchment itself, using a stratified sampling approach to cover a range of locations and community types. A random sample was taken from each stratified sampling unit, using a drop-off/pick-up method with the assistance of local service clubs. Surveyors

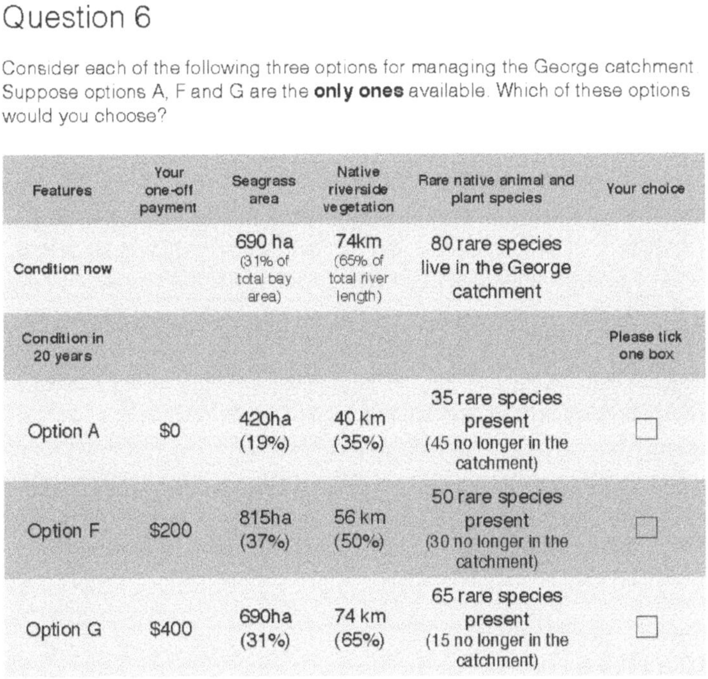

Question 6

Consider each of the following three options for managing the George catchment. Suppose options A, F and G are the **only ones** available. Which of these options would you choose?

Features	Your one-off payment	Seagrass area	Native riverside vegetation	Rare native animal and plant species	Your choice
Condition now		690 ha (31% of total bay area)	74km (65% of total river length)	80 rare species live in the George catchment	
Condition in 20 years					Please tick one box
Option A	$0	420ha (19%)	40 km (35%)	35 rare species present (45 no longer in the catchment)	☐
Option F	$200	815ha (37%)	56 km (50%)	50 rare species present (30 no longer in the catchment)	☐
Option G	$400	690ha (31%)	74 km (65%)	65 rare species present (15 no longer in the catchment)	☐

Figure 8.5: Example choice question in the George catchment experiment.

received training sessions and detailed instructions on the sampling locations and procedures. The questionnaires were collected between November 2008 and March 2009, yielding a total of 933 surveys. Descriptive statistics of the sample and the results of the econometric analysis of the survey data are available in Kragt (2010) and Kragt and Bennett (2011). The most relevant outcomes of the economic analysis for the integrated modelling are the value estimates for seagrass, riverside vegetation and rare species (Table 8.1). These estimates are expressed as the marginal willingness to pay (WTP) per household. In this way, the choice experiment data provide information about the non-use value of marginal changes in George catchment environmental conditions.

THE INTEGRATED MODEL

A major challenge in our study was the integration of knowledge from different sources about changes in catchment systems. A process-based model provided predictions of water quality

Table 8.1: Marginal willingness to pay estimates ($) for the George catchment environmental attributes

Attribute	Measurement unit	Mean	Uncertainty range
Seagrass (ha)	$/ha	0.17	(0.05–0.41)
Riverside vegetation	$/km	5.39	(2.23–12.5)
Rare native species	$/no. of species	11.7	(2.02–27.0)

changes, literature values and expert judgements were used to predict changes in ecosystem variables, and choice experiment survey data provided information about non-use value impacts. These different data sources needed to be combined into a logically consistent modelling framework. A further challenge was the representation of knowledge uncertainty about biophysical and socio-economic systems and the interactions between them. We used BN modelling to combine different data sources and represent uncertainties in an integrated bio-economic BN model. Although BNs are widely used to support catchment NRM (Bromley *et al.* 2005; McCann *et al.* 2006; Castelletti and Soncini-Sessa 2007), there are few BN applications that focus on economic impacts of environmental changes (see Barton *et al.* 2008 for an exception).

There was insufficient information about all the variables and processes to develop the conceptual model (Fig. 8.2) into a fully functioning BN for the whole George catchment system. Research efforts therefore focused on the native vegetation subsection of the conceptual model. We developed a BN that integrates the costs of management actions (streambank engineering, establishing riparian buffer zones, changing catchment land use and weed management) with predictions of river water quality (flows and total suspended sediment, phosphorus and nitrogen loads), native riparian vegetation length and non-market benefits (Fig. 8.6). For detailed explanation of the model variables, see Kragt (2010) or Kragt *et al.* (2010).

BN PROBABILITIES

Our BN model incorporates different sources of uncertainty, represented by the probability distributions that define the linkages between variables (Kragt 2009a). We used different techniques to generate these probability distributions. Uncertainties about the costs of changing NRM arise from, for example, knowledge gaps about the returns to land use, the types of materials used and the labour time involved in implementing and maintaining engineering works or riparian buffer zones. We represent these uncertainties in the BN model by estimating a range of likely costs rather than a single value (Kragt *et al.* 2010).

Monte Carlo simulations of the George-CatchMODS model provided conditional probability distributions for the water quality variables. Our simulations combined different scenarios of land use changes with varying lengths of streambank engineering works and riparian buffers, to predict the impacts on flow, sediment and nutrient loads. Uncertainties in the predictions arise from variations in the model parameters and were specified as an uncertainty bound around the deterministic predictions from the George-CatchMODS model. The results from these simulations were used to populate the probability tables in the water quality component of the BN.

The impacts of NRM actions on native riparian vegetation were predicted based on information collected through literature reviews, Tasmanian digital vegetation mapping (DPIW 2005a, 2005b) and expert reviews. Using the available monitoring data and the model structure depicted in Figure 8.6, experts were pressed to define their best-guess estimates about the proportion of 'natural' vegetation in the riparian zone under each land use, revegetation and weed management scenario, as well as their level of confidence about these estimates (Kragt *et al.* 2010). This provided a range of possible vegetation scenarios that were used to define probability distributions.

The non-use values of native riparian vegetation in the George catchment were estimated in the choice experiment. The results indicated that the Tasmanian survey respondents are, on

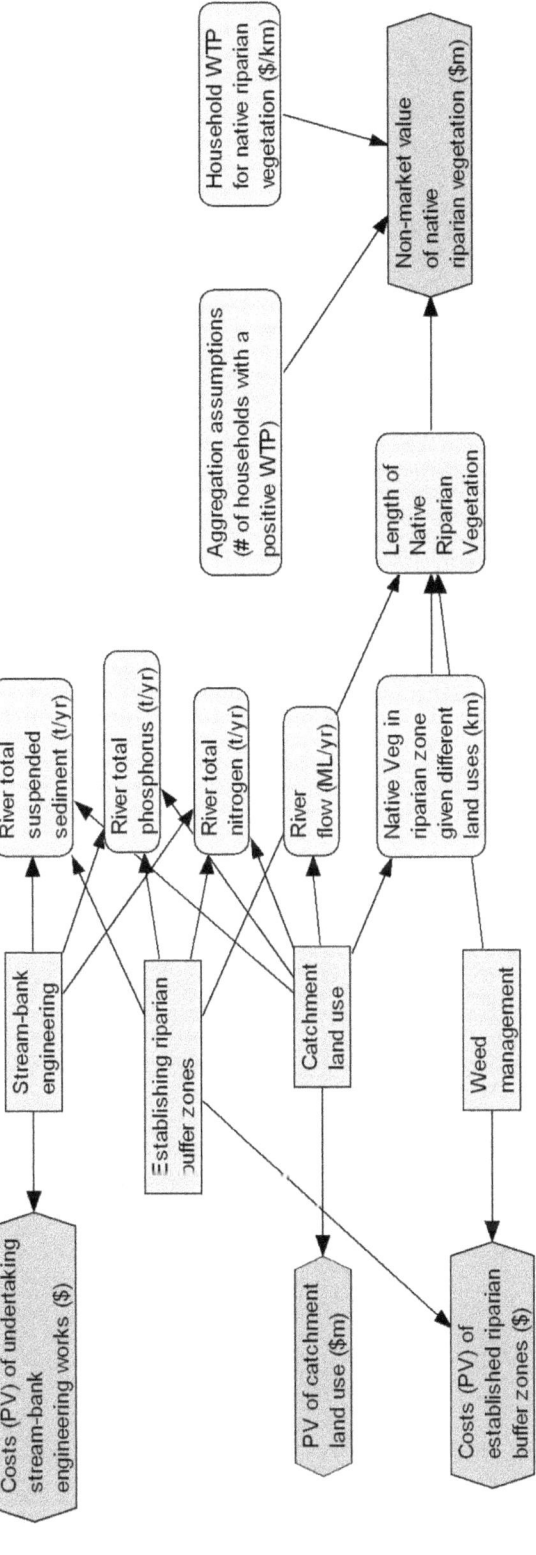

Figure 8.6: An integrated bio-economic model of water quality, native riparian vegetation and costs and benefits in the George catchment.

average, willing to pay $5.39 for every kilometre increase in native riparian vegetation, compared to the predicted worst case scenario (= 40 km of native riparian vegetation; Kragt and Bennett 2011). The econometric analysis also provided information about the uncertainty range in the WTP distribution (Table 8.1), which we used to calculate probabilities in the BN (assuming a normal distribution for each variable).

MODELLING RESULTS

The biophysical models predict the state of the environment under different management scenarios. This contrasts with the economic predictions in a cost–benefit analysis, which are based on marginal value changes. To link the biophysical modelling to economic valuation, changes (rather than states) in environmental conditions needed to be predicted.

In the integrated BN, we achieved this by including two predictions of the biophysical models (Fig. 8.7). The top section of the model predicts environmental conditions in the George catchment under a 'status quo' scenario (before any management changes). The bottom section predicts environmental conditions that result from implementing new management

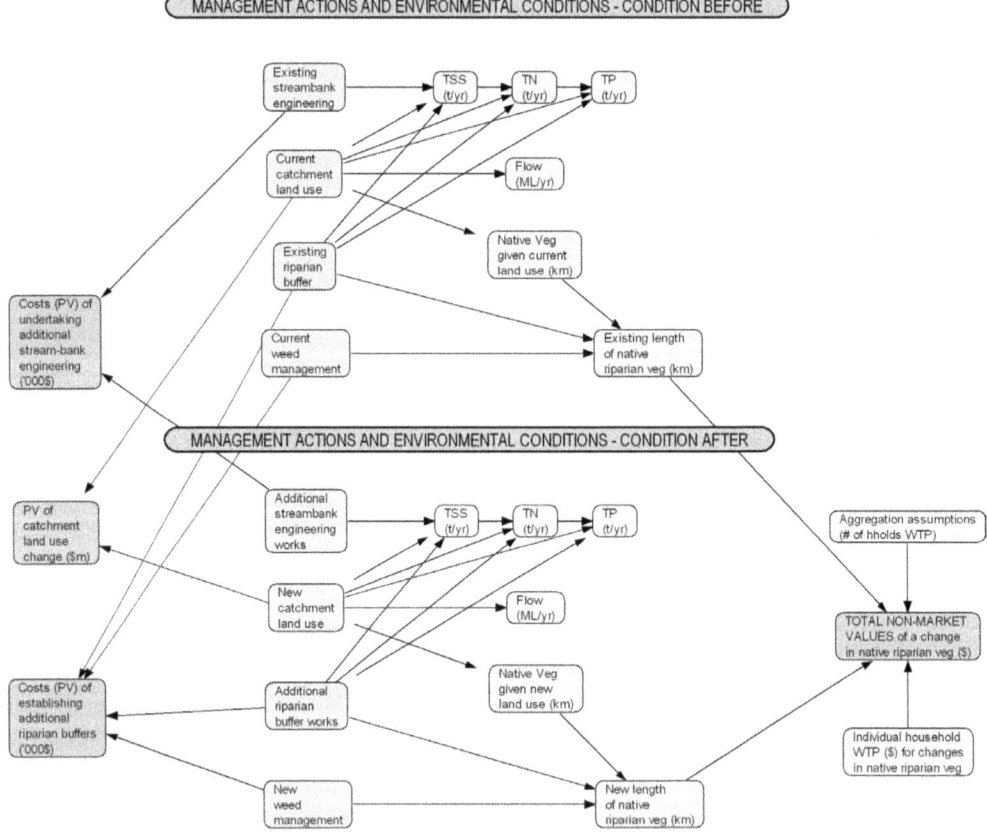

Figure 8.7: Final Bayesian network to link biophysical modelling and economic valuation in the George catchment (see Kragt *et al.* 2010 for more detail).

actions. Economic values are included by calculating the change in management costs (on the left) and the change in non-use values (on the right). The integrated model can be used to assess the impacts of NRM actions on a range of indicators, including water quality parameters, native riparian vegetation condition and non-use environmental values. Including the management costs of NRM actions as well as non-use benefits allows a cost–benefit analysis to determine which management investments deliver the greatest net returns to society. Using a BN modelling approach accounts for knowledge uncertainty in the input data and allows an analysis of the probability that one of the output indicators is in a certain state, given the management interventions. Kragt (2010) and Kragt *et al.* (2010) show how the BN model can be used to evaluate alternative policy scenarios, as well as results of model sensitivity analyses.

CONCLUSIONS

A major focus of our research was to assess the impacts of catchment NRM actions in the George catchment, Tasmania, on biophysical and socio-economic systems in an integrated manner. We based the economic analysis of direct management costs and non-use value impacts on biophysical model predictions. An iterative model development process which included consultation with academic experts, policy-makers and community stakeholders ensured that multiple stakeholder perspectives were accounted for in the final integrated model. The biophysical modelling was integrated with economic information about the costs and benefits of NRM actions in a BN modelling framework. The structure of our model allows, in theory, for a cost–benefit analysis of alternative NRM investment strategies and can thus provide a tool for policy-makers to assess the net social benefits of their decisions. However, the model development thus far is based on limited information about management costs and ecosystem changes in the George catchment. Because of the limited number of data sources and the high levels of knowledge uncertainty, we recommend further data collection before using the model's predictions as a reliable basis for decision-making.

Several challenges were encountered during this study. The most important obstacle to model development were the difficulties in reaching consensus between stakeholders (scientists, economists and community members) about the processes and indicators that were important in the George catchment. Using influence diagrams to visualise stakeholders' understanding of the system was very useful in gaining a shared understanding about the different stakeholder perspectives. An iterative process that involved numerous workshops led by experienced facilitators was required to reach agreement on the most important decision variables and environmental indicators in the George catchment.

We encountered a major divergence between the definition of various indicators by scientists, economists and members of the public. The detailed nature of scientific indicators generally made them incomprehensible to other stakeholders. This included not only the type of indicators but also the way in which they were described (e.g. riparian versus riverside). Scientists were hard-pressed to express their expert knowledge in a way that could be understood by economists and survey respondents.

We faced similar challenges using the BN modelling approach. The stakeholders involved in the model development process found it difficult to express their expert knowledge as probabilistic prediction, particularly without full knowledge of all the processes in the catchment. Where economists think in terms of *ceteris paribus* ('nothing else changes'), scientific experts typically find it hard to think of one component without considering the system as a whole ('it all depends').

A final discrepancy between biophysical and economic models lies in the nature of their predictions. Economists are interested in predicting the values associated with a change in environmental conditions from state A to state B. However, biophysical models provide predictions of environmental conditions given a certain combination of (management) inputs. To allow the integration of scientific knowledge with economic analyses, biophysical models need to allow predictions of marginal changes.

This study and the lessons from the process provide valuable insights for future projects that aim to link biophysical processes and economic values. There is still much scope for mutual learning and integration of scientific and economic knowledge. An important lesson is the arduous but vital role of an 'integrator' to bring together academic disciplines from disparate backgrounds. A model developer with experience in science, economics and modelling served as the knowledge broker in this project. Facilitating communication and knowledge exchange between stakeholders and translating the information into a final modelling tool remain huge challenges for integrated modellers.

REFERENCES

Alpízar F, Carlsson F and Martinsson P (2001) Using choice experiments for non-market valuation. *Economic Issues* **8**, 83–110.

Barton DN, Saloranta T, Moe SJ, Eggestad HO and Kuikka S (2008) Bayesian belief networks as a meta-modelling tool in integrated river basin management: pros and cons in evaluating nutrient abatement decisions under uncertainty in a Norwegian river basin. *Ecological Economics* **66**, 91–104.

Bennett J (2005) Australasian environmental economics: contributions, conflicts and 'cop-outs'. *Australian Journal of Agricultural and Resource Economics* **49**, 243–261.

Bennett J and Blamey R (2001) *The Choice Modelling Approach to Environmental Valuation*. Edward Elgar Publishing, Cheltenham, UK.

Bromley J, Jackson NA, Clymer OJ, Giacomello AM and Jensen FV (2005) The use of Hugin to develop Bayesian networks as an aid to integrated water resource planning. *Environmental Modelling and Software* **20**, 231–242.

Brouwer R, Turner RK, Georgiou S and van den Bergh JCJM (2003) Integrated assessment as a decision support tool. In *Managing Wetlands: An Ecological Economics Approach*. (Eds RK Turner, JCJM van den Bergh and R Brouwer) pp. 19–40. Edward Elgar Publishing, Cheltenham, UK.

BRS (2003) *Land Use, Tasmania. Version 5*. Bureau of Rural Sciences, Dept of Agriculture, Fisheries and Forestry, Canberra.

Bureau of Meteorology (2007) 'Climate statistics for Australian locations'. Australian Government, Bureau of Meteorology, Melbourne. <http://www.bom.gov.au/climate/averages/tables/ca_tas_names.shtml>.

Castelletti A and Soncini-Sessa R (2007) Bayesian Networks and participatory modelling in water resource management. *Environmental Modelling and Software* **22**, 1075–1088.

Croke B, Ticehurst J, Letcher R, Norton J, Newham L and Jakeman A (2007) Integrated assessment of water resources: Australian experiences. *Water Resources Management* **21**, 351–373.

Davies PE, Long J, Brown M, Dunn H, Heffner D and Knight R (2005) The Tasmanian Conservation of Freshwater Ecosystem Values (CFEV) framework: developing a conservation and management system for rivers. *Freshwater Protected Areas Conference 2004.* 27–28 September 2004, Sydney. pp. 45–50. IRN and WWF-Australia.

Department of Environmental Affairs and Tourism (2000) 'State of the environment report'. Division of Water, Environment and Forestry Technology, Pretoria, South Africa. <http://www.ngo.grida.no/soesa/nsoer/general/glossary.htm>.

De Ridder W, Turnpenny J, Nilsson M and Von Raggamy A (2007) A framework for tool selection and use in integrated assessment for sustainable development. *Journal of Environmental Assessment Policy and Management* **9**, 423–441.

DPIW (2005a) 'The conservation of freshwater ecosystem values data layers'. Dept of Primary Industries and Water, Water Assessment Branch, Hobart.

DPIW (2005b) *TASVEG: The Tasmanian Vegetation Map.* Dept of Primary Industries and Water, Information and Land Services Division, Hobart.

DPIW (2007) 'Annual waterways monitoring teports 2006: George vatchment'. Department of Primary Industries and Water, Hobart.

Drewry JJ, Newham LTH, Greene RSB, Jakeman AJ and Croke BFW (2005) An approach to assess and manage nutrient loads in two coastal catchments of the Eurobodalla region, NSW, Australia. In *MODSIM 2005: International Congress on Modelling and Simulation.* 12–15 December 2005, Melbourne. (Eds A Zerger and RM Argent). Modelling and Simulation Society of Australia and New Zealand.

Hanley N and Barbier EB (2009) *Pricing Nature: Cost–benefit Analysis and Environmental Policy.* Edward Elgar Publishing, Cheltenham, UK.

Jakeman AJ, Letcher RA and Norton JP (2006) Ten iterative steps in development and evaluation of environmental models. *Environmental Modelling and Software* **21**, 602–614.

Kragt ME (2009a) 'A beginners' guide to Bayesian Network Modelling for integrated catchment management'. Technical Report no. 9. Landscape Logic, Hobart.

Kragt ME (2009b) 'Developing a non-market valuation survey of natural resource condition'. Technical Report no. 8. Landscape Logic, Hobart.

Kragt ME (2010) An integrated assessment approach to linking biophysical modelling and economic valuation. PhD thesis. Australian National University, Canberra.

Kragt ME and Bennett J (2009) Integrated hydro-economic modelling: challenges and experiences in an Australian catchment. In *17th Annual Conference of the European Association of Environmental and Resource Economists. Pre-conference on Hydro-economic Modelling.* 24 June 2009. Amsterdam, The Netherlands.

Kragt ME and Bennett J (2011) Using choice experiments to value catchment and estuary health in Tasmania with individual preference heterogeneity. *Australian Journal of Agricultural and Resource Economics* **55**, 159–179.

Kragt ME and Newham LTH (2009) 'Developing a water quality model for the George catchment, Tasmania'. Technical Report no.16. Landscape Logic, Hobart.

Kragt ME, Bennett JW and Jakeman AJ (2010) An integrated assessment approach to linking biophysical modelling and economic valuation tools. In *iEMSs2010: International Congress on Environmental Modelling and Software.* 5–9 July 2010, Ottawa, Canada.

Kragt ME, Newham LTH, Bennett J and Jakeman AJ (2011) An integrated approach to linking economic valuation and catchment modelling. *Environmental Modelling and Software* **26**, 92–102.

Letcher RA and Giupponi C (2005) Policies and tools for sustainable water management in the European Union. *Environmental Modelling and Software* **20**, 93–98.

McCann RK, Marcot BG and Ellis R (2006) Bayesian belief networks: applications in ecology and natural resource management. *Canadian Journal of Forest Research* **36**, 3053–3062.

Newham LTH, Letcher RA, Jakeman AJ and Kobayashi T (2004) A framework for integrated hydrologic, sediment and nutrient export modelling for catchment-scale management. *Environmental Modelling and Software* **19**, 1029–1038.

NWI (2004) 'Intergovernmental agreement on a National Water Initiative'. COAG, Council of Australian Governments, Canberra.

TIAS (2009) The Integrated Assessment Society (TIAS). Osnabruck. <http://www.tias-web.info>.

Turner RK, van den Bergh JCJM and Brouwer R (2003) *Managing Wetlands: An Ecological Economics Approach*. Edward Elgar Publishing, Cheltenham, UK.

Vigiak O, Newham LTH, Whitford J, Melland A and Borselli L (2009) Comparison of landscape approaches to define spatial patterns of hillslope-scale sediment delivery ratio. In *MODSIM 2009: International Congress on Modelling and Simulation*. 13–17 July 2009, Cairns. Modelling and Simulation Society of Australia and New Zealand.

Ward FA (2009) Economics in integrated water management. *Environmental Modelling and Software* **24**, 948–958.

9

Lessons from studying water quality in agricultural catchments

Ted Lefroy, Aniela Grun, Anthony Jakeman and

James McKee

An evolutionary/ecological worldview encourages us to draw our boundaries of consideration to overlap the boundaries of causation (Jackson 2002).

The research described in the seven chapters in Part I varied in its scale of inquiry (from site to property to catchment to region), time period under consideration (historic and contemporary, hours to decades) and combination of methods employed (data mining, spatial analysis, direct measurement, experimentation, social survey and modelling). What it had in common were two objectives:

1   to contribute to a better understanding of the causal links between land use and water quality in rivers and estuaries;
2   to identify land management practices likely to improve water quality or prevent further decline.

The authors shared a commitment to integrated and participatory research methods (e.g. Bousquet and Voinov 2010). That is, the researchers attempted to identify and combine knowledge from a range of scientific disciplines with the experience of managers and other experts to address key research gaps and improve the relevance and uptake of research findings (Fig. 9.1).

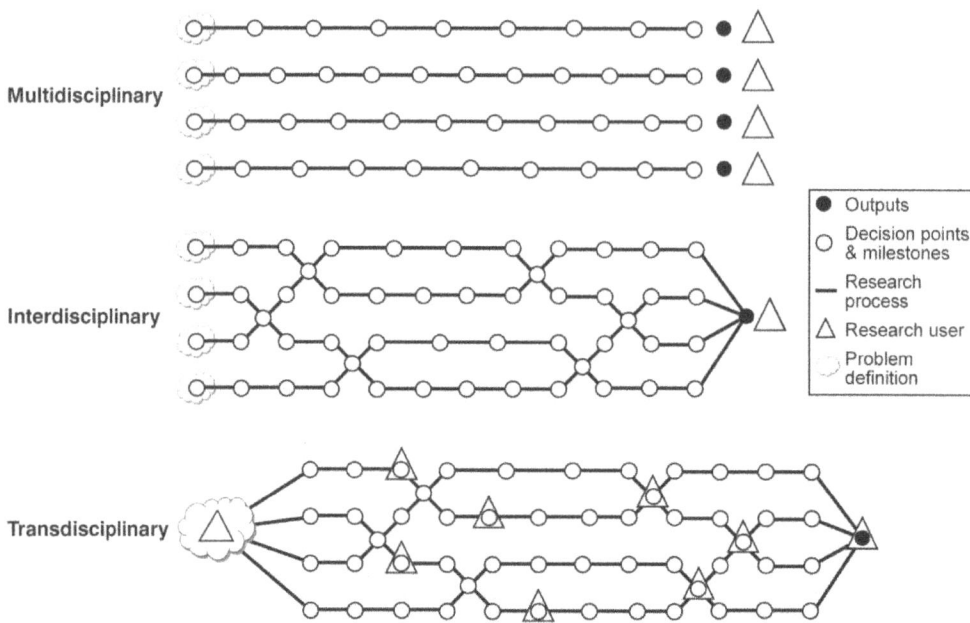

Figure 9.1: Schematic representation of research processes that feature different degrees of integration and participation where multi-disciplinary research is defined as more than one discipline under one roof, interdisciplinary research as integrating unrelated disciplines in ways that cross boundaries to solve problems, and transdisciplinary research as integrating unrelated disciplines in participation with user groups to solve problems (Klein 2008; Roux *et al.* 2010).

This chapter briefly reflects on the extent to which integration and participation were achieved, the obstacles faced and lessons for future research of this nature.

BREAKING DOWN AND BUILDING UP AGAIN

Questions about water quality posed at landscape scale confront the complexities of large spatial scales, long time-lags between intervention and response, multiple confounding human and natural influences, non-linear processes and high levels of natural variability in processes such as rainfall, runoff and stream flow. Identifying relationships between natural and human drivers of water quality in such complex environmental systems is made more difficult by a lack of data, particularly with respect to stream flow and even more with respect to water quality response. Data tend to be inadequate in spatial and temporal definition, quality and period of coverage.

These complexities force researchers to break questions down into tractable but conceptually linked components where they can exert some degree of control over variability. The challenge of integration is then to develop a conceptual model of how a landscape functions and is likely to respond to intervention. The modelling process is not simply reductionism in reverse (top-down) as it demands that researchers identify the key driving variables, the response functions that govern them and their interrelationships (bottom-up). Typically, the gaps between the tractable component pieces of knowledge is too great to achieve this, hence researchers and managers rely on experience, referred to variously as 'expert opinion' (Chapters

3, 4, 5, 6, 7 and 8) and 'gut feel' (Chapter 2). This reinforces the view expressed in Chapter 7 that, given the uncertainty at the intersection of science and management, no one group has a monopoly on knowledge. This is an argument supporting investment in integration to develop a systems view of the issues under scrutiny, that can be updated as ideas are challenged and new knowledge and perspectives develop.

UNCERTAINTY, COMPLEXITY AND THE DIVERSITY OF KNOWLEDGE

Early in the Landscape Logic project, we identified Bayesian network (BN) models as a means of tackling the 'apples and oranges' problem of different sources, types and quality of data across and within our system components. BN models were considered appropriate as they can explicitly accommodate the uncertainty associated with different datasets and their graphical format lends them to participatory research (Ticehurst *et al.* 2011). Influence diagrams, the first stage of BN modelling, were used by all research teams and their utility was acknowledged by managers. These diagrams allowed the development of shared hypotheses about how systems worked and forced teams to clearly articulate and converge on their environmental goal (especially endpoints of predictive interest), their hypotheses about system interactions and the structural level of interactions to be represented (Chapter 18). In adaptive management, whether those shared hypotheses are right or wrong is less important than whether they can be expressed, understood and tested. Chapter 3 provides a good example of the application of Bayesian modelling in natural resource management, culminating in a working quantitative model that produced new insights into the processes influencing river ecological condition and the likely response to intervention. This was a reflection of both the available data and the ability, in that instance, to break a large complex question down to tractable components that could be approached as a series of separate studies.

LONG TIME-LAGS

The long time-lags between intervention and response meant that three of the seven chapters were retrospective in nature, relying heavily on historic datasets on land use, water quality, river ecological condition and estuarine condition (Chapters 2, 3 and 4). Two chapters were based on contemporary studies that assessed the performance of a novel land management practice (Chapter 5) and relationships between social and economic characteristics of land ownership and the adoption of management practices (Chapter 6). The remaining two used a mixture of contemporary surveys and historic data to, in one case, develop a generic diagnostic method for identifying the source area, origin and timing of stream pollutants (Chapter 7) and, in the other, to ascribe economic values to resource management outcomes (Chapter 8).

INVESTIGATING AT MULTIPLE SCALES

The research discussed in four of the chapters required observations at multiple scales in order to infer causal links between land use and water quality. Landscape-scale investigation involving multiple catchments enabled generalisations about the relationship between land use and nutrient generation. They indicated that nitrogen and phosphorus loads in irrigated dairy pastures in Tasmanian landscapes are at the higher end of recorded values worldwide (Chapter 2) and identified a new land use disturbance threshold for ecological river condition (Chapter 3). Property-scale investigation provided improved understanding of the current management of riparian land and showed correlations between the occupation, social networks and income source of land managers and their adoption of different practices (Chapter 6). Detailed site-scale experiments and monitoring were necessary to infer a causal connection between land use in the upstream catchment and biophysical process in rivers (Chapters 3, 5 and 7) and to identify a causal connection between fertiliser application and irrigation practices on farms and rates of water and nutrient leaching under irrigated cropping (Chapter 2).

INTEGRATED ASSESSMENT

The power of integrated assessments such as these is that investigation at multiple scales can be combined to improve understanding and suggest remedies. This was possible in the case of ecological river condition (Chapters 3 and 6). The combined catchment- and site-scale studies pointed to fine sediment as a whole-of-catchment contributor to declining river condition and to high levels of shade (80%) as a major factor controlling algal growth (Chapter 3). When these findings were combined with property-level data from the survey of riparian land owners (Chapter 6), water quality managers were provided with the following critical information about potential remediation strategies and target audiences:

- that whole-of-landscape sediment management and increased levels of shading are likely to be necessary for remediation;
- that a realistic time-line for response to riparian intervention is in the order of 30–40 years, the time it will take for riparian revegetation to achieve threshold levels of shade;
- that the majority of land owners (61%) at the time of the survey were not commercial farmers and therefore may not respond to conventional forms of communication and incentives;
- that a significant proportion of land owners are reluctant to replace exotic deciduous riparian vegetation (willows, *Salix fragilis*) with endemic evergreens, and hence increase shading.

A factor that aided the integrated assessment of river condition across scales was the ability of the river health team to design gradient studies in response to significant correlations in the initial multi-catchment analysis, which moved their conclusions from correlation to strong inference (Chapter 3). The ability to integrate knowledge across multiple scales was more limited in the other historically based studies (Chapters 2 and 4), largely due to the inferior quality and consistency of the original datasets. In the case of the land use and water quality study (Chapter 2), the reliance on monthly gauging station data with very little event-based monitoring severely limited the ability to generalise about the relationship between land use and concentrations of nitrogen, phosphorus and sediment. The response of that team was to instigate a series of smaller-scale studies to examine the performance of different land management practices. However, the ability to link these to form generalisations about larger-scale consequences was severely limited by the temporal scale and design of the available water quality datasets. In the case of estuarine condition (Chapter 4), the differences in timing, location and method of data collection for the 34 estuaries in the study limited interpretation. The response of that team was to move up in scale. By stepping back from the fine-scaled and very noisy datasets, they applied first principles to develop a risk assessment framework based on residence time, derived from readily available data on tides, bathymetry, river flow and shape of the estuary mouth. This approach has general application to targeted monitoring of the estuaries (and locations within estuaries) most at risk.

These examples suggest that inferring cause in water quality and identifying effective interventions requires the integration of knowledge from three scales: landscape scale to establish correlations between land use and water quality (Chapters 2, 3 and 7), site scale to assess the performance of different land management practices (Chapters 2, 3 and 5) and property scale to locate and date interventions and understand the social and economic determinants of adoption (Chapters 6). The seven chapters involved some 20 individual studies in the form of data mining, spatial analysis, surveys and experiments across these three scales of inquiry. In hindsight, better planning and coordination could have increased the degree of vertical integration between scales by co-locating more studies and ensuring use of compatible methods. This is a major challenge for interdisciplinary and transdisciplinary research (Fig. 9.1) where individual studies are often, partially of necessity, designed sequentially in response to the results of earlier and often larger-scale inquiry.

ENGAGING RESEARCH USERS

The opportunity to participate in a large collaborative research project gave the three regional NRM organisations in Tasmania a welcome opportunity for deeper learning than is normally possible through project-based research or exposure to science at conferences where there is little opportunity to influence the research process. Landscape Logic provided the opportunity to explore locally critical questions of landscape function in a rigorous fashion. Participation across the three regions was varied and some regions weren't able to participate to the extent they had anticipated. Engagement in the scoping phase at the project's inception proved more difficult than anticipated as the NRM regions did not all have established modes of engagement in participatory research, relying on the ability and willingness of individuals. As organisations with a primary focus on facilitating environmental management through community engagement, their capacity to maintain strategic and consistent involvement in the research varied over time. Engagement was affected by changes in government policy and funding directions which changed their program focus and staffing. The technical nature of the research proposals that emerged from the initial scoping sessions also limited the ability of some to engage. Regional NRM staff indicated that the quality of communication from the research projects improved noticeably throughout its life, but greater emphasis on clarity in the formative stage may have assisted stronger 'buy-in'. Clearer relationships between the range of different research activities and how the managers might incorporate research findings into their regional plans and strategies was difficult to gain in the early stages. By the end of the project it was apparent that the skills and capacity within the organisations meant that model outputs and scenarios were more likely to be taken up than the models themselves, requiring a further level of interpretation and consultation between the researchers and managers, expressed through fact-sheets and other high-level summaries. However, as the project progressed, one aspect of the research process itself was adopted by some managers. Influence or causal loop diagrams were incorporated into their planning to share understanding about how landscapes work, identify common goals over which they were most likely to have influence and assess the likely feasibility of interventions.

Three factors were identified as contributing to the success of the collaboration with the three NRM regions: having a dedicated knowledge broker to act as a go-between, being able to engage with researchers at different levels from one-on-one engagement in the field to hands-on training in models and decision frameworks, and having a website to access results and products. Having a range of products on a given topic, from detailed BN and other models to technical reports and fact-sheets, meant that managers could select the level of detail most relevant to their knowledge and interests. The effort made by the research teams to seek comment on the design of case studies and scenarios was important to the relevance of the outputs, as was the emphasis placed in the last year on jointly identifying the implications of key findings for management.

CONCLUSIONS

At the heart of the two objectives (inferring causality between land use and water quality, and identifying effective interventions) is the issue of scale, particularly the ability to transfer knowledge between scales. Inferring causality between land use and river condition required demonstration of relationships between ecological variables at site scale and measures of human activity at landscape scale. Identifying effective interventions required investigation at property scale to characterise the performance of different land management practices within a land use, and social research to understand the demographics of land ownership and the economic and social determinants of adoption. This type of integrated assessment requires integration in two dimensions, vertically between spatial scales and horizontally across disciplines, to establish the relative importance of natural and human influences on the chosen variable.

The extent to which the two objectives were achieved depended on the appropriateness of existing long-term datasets to the questions, the ability to identify multiple lines of evidence to overcome high levels of uncertainty and the ability to design and undertake sequential investigations in response to each set of results. A weakness, apparent in retrospective approaches, is relying on surveillance monitoring where data are rarely collected at sufficient spatial and temporal scales or with adequate replication to answer specific questions about intervention and response (Lindenmayer and Likens 2010; Field *et al* 2007).

While each chapter in Part I adds incrementally to our knowledge about the biophysical and social processes that influence water quality, the extent to which integration was achieved across scales and disciplines was inevitably limited. This experience suggests that this type of integrated assessment involves two distinct challenges: technical integration and social integration. Technical integration requires access to suitable datasets, conceptual frameworks and modelling platforms capable of incorporating knowledge from different sources, disciplines and scales. Social integration requires planning and coordination to ensure sufficient compatibility between individual studies to achieve integration between scales and across disciplines. It also requires effective leadership, to maintain engagement between research teams and end users and to communicate the common purpose. Many participants in this project were new to integration, but it was evident that the whole team benefited from an experience which they can apply to future research. Worldwide, researchers are beginning to learn how to undertake integrated assessments which are by definition necessary for wicked problems where uncertainty and conflicts are rife. As noted by Jakeman and Letcher (2003), analysis frameworks, including Bayesian networks, for integrated assessment and other tools have come of age but there is much that is problem-specific because issues, scales, models and stakeholders vary. By continuing to practise integrated assessment on real world problems in participation with managers, this metadiscipline will mature as lessons accrue and learning is systematised.

REFERENCES

Bousquet F and Voinov AA (2010) Modelling with stakeholders. *Environmental Modelling and Software* **25**, 1268–1281.

Field SA, O'Connor PJ, Tyre AJ and Possingham HP (2007) Making monitoring meaningful. *Austral Ecology* **32**, 485–491.

Jackson W (2002) Poverty and agricultural policies. *Population and Environment* **24**, 55–67.

Jakeman AJ and Letcher RA (2003) Integrated assessment and modelling: features, principles and examples for catchment management. *Environmental Modelling and Software* **18**, 491–501.

Klein JT (2008) Education. In *Handbook of Transdisciplinary Research.* (Eds G Hirsch Hadorn, H Hoffmann Riem, S Biber-Klemm, W Grossenbacher-Mansuy, D Joye, C Pohl, U Wiesmann and E Zemp) pp. 399–410. Springer, Berlin.

Lindenmayer DB and Likens GE (2010) *Effective Ecological Monitoring.* CSIRO Publishing, Melbourne.

Roux DJ, Stirzaker RJ, Breen CM, Lefroy EC and Cresswell HP (2010) Framework for participative reflection on the accomplishment of transdisciplinary research programs. *Environmental Science and Policy* **13**, 733–741.

Ticehurst JL, Curtis A and Merritt WS (2011) Using Bayesian Networks to complement conventional analyses to explore landholder management of native vegetation. *Environmental Modelling and Software* **26**, 52–65.

Part II

VEGETATION CHANGE IN RURAL LANDSCAPES

10

Measuring change in vegetation extent at regional and property scales

Garreth Kyle, David Duncan and Graeme Newell

SUMMARY

This chapter describes the ecological component of a multi-disciplinary investigation of change in native vegetation extent at landscape and property scales in northern Victoria, Australia. We quantified change in mature native tree cover at landscape level in three study areas covering ~500 000 ha using aerial photography from 1946–2008 and conducted participatory workshops to attribute probable causes to mapped change. We then mapped the extent and type of all forms of vegetation management on a subset of 71 properties across the case study areas based on interviews with landholders to identify the year and funding source of those works. Further work at the site scale (Chapter 12) examined the trajectory of ecological change that follows funded intervention. To our knowledge, no previous studies have attempted an analysis of natural resource management (NRM) effectiveness spanning these three scales.

We demonstrated how the combined forces of socio-economic factors and regulation slowed then largely ended the clearing of woody vegetation by around 1990, and how more recently NRM incentives have laid the foundation for a modest increase in landscape-level cover of just under 2% within the last two decades. Spontaneous recolonisation or regeneration of native woody vegetation was by far the biggest contributor to the likely increase in extent, contributing far more area than revegetation projects.

Our landholder-scale mapping and interviews revealed that, by area, joint public–private funding of revegetation, remnant fencing and restoration projects outweighed wholly private works by four to one over the entire dataset, revising a key reporting assumption about area of impact. The annual total area of wholly privately funded work has increased slightly over time but has not kept pace with the large increase in public investment that occurred through the Australian government's Natural Heritage Trust program on a site-level basis. However, the total contribution from private individuals to the subsidised projects means that overall they matched this major increase in public investment. Our approach yielded data that placed the impact of NRM investment in the context of 160 years of post-European land use, and has revised understanding of how public investment has leveraged private landholder activity in native vegetation protection and revegetation.

INTRODUCTION

Australian governments at all levels share the goal of reversing the long-term decline in the extent and condition of Australia's native vegetation (Dept of Natural Resources and Environment 2002; North Central CMA 2005; Australian State of the Environment Committee *et al.* 2006). In 1989, the Australian government responded to the growing public appetite for conservation action on private land and pledged to plant one billion trees to combat land degradation. Over subsequent decades, public investment in NRM has evolved from a focus on plantings of a few tree species to biodiverse plantings of local endemics. The scale of works has increased from single rows or small isolated plantings to larger strategic programs, with an additional emphasis on replanting and protecting the structure and composition of existing native vegetation (habitat quality; Parkes *et al.* 2003). The delivery model has also changed. Where previously only government departments and a few NGOs operated, there are now 56 regional catchment management organisations nationally, collaborating with a broad range of agencies or other organisations to deliver NRM investments. With the increase in investment has come greater scrutiny of the extent to which the landscape has changed in response to the increased effort and investment, with reviews concluding that it is not possible to judge the extent to which stated environmental outcomes have been achieved (Australian National Audit Office 2004, 2008).

It is relatively easy to quantify the material outputs of investments (area of remnant fenced, area of revegetation, number of trees planted) and indeed these details are routinely reported (North East CMA 2006; Wimmera CMA 2007; Goulburn Broken CMA 2010), but the ecological outcomes of these activities are more difficult to ascertain. Not surprisingly, the uncertainties compound as we move from considering a single investment site to cumulative changes at a whole-of-region level. Net landscape change in native vegetation extent and quality (Fig. 10.1) is influenced by a complex mixture of socio-economic and biophysical drivers operating at different temporal and spatial scales (Duncan and Wintle 2008; DSE 2008). Only a modest number of these variables are within the influence of natural resource managers, therefore an analysis of NRM effectiveness must consider the variables over which they have no control. Where private land tenure is dominant, such as in our project, we considered there would be three critical scales of inquiry required to compile a picture of the effectiveness of NRM investment in native vegetation change. These scales were landscape, landholding and site (Fig. 10.1) and there were distinct questions applicable to each scale.

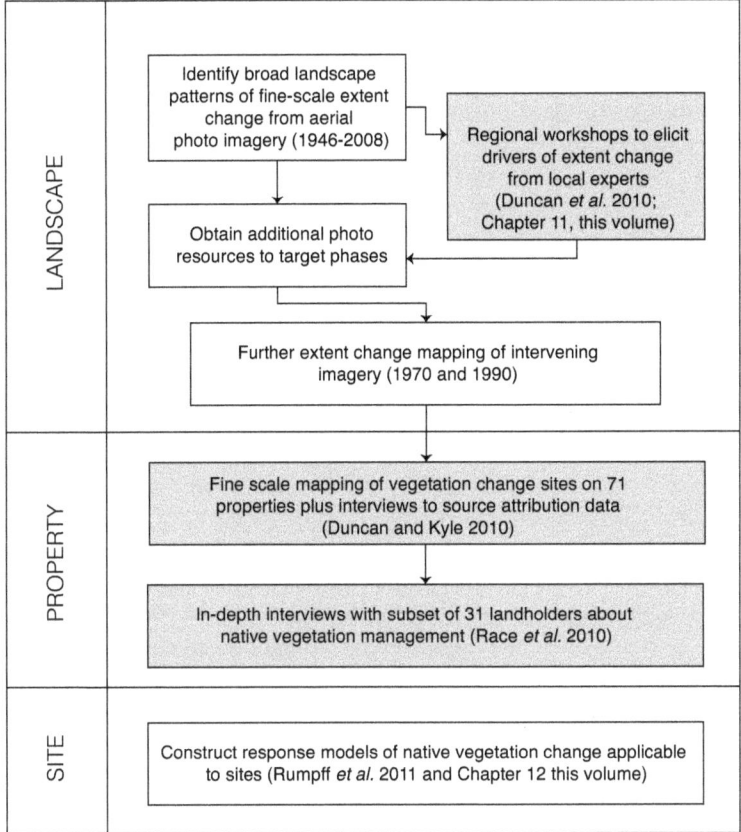

Figure 10.1: Elements of the study spanning landscape, property and site scales. Shaded boxes indicate particular foci of collaboration with social research; the site-scale box included a major collaboration with ecologists and modellers from the Adaptive Environmental Decision Analysis (AEDA) Hub within the Commonwealth Environmental Research Facilities program.

Beginning with the landscape scale, we wanted to quantify the amount of change in the cover of native vegetation and identify the major drivers of that change. Second, at the property scale, we wanted to know how vegetation change was influenced by decisions made by private landholders. These questions were:

- Who was engaging in restorative works on their properties?
- What kind of activities were they engaging in?
- How much were they spending?
- Who was resourcing those activities?

Third, at the scale of site investments such as a revegetation site or a remnant protected from stock grazing, we wanted to know the trajectory of ecological change that follows the funded action. To our knowledge, no previous studies have attempted an analysis of NRM effectiveness spanning these three scales.

To adequately address these questions, our activities required an approach that was both multi-scale (in space and time) and multi-disciplinary. While the combination of spatial analysis, ecology and social research that we employed presented a number of challenges, it also produced a rich and comprehensive analysis. In this chapter we review the methodology that we used to examine change in the extent of native vegetation cover at the landscape and property scales. Our approach to site-level analysis is described in Chapter 12. We highlight key decision points in our project, the major findings, their implications and the overall lessons about research of this nature.

SELECTION OF CASE STUDY AREAS

We chose the location of our case study areas in consultation with our partner catchment management authorities (CMAs): North Central, Goulburn Broken and North East. To ensure broad relevance of our results, we sought areas that were representative of the dominant land use, vegetation and bioregion classes of the CMAs (Fig. 10.2). We had to decide the extent to which we would seek a landscape with contrasting or similar land use and land management histories. We chose the latter, reasoning that contrasting landscapes would produce too much variation and impede the construction of a cohesive narrative.

Each CMA was looking for new data and insights for their regions, therefore we chose a single case study area within each of the three CMAs: Muckleford in North Central, Longwood–Violet Town in Goulburn Broken and Chiltern–Springhurst in North East (Fig. 10.2). The case study areas were situated within the fragmented portion of the landscape that lay between the intact forest landscapes and the highly modified landscapes dominated by intensive agriculture. One of the most obvious differences between the case study areas was the extent of remnant vegetation. The Longwood–Violet Town case study area had a little over 4% cover of mature native trees, while Muckleford and Chiltern–Springhurst had about 37% and 34% respectively (DSE 2005a). Despite differences in the extent of native trees and a higher preva-lence of agriculture in Longwood–Violet Town, all the case study areas fell within the transi-tion and amenity landscapes described by Barr *et al.* (2005). Accordingly, we assumed the socio-economic character would be broadly similar across all three regions. The areas repre-sented a diversity of land uses and had a history of land use change, good records of NRM investment and reliable spatial data, in particular historical aerial photographs.

LANDSCAPE-SCALE METHODS AND KEY RESULTS

We used aerial photographs from 1946–2008 to map change in native tree extent at the landscape level. Aerial photographs were preferred as the primary resource as they provided superior historical depth than satellite imagery, and offered the highest resolution of the remote sensed data available to us. Previous automated analysis of change in vegetation cover within Victoria using Landsat imagery (1989–2005) failed to detect change on the scale of the NRM investment we were interested in, due to the coarse resolution of the imagery (DSE 2005b). Indeed, the narrow elongated shape of many parcels of revegetation presents a problem for some automated detection methods, which often fail to map narrow shapes that are incon-gruent in size and shape with larger square pixels (Chapter 14; Lechner *et al.* 2009). Despite the superior resolution of much of the aerial photography we were nonetheless limited to mapping change in tree extent, as it was not possible to accurately identify zones of native shrubs and grasslands.

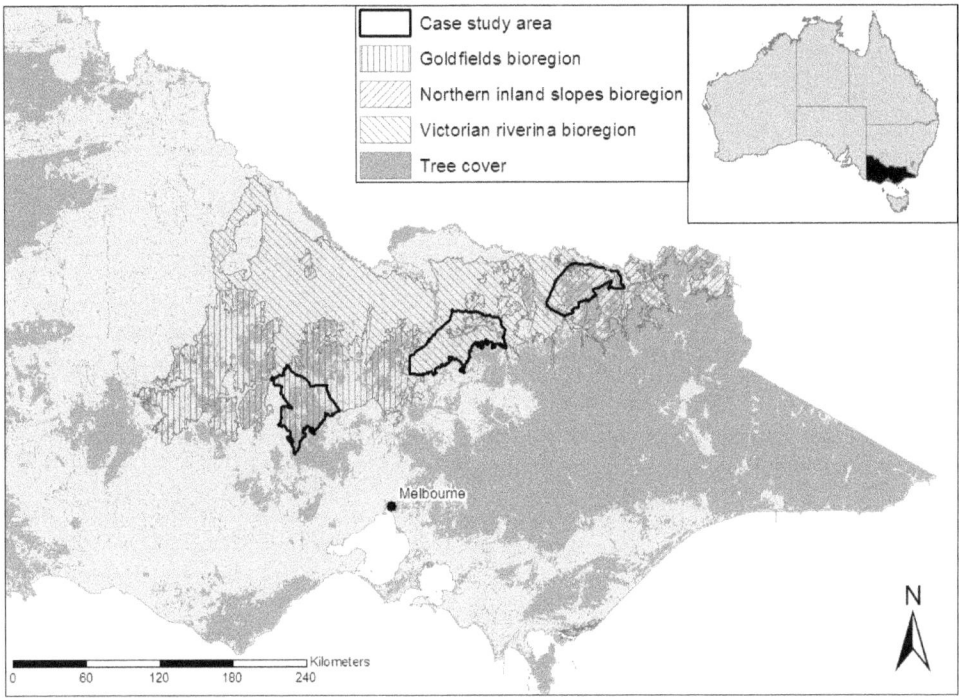

Figure 10.2: The Victorian case study areas, dominant bioregions and native tree cover.

We manually mapped change in tree extent over our case study areas using a lattice subsampling technique. The lattice of 1.5 km diameter circles, centred at 3 km, accounted for a little over 19% of the three case study areas (569 circles in total). These circles were not the minimum sampling unit but a window through which the landscape was surveyed. Within each circle we identified and mapped change in mature tree cover by comparing 1946 to 2004–08. Change polygons were attributed to one of three categories: loss, gain and scattered loss (Table 10.1). Using the most recent aerial photography (2004–08) we also identified and mapped zones of gain in immature tree cover as revegetation or regeneration (Table 10.1). Based on investigation into the potential drivers of change in tree cover, we identified two intervening time slices that we felt would provide an important inflection point for our mapped datasets: 1990 and 1970 (Duncan *et al.* 2010). We compared imagery from these time slices to the mapped layers of change from 1946–2008 and the 2008 aerial imagery, and attributed change to one of three time periods: 1946–70, 1971–90 and 1991–2008. All zones of vegetation change were individually identified and mapped by an operator using a GIS, then re-checked by a second operator. Where possible inconsistencies in the mapping process appeared, the site was visited in the field where possible, or a third operator consulted.

Much of the clearing within our three case study areas occurred between European settlement in the 1800s and 1946 when the aerial photography record begins. By that time the extent of native trees in the three regions had been reduced to about 5% in Longwood–Violet Town, and had returned to 38% in Muckleford and 35% in Chiltern–Springhurst following heavy clearing (Fig. 10.3). After 1946 the variation in loss and gain was relatively subdued, with the area of loss of solid and scattered canopy greatest from 1946–70 and the overall rate of loss

Table 10.1: Categories and definitions used in mapping historical and contemporary change in native tree cover from 1946–47 to 2004–08

Vegetation cover type	Description
Gain	A gain in continuous canopy of mature trees
Loss	A loss in continuous canopy of mature trees
Scattered loss	The loss of several remnant or 'paddock' native trees. These were mapped as patches of three or more trees. The average density of each patch was 10.4 trees/ha (SE. 0.44)
Spontaneous regeneration	Immature tree canopy that resulted from native regeneration
Revegetation	Immature native tree canopy that resulted from active revegetation works

decreasing throughout the study period. Gain in solid canopy was a little more variable, with the peak in canopy gain occurring from 1970–90. From 1946–2008 there was a net loss in mature solid canopy (Fig. 10.3, inset graph). This would be substantially greater if scattered loss was also considered. The consideration of scattered loss is complicated by values associated with whether the understorey is native or exotic and by the density of the trees within the polygons. This latter point is important as polygons of scattered loss include a sometimes extensive matrix of grasses, shrubs or crops, potentially inflating the area of loss. The average

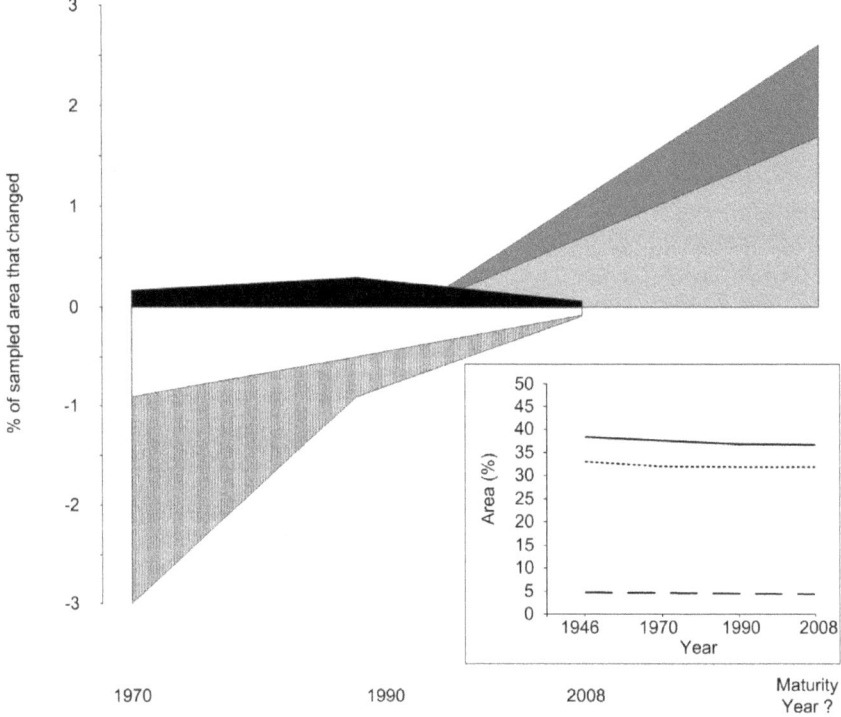

Figure 10.3: Historical vegetation change from 1946–2008 as a percentage of the three case study areas. Change categories for mature woody cover are gain in solid canopy (black), loss in solid canopy (white) and loss of scattered tree areas (stippled). Currently immature categories of revegetation (dark grey) and regeneration (light grey) are shown with an uncertain origin from 1990–2000, and their contribution is projected into the future in recognition of time-to-maturity. The inset graph represents net change in solid canopy within the three case study areas: Muckleford (solid line), Chiltern–Springhurst (dotted line) and Longwood–Violet Town (dashed line).

Table 10.2: Area of revegetation and regeneration mapped at the landscape scale

Region	Vegetation type	Total area (ha)	% of sampled area
Longwood–Violet Town	Spontaneous regeneration	635.17	1.73
	Revegetation	384.11	1.05
Muckleford	Spontaneous regeneration	732.55	2.24
	Revegetation	236.00	0.72
Chiltern–Springhurst	Spontaneous regeneration	293.70	1.11
	Revegetation	252.19	0.95

density of trees in polygons of scattered loss was 10.4 per hectare, which sits at the lower end of the 10–20 per hectare estimated for south-eastern Australian woodlands from early 1800s surveyor plans (Vesk and MacNally 2006).

In comparison to the small amount of gain documented from 1946–2008 the area of immature native trees identified using aerial photographs from 2004–08 was relatively large. The total area of immature native trees, including revegetation and spontaneous regeneration, across the three case study areas was 2533 ha (approximately 2.9% of the survey area). The area of spontaneous regeneration was almost twice that of revegetation being 1.9% of the sampled region compared to 1%. The highest incidence of spontaneous regeneration occurred in Longwood–Violet Town, followed by Muckleford then Chiltern–Springhurst, while the highest incidence of revegetation occurred in Muckleford followed by Chiltern–Springhurst then Longwood–Violet Town (Table 10.2).

Assuming that the current extent of immature vegetation reaches maturity and that the decreasing rate of canopy loss is not reversed, from 1946 until around 2030 there will have been a net gain in mature canopy (Fig. 10.3). Excluding scattered loss, this will vary from 3.1% in Longwood–Violet Town to 1.1% in Muckleford and 0.4% in Chiltern–Springhurst. While this represents a positive outcome following the net decrease in native cover documented up to 2008, these figures should be used with caution. In terms of capital expenditure, spontaneous regeneration represents an inexpensive and therefore attractive alternative to revegetation (Schirmer and Field 2000). There are a number of other factors that should be considered, including the opportunity costs of interventions (Dorrough *et al.* 2008), the influence of incentive schemes associated with revegetation or remnant management (Duncan and Kyle 2010), landholder perception of regeneration (Jay 2005), the composition of local seedbanks and the influence of environmental change on the regeneration of native species. These factors are likely to have a considerable influence on the quality and long-term viability of patches of regeneration.

IDENTIFYING CAUSES OF WOODED VEGETATION CHANGE

We investigated the drivers of long-term tree cover change using a series of regional workshops (Duncan *et al.* 2010; Chapter 11, this volume). We summarised the information gained from each workshop in a time-line (Fig. 11.3, Chapter 11, this volume), a map (Fig. 10.4) and reports (Race *et al.* 2009a, b, c). The regional workshops also informed further research, in particular the selection of intervening time periods for aerial photographs, and gauged the accuracy of our preliminary mapping using the knowledge of local landholders.

The regional workshops allowed us to delve beyond the biophysical drivers of change to understand the socio-economic factors that influenced vegetation change within our case study areas. Many of the major drivers of change were influential in all three case study areas. In the decades preceding 1970, drivers of vegetation change included the release and spread of myxomatosis (and associated decline in the effects of rabbit browsing), the increase and subsequent decrease in wool prices and the decline of hilltop grazing. From 1970–90 important drivers included the removal of public lands from grazing, changing land use including an increase in the number of lifestyle landholders, the inception of Landcare and legislation introduced in 1989 that placed considerable limitations on the clearing of native vegetation. Many of the latter drivers still influence the extent and quality of the native vegetation.

To examine spatial drivers of spontaneous regeneration and revegetation we used boosted regression tree modelling (Elith *et al.* 2008). Although many factors cited by workshop participants could not be effectively drawn into a regression modelling context, we felt that recent phenomena of revegetation and regeneration could be tested using plausible models constructed from available surrogates of some of these factors represented as spatial data. The results of the modelling were largely intuitive, in that regeneration was strongly associated with existing native canopy cover and to a lesser extent by moderate to highly sloping land. However, the explanatory power of the models was low, indicating that factors other than those we examined play an important role.

The drivers of vegetation change have been complex, operating at a range of spatial and temporal scales. While it may be possible to influence some of these drivers directly through active intervention or policy, there is another subset, such as climatic variation and commodity prices, that are not as easily influenced, if at all. The agencies tasked with the management and promotion of native vegetation should recognise their limited ability to control some drivers and be ready to make the most of any positive situations that occur. As well, agencies should be wary of overreporting on native vegetation gains. On the surface, the net gain based on currently immature vegetation appears positive; however, over half the immature vegetation gain included in this calculation exists as spontaneous regeneration. Further research is needed to investigate the biodiversity values of spontaneous regeneration compared to revegetation and remnant native vegetation, and to improve our understanding of the biophysical and social factors that affect the perseverance of spontaneous regeneration.

PROPERTY-SCALE METHODS AND RESULTS

At the property scale we were interested in finding out what kind of landholders were engaging in native vegetation protection and revegetation, how much they were doing and the extent to which it was publicly supported. To address these questions we mapped polygons of native vegetation works in 71 landholdings across the three case study areas (Duncan and Kyle 2010). We found more than 400 individual projects, accounting for 10% of the combined land area (32 360 ha). Mapped vegetation units were classified into four categories:

- revegetation, which included fencing, weed control and planting tubestock or direct seeding;
- active remnant restoration, which involved the active management of existing remnants for biodiversity reasons. Management included pest plant and animal control, revegetation and fencing;

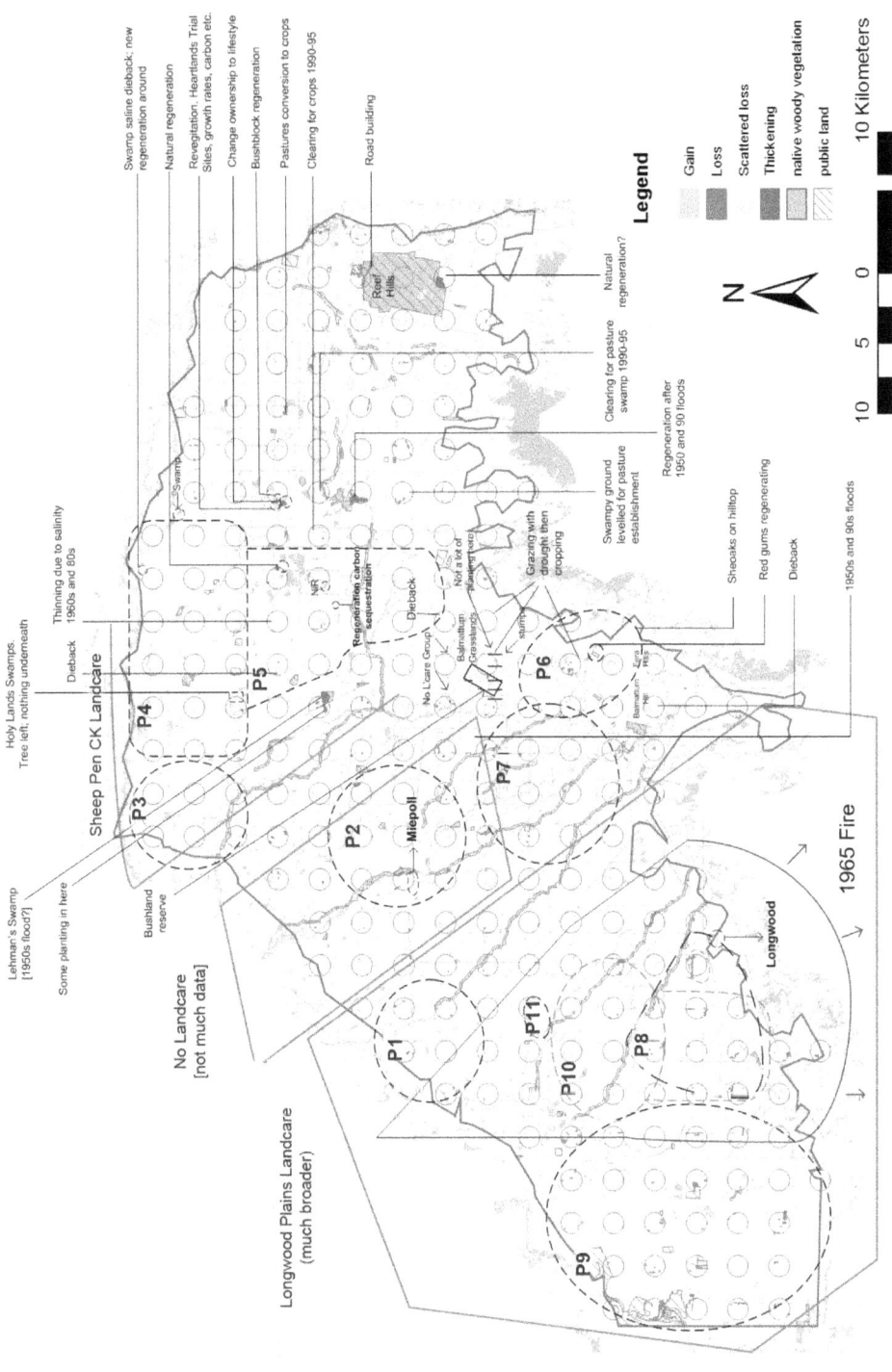

Figure 10.4: Mapped vegetation change in the Longwood–Violet Town region presented at the regional landscape history workshops with annotations from participants on likely local influences.

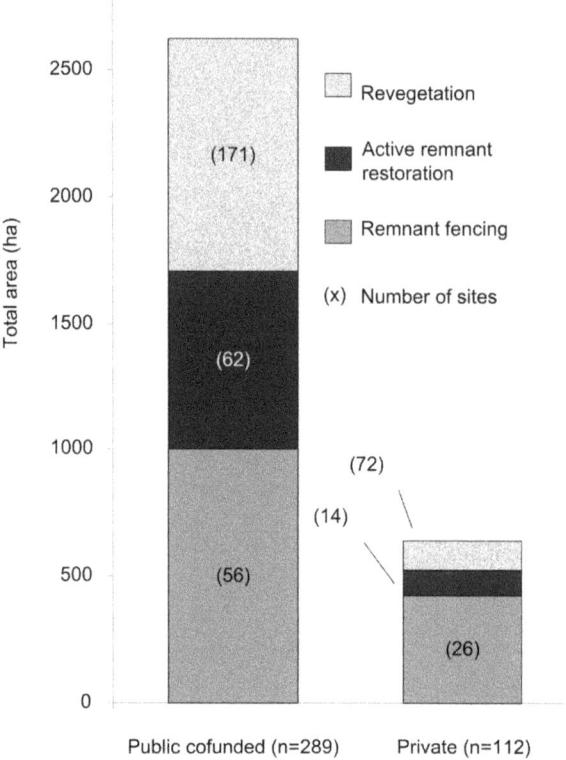

Figure 10.5: Total publicly co-funded and privately funded contributions to new revegetation, remnant fencing and active remnant restoration sites for 71 landholdings (Duncan and Kyle 2010).

- remnant fencing, which involved the fencing of existing patches of native vegetation often for stock control purposes rather than biodiversity objectives.

For each polygon, the landholders were also asked to nominate the year, source of funding and specific actions undertaken.

Our data enabled a direct test of an assumption used in native vegetation reporting in Victoria, known as the 'two-times assumption' (Brunt and McLennan 2006): that for each hectare of co-funded work, another hectare is undertaken privately without direct public contribution. In contrast, we found 4 ha of co-funded works for every hectare of wholly privately funded works (Fig. 10.5). In fact, it is misguided to consider the annual contribution of co-funded and privately funded works to native vegetation extent as a ratio. Although private and co-funded area were positively correlated (Fig. 10.6), privately funded works maintained a stable average despite co-funded works increasing dramatically with the advent of the Natural Heritage Trust in 1998. Thus, we were able to definitively disprove the two-times assumption, which has been used in reporting against resource condition targets by CMAs (e.g. Goulburn Broken CMA 2010) and in net gain accounting by the Victorian government (e.g. DSE 2008). In comparison with entirely privately funded works, publicly subsidised works are more than twice as likely to be located in zones of higher conservation significance (Duncan and Kyle 2010). Last, we were able to validate a landholder typology proposed by Race *et al.* (2010) using

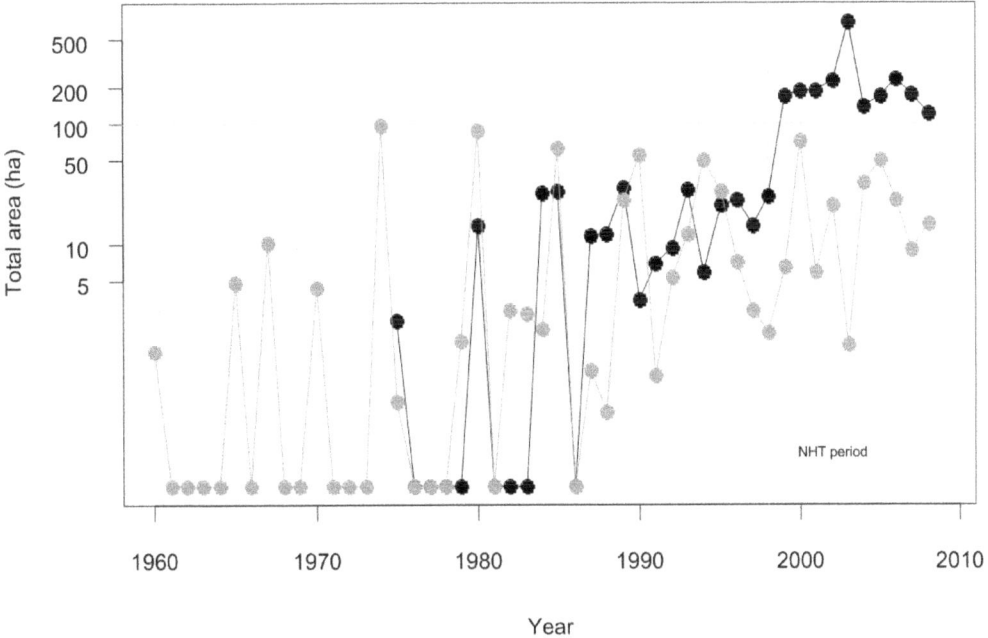

Figure 10.6: Total area (ha) of privately funded (●) and publicly co-funded (●) revegetation and regeneration (Duncan and Kyle 2010). The y-axis has the natural log of total area presented with back-transformed labels. Reference lines are given to indicate 100 ha on the y-axis, and the commencement of National Heritage funding on the time axis.

empirical landholder and property data, which suggests that the typology may have some utility as a lens for considering NRM engagement.

LINKING LANDSCAPE- AND PROPERTY-SCALE DATA

Although we endeavoured to control the bias in our landholder sample so that we could relate the findings back to the broader case study areas, we still needed to quantify residual bias. The systematic survey methods that we employed in the landscape-level survey meant we could presume this was a representative sample of the landscape. Comparisons between the property-scale mapping and the landscape lattice mapping were not initially possible, as classifications we used in aerial photograh interpretation were not entirely congruent with those used by landholders. To overcome this we used aerial photographs from 2004–08 to map instances of revegetation and regeneration for all properties irrespective of the categories previously used in the property scale mapping. We then directly compared the proportion of revegetation and regeneration at the property scale with the equivalent data at the landscape scale (Fig. 10.7). This comparison revealed that, while we had a representative subsample in terms of regeneration, our properties contained a considerably higher proportional cover of revegetation than was typical within the case study areas.

REFLECTIONS

This project was both spatially and temporally complex. To gain an appreciation of the influence of socio-economic factors on vegetation change at this scale required a different set of

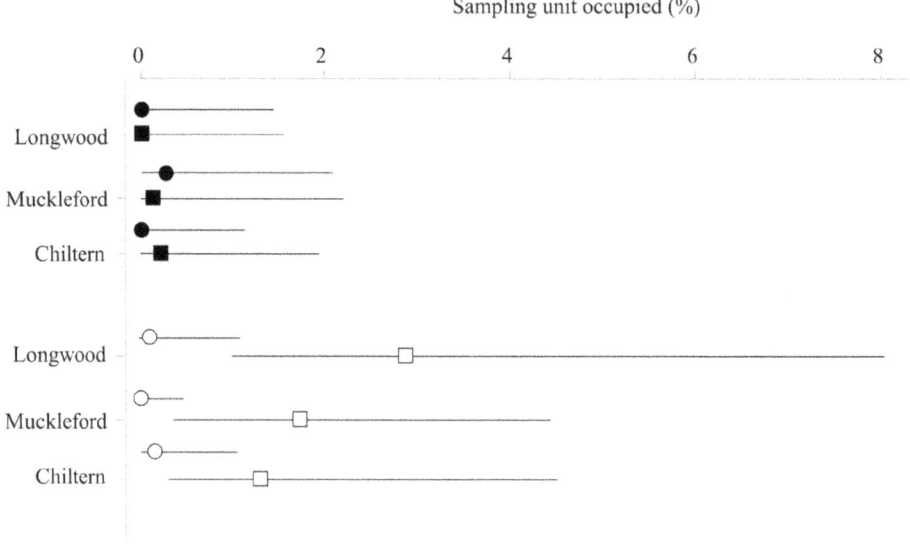

Figure 10.7: Comparison of median percentage of sampling units occupied by revegetation (open) and regeneration (closed) at the property (squares) and landscape (circles) scales. Bars represent 25th and 75th quartiles.

skills than those available to the ecologists within our research group. Our collaboration with social researchers was vital, and it was important that it was forged early through the three regional landscape history workshops. In addition to providing an opportunity to conduct real field work together, these workshops allowed us to rapidly gain insight into the drivers of change and the relative importance of these drivers, and to become familiar with the region. In hindsight, we underestimated the potential of the regional workshops to provide spatially and temporally specific information, which proved invaluable to the progression of the project (Duncan *et al.* 2010).

THE USE OF AERIAL PHOTOGRAPHY

We initially chose to pursue aerial photograph interpretation as the main source of new data for this project because of its superior resolution (Fensham *et al.* 2002) and its historical depth – back to the 1940s, when the technology came of age (Rango *et al.* 2008). In recent decades, satellite imagery has increasingly been used and advocated for environmental monitoring and mapping of vegetation communities, as evidenced by peer review journals such as *Remote Sensing and the Environment*. Satellite and airborne data have relative technical strengths and weaknesses, which are discussed elsewhere (Wyatt 2000; Farmer 2008). With specific reference to the inconsistency of aerial photographs in image resolution, contrast and spatial accuracy prevented the effective use of automated mapping techniques (Fensham and Fairfax 2002; Bowman *et al.* 2001; Wyatt 2000). The consequent reliance upon operator judgement rather than automation constrained the amount of area that we could investigate and limited the direct transferability of our approach to other case study areas.

Despite these reservations, at the scale of our case studies and for our objectives, we are convinced aerial photography was the correct choice. Debates about aerial photography versus satellite data typically focus on relative economy and technical merits. We suspect one largely overlooked benefit of aerial photography is the ability of photographic imagery to facilitate participatory research (e.g. Duncan *et al.* 2010a). Most people are accustomed to looking at and interpreting photographs and thus have an intuitive understanding of the factors that limit interpretation, such as tone, resolution and scale. We suspect that this power to engage non-specialists will mean that aerial photograph interpretation will continue to play an important role in landscape change analysis even if it reaches technical obsolescence and is eclipsed by more readily available satellite imagery.

Just as the choice of aerial photography influenced the transportability of our methodology beyond our case study areas, the nature of the case study areas themselves will influence the degree to which our findings could be extrapolated within the partner CMAs and to similar landscapes within south-eastern Australia. Given the importance of land use history, current land use and management patterns and socio-economic characteristics, we urge that caution and consideration be applied to extrapolation of our findings.

CONCLUSIONS

This multi-scale, multi-disciplinary project was challenging but ultimately yielded a number of important findings relevant to our project partners. Working with social researchers and regional agency partners in particular was highly successful and greatly enhanced our work, allowing us to explore and contextualise the broad range of drivers that influence native vegetation. We have illustrated how the combined forces of socio-economic factors and regulation slowed then largely ended the clearing of woody vegetation, and that more recently NRM incentives have laid the foundation for a modest increase in native vegetation cover. In our case study areas, spontaneous regeneration is likely to be the largest contributor to future gain in woody vegetation extent. The long-term perseverance and ecological significance of this vegetation type needs to be better understood. Our landholder-scale mapping and interviews enabled us to understand the contribution of different kinds of remnant fencing, restoration and revegetation projects to future woody native vegetation, and the relative contribution of voluntary action. The average yearly total of wholly privately funded work has increased slightly over time but did not keep pace with the rapid increase in public investment that occurred through the National Heritage program. However, the total co-investment contribution from private individuals to subsidised projects means that overall they matched the quantum leap in public investment.

While the patterns we found illustrate the impact of NRM investment in our case study areas, they may not have strong predictive value outside these areas, or even within these areas into the future. If new markets emerge to support enterprises such as carbon farming, the current farm-scale economics could be radically altered within a short space of time. We hope that the process of reflection shown in this chapter will give readers a more complete understanding of the background to our key results. Similarly, we hope that fellow researchers will have more information on which to judge the aspects of our methodology that are worthy of adoption and those that should be avoided, depending on their specific questions.

REFERENCES

Australian National Audit Office (2004) 'The administration of the National Action Plan for Salinity and Water Quality'. Australian National Audit Office, Canberra.

Australian National Audit Office (2008) 'Regional delivery model for the Natural Heritage Trust and the National Action Plan for Salinity and Water Quality'. ANAO Audit Report no. 21, 2007–08. Australian National Audit Office, Canberra.

Australian State of the Environment Committee, Beeton RJS and Dept of Environment and Heritage (2006) 'Australia State of the Environment 2006: independent report to the Australian Government Minister for the Environment and Heritage'. Dept of Environment and Heritage, Canberra.

Barr N, Wilkinson R and Karunaratne K (2005) 'Understanding rural Victoria'. Dept of Primary Industries, Bendigo.

Bowman DMJS, Walsh A and Milne DJ (2001) Forest expansion and grassland contraction within a savanna matrix between 1941 and 1994 at Litchfield park in the Australian monsoon tropics. *Global Ecology and Biogeography* **10**, 535–548.

Brunt K and McLennan R (2006) 'Biodiversity Monitoring Action Plan'. Goulburn Broken Catchment Management Authority, Shepparton.

Dept of Natural Resources and Environment (2002) 'Victoria's native vegetation management. A framework for action'. Dept of Natural Resources and Environment, Melbourne.

Dept of Sustainability and Environment (DSE) (2005a) 'Native vegetation extent 2005'. Spatial layer from the spatial corporate data library. Dept of Sustainability, Melbourne.

Dept of Sustainability and Environment (DSE) (2005b) 'Pre-1750 ecological vegetation classes'. Spatial layer from the spatial corporate data library. Dept of Sustainability, Melbourne.

Dept of Sustainability and Environment (DSE) (2008) 'Native vegetation net gain accounting first approximation report'. Dept of Sustainability and Environment, Melbourne.

Dorrough J, Vesk P and Moll J (2008) Integrating ecological uncertainty and farm-scale economics when planning restoration. *Journal of Applied Ecology* **45**, 288.

Duncan D and Kyle G (2010) 'Management of native vegetation on private land in northern Victoria: the magnitude, motivation and resourcing of revegetation and restoration'. Technical Report no. 23. Landscape Logic, Hobart.

Duncan DH and Wintle BA (2008) Towards adaptive management of native vegetation in regional landscapes. In *Landscape Analysis and Visualisation: Spatial Models for Natural Resource Management and Planning.* (Eds C Pettit, W Cartwright, I Bishop, K Lowell, D Pullar and D Duncan) pp. 159–182. Springer-Verlag, Berlin.

Duncan DH, Kyle G and Race D (2010) Combining facilitated dialogue and spatial data analysis to compile landscape history. *Environmental Conservation* **37**, 432–441.

Elith J, Leathwick JR and Hastie T (2008) A working guide to boosted regression trees. *Journal of Animal Ecology* **77**, 802–813.

Farmer EA (2008) A critical evaluation of remote sensing based land cover mapping methodologies. PhD thesis. School of Applied Sciences, Cranfield University, UK.

Fensham RJ and Fairfax RJ (2002) Aerial photography for assessing vegetation change: a review of applications and the relevance of findings for Australian vegetation history. *Australian Journal of Botany* **50**, 415–429.

Fensham RJ, Fairfax RJ, Holman JE and Whitehead PJ (2002) Quantitative assessment of vegetation structural attributes from aerial photography. *International Journal of Remote Sensing* **23**, 2293–2317.

Goulburn Broken CMA (2010) 'Annual Report 2009–2010'. Goulburn Broken Catchment Management Authority, Shepparton, Victoria.

Jay M (2005) Remnants of the Waikato: native forest survival in a production landscape. *New Zealand Geographer* **61**, 14–28.

Lechner A, Stein A, Jones S and Ferwerda J (2009) Remote sensing of small and linear features: quantifying the effects of patch size and length, grid position and detectability on land cover mapping. *Remote Sensing of Environment* **113**, 2194–2204.

North Central CMA (2005) 'North Central Native Vegetation Plan'. North Central Catchment Management Authority, Huntly, Victoria.

North East CMA (2006) 'Annual Report 2006/2007'. North East Catchment Management Authority, Wodonga, Victoria.

Parkes D, Newell N and Cheal D (2003) Assessing the quality of native vegetation: the 'habitat hectares' approach. *Ecological Management and Restoration* **4**(Supplement), S29–S38.

Race D, Duncan DH, Kyle G and Merritt W (2009a) 'Landscape history and vegetation change in the Chiltern–Springhurst region of Victoria'. Landscape Logic, Hobart. <www.landscapelogic.org.au/publications/Brochures/Chiltern_Springhurst_brochure.pdf>.

Race D, Duncan DH, Kyle G, Merritt W and Park G (2009b) 'Landscape history and vegetation change in the Muckleford region of Victoria'. Landscape Logic, Hobart. <www.landscapelogic.org.au/publications/Brochures/Muckleford_brochure.pdf>.

Race D, Duncan DH, Kyle G and Merritt W (2009c) 'Landscape history and vegetation change in the Violet Town–Longwood region of Victoria'. Landscape Logic, Hobart. <www.landscapelogic.org.au/publications/Brochures/Violet_Town–Longford_brochure.pdf>.

Race D, Sample R, Curtis A and McDonald S (2010) 'Management of native vegetation on private land: perspectives of landholders and NRM program managers in northern Victoria'. Technical Report no. 24. Landscape Logic, Hobart.

Rango A, Laliberte A and Winters C (2008) Role of aerial photos in compiling a long-term remote sensing data set. *Journal of Applied Remote Sensing* **2**, 1–21.

Rumpff L, Duncan DH, Vesk PA, Keith DA and Wintle B (2011) State-and-transition modelling for adaptive management of native woodlands. *Biological Conservation* **144**(4), 1175–1178.

Schirmer J and Field J (2000) 'The cost of revegetation. Final report'. Australian National University Forestry and Greening Australia, Canberra.

Vesk PA and MacNally R (2006) The clock is ticking: revegetation and habitat for birds and arboreal mammals in rural landscapes of southern Australia. *Agriculture, Ecosystems and the Environment* **112**, 356–366.

Wimmera CMA (2010) 'Annual Report 2009–2010'. Wimmera Catchment Management Authority, Horsham, Victoria.

Wyatt BK (2000) Vegetation mapping from ground air and space: competitive or complementary techniques. In *Vegetation Mapping from Patch to Planet*. (Eds R Alexander and A Millington) pp. 3–15. John Wiley and Sons, Chichester.

Exploring landscape history through integrated participatory research: experiences from Victoria

Digby Race, Allan Curtis, Garreth Kyle, Wendy Merritt and Geoff Park

SUMMARY

The extent of native vegetation in the landscape is the function of many contemporary and historical factors. While the contemporary influences on native vegetation are relatively well understood, the historical perspective is often ignored or poorly understood. In part this is due to the difficulty of obtaining reliable data, but it is also because the collection of such data requires the combination of disciplines and perspectives. This chapter documents the process and results of a multi-disciplinary investigation into the historical and contemporary drivers of landscape change. This process brought together researchers in rural sociology, vegetation ecology and environmental modelling, as well as natural resource management (NRM) practitioners. Together, we sought to construct a picture of historical change in native vegetation within our three case study areas by mapping the change in native vegetation using aerial photographs from 1946–2008, engaging a selective sample of landholders in a series of landscape history workshops, conducting site inspections with landholders to explore specific native vegetation management actions, and in-depth interviews with landholders that explored the drivers of changes in landholder management of native vegetation.

Over the 60-year period there were substantial changes in farming practices and viability, community lifestyles and perceptions of native vegetation, all with a profound impact on native vegetation cover. The engagement of long-term resident landholders was invaluable for documenting activities, events and land use trends that often pre-dated the existing 'corporate memory' of NRM organisations. However, our approach also encountered some logistical and methodological challenges. These included effective communication within the research team that comprised different disciplines, engaging people from different organisations, and creating a shared understanding between researchers and landholders.

INTRODUCTION

It is difficult to understate the impact that humans have had on Australia's native vegetation since European settlement in the early 19th century. In recent decades, in recognition of the integral role of native vegetation in a healthy landscape, state and federal governments have invested considerable funds in native vegetation works in an attempt to mitigate some of this impact. For example, since the late 1990s, $2.8 billion has been spent via the Natural Heritage Trust on improving Australia's natural environment (ANAO 2008). A substantial portion of this money has been used to encourage landholders to increase, or at least retain, native vegetation on their land. Unfortunately, as a result of the complexity of natural systems and the lack of adequate monitoring (Duncan and Wintle 2008), the outcome of this spending in terms of vegetation extent and quality is largely uncertain (ANAO 2008). An improved understanding of the outcomes of investment in native vegetation is essential to improving future management interventions.

Based on historical accounts and vegetation modelling, we have a broad understanding of the nature of pre-European vegetation in Victoria, as well as the transformation that occurred following the arrival of Europeans. Within much of northern and central Victoria the most significant period of change occurred immediately after European colonisation in the early to mid 1800s. The predominant drivers of the initial wave of vegetation clearance were the discovery of alluvial gold and development of agriculture. Since the beginning of the 1900s the quality and extent of native vegetation has continued to change, though not in a consistent manner. Understanding the magnitude and drivers of these changes is an important step in designing and implementing effective management interventions.

Despite their importance, relatively little is understood about the historical human-mediated drivers of change in native vegetation, making it difficult to predict and manage the trajectory of the extent of native vegetation. Identifying these drivers and their impacts at the landscape scale is a complex and difficult task. The 'driver–pressure–state–impact–response' framework, originally developed by the OECD, is a conceptual model that explains the causal links and feedbacks associated with major influences that affect landholders' land use. It illustrates this potential complexity (Fig. 11.1).

A single outcome can be influenced by multiple drivers operating at different locations, times and dimensions (Braimoh 2009; Seabrook *et al.* 2006). The impact of a single driver may be amplified by other drivers or, conversely, moderated by opposing drivers (Brook *et al.* 2008), while the legacy of some drivers can persist long after the source of their influence has ceased

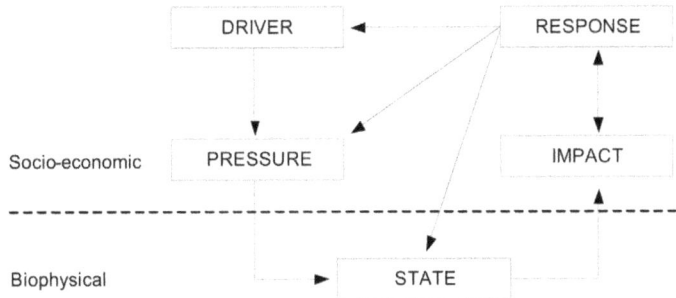

Figure 11.1: The driver–pressure–state–impact–response framework.

(Mottet *et al.* 2006). Attempting to understand the interaction of socio-economic and biophysical drivers on change in native vegetation extent presents a further layer of complexity. Several comprehensive studies have reviewed the drivers of environmental change in Australia (Cork and Delaney 2005), and specifically in Victoria (Martin and Werren 2009; Crosthwaite *et al.* 2004). These have included consideration of the scale of the challenge to create sustainable landscapes and the implications of predicted trends for NRM. The need to achieve a reducing rate of decline in native vegetation as well as an increase in desired landscape attributes is important as, despite investment to date, a recent assessment found that most catchments in Victoria were showing continuing signs of decline in land, water and biodiversity (DSE 2009).

Victoria's rural landscapes are as much social constructs as they are biophysical entities, with the values, aspirations and capabilities of local communities deeply etched in contemporary land use and condition of native vegetation. As such, exploring the convergence of social and ecological systems in rural landscapes invariably requires understanding the knowledge held by the local community (Alessa *et al.* 2007). Also, over recent decades there has been a growing consciousness that the process of acquiring and interpreting information about rural communities can also be a process of community empowerment (when rural communities are actively engaged as partners with researchers in data collection and analysis). Robert Chambers (Chambers 1983, 1997), among others, popularised the concept of 'rapid rural appraisal' (engaging the people who are the subject of the research in data analysis and problem-solving stages) in rural development (international aid) projects in the 1980s. Much has been written about the application and value of participatory processes with rural communities (Aslin and Brown 2004), such as yielding high-quality information at a low cost (Anadon *et al.* 2009). It is now a well-entrenched tenet of regional NRM in Australia.

For these reasons we adopted a participatory approach that directly engaged local community representatives and researchers from different disciplines in the analysis and documentation of landscape history. Through this approach, local people with detailed knowledge and experience of the changes within the local landscape could help us:

- avoid expensive and time-consuming approaches that require information from disparate sources to be pieced together;
- offset imperfect empirical data in publicly accessible sources;
- identify influences with a very localised impact and legacy;
- record historical activities and events that may retain a powerful influence among landholders, such as shaping their land use practices (bushfires, floods), that may not be fully explained or evident in agency policy documents or management plans;

- avoid perpetuating or entrenching an agency-centric or discipline-centric view of landscape history, by balancing these views with those of local landholders;
- engage local people and scientists in a shared 'conversation' (using a common language) that can inform each other's perspective – valuable when a cooperative approach is needed for NRM.

This chapter describes the process and results of an interdisciplinary collaboration that examined the historical drivers of change in extent of native vegetation at the landscape scale. Our work involved researchers with backgrounds in rural sociology, vegetation ecology and environmental modelling as well as NRM agency staff. We mapped change in native vegetation using aerial photographs from 1946–2008 and used a series of regional workshops to validate the findings from the mapping process, identify potential drivers of vegetation change and guide subsequent research. Our approach differed from previous studies in that it was interdisciplinary throughout the problem definition, data collection and analysis stages and reviewed influences on land use change over a lengthy period (60 years).

LANDSCAPE HISTORY APPROACH

BACKGROUND

We selected one case study area within each of our three partner catchment management authorities (CMAs): North Central, Goulburn Broken and North East (Fig. 11.2). The selection of the case study areas considered land uses, investment in native vegetation works and our confidence in distinguishing between natural regeneration due to land use changes and the presence of native vegetation remnants (see Chapter 10, this volume). The socio-economic characters of these areas have been broadly characterised as 'rural amenity' and 'rural transitional' landscapes (Barr *et al.* 2005). Traditional farming practices (e.g. wool and beef production) are decreasing in area and intensity while rural residential and alternate farm enterprises are increasingly popular, resulting in an increase in rural land values beyond the perceived agronomic value for traditional farm enterprises.

CONNECTING SCIENTIFIC AND COMMUNITY KNOWLEDGE

The regional workshops that are the focus of this chapter were primarily designed and implemented to inform other research carried out by Landscape Logic. The following section provides an overview of the methods associated with this work, with more detailed descriptions provided elsewhere (Duncan *et al.* 2010; Duncan and Kyle 2010).

We used aerial photographs from 1946–2008 to map change in the extent of mature native trees. Change was attributed to one of five categories: loss in solid canopy (loss); gain in solid canopy (gain); increase in the density of existing tree cover (thickening); decrease in the density of existing canopy (thinning); and loss of a group (>3) of remnant or 'paddock' trees (scattered loss). The spatial data depicting change in native tree canopy were presented in a series of regional landholder workshops, one in each case study area. The study period of 1946–2008 reflects the earliest aerial photographs available in the region. We focused on tree cover as the resolution of some of the available aerial photographs was insufficient to accurately identify shrubs, forbs and grasses.

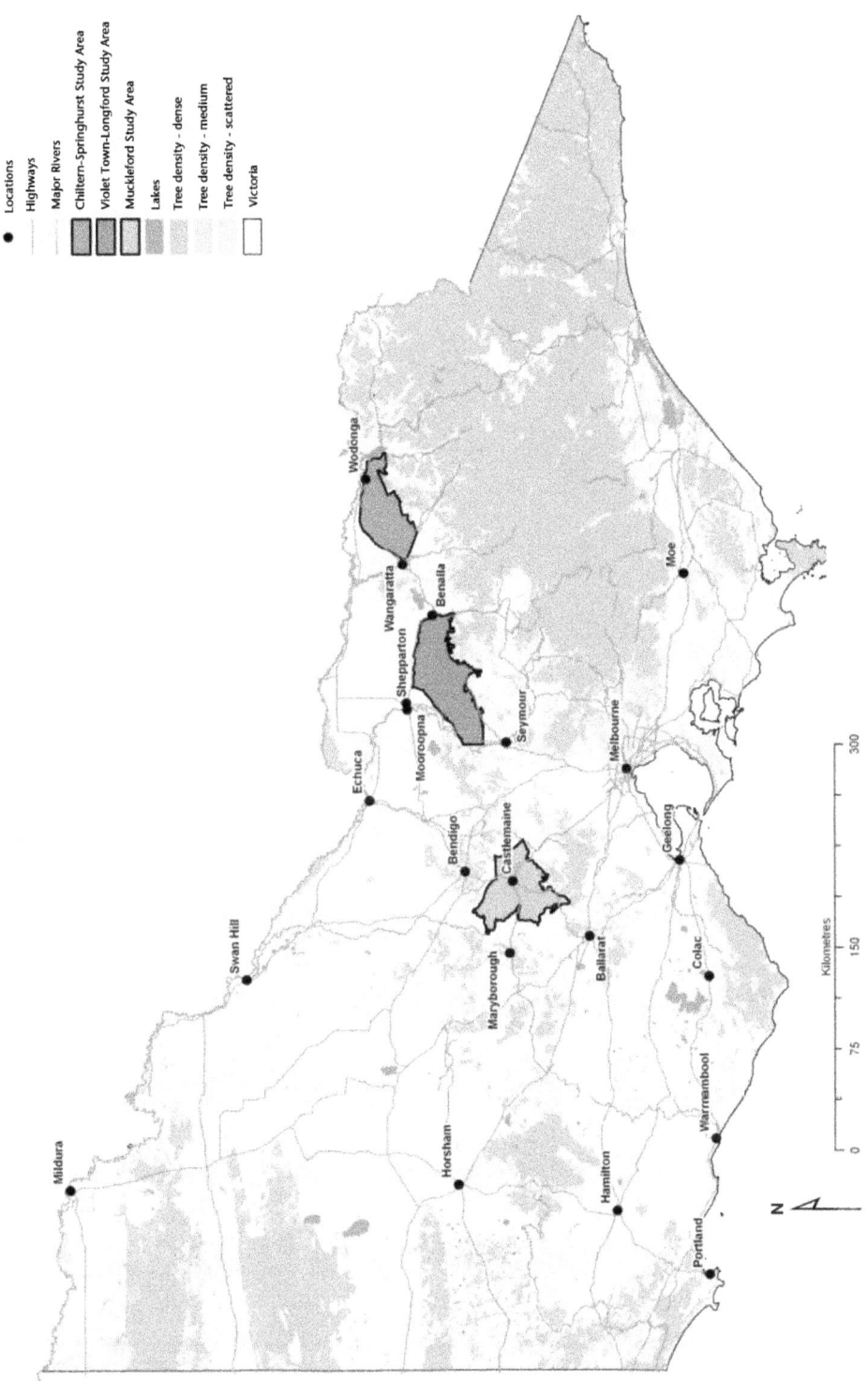

Figure 11.2: Location of study areas in Victoria.

The workshops ran for about two and half hours. The research team gave a short presentation outlining the purpose of Landscape Logic and the Victorian vegetation research and presented the spatial data. The results of the mapping process were discussed among researchers and participants, with the latter encouraged to annotate hard-copy maps of tree cover change, showing the areas with which they were familiar and commenting on polygons of change. Following this was a period of facilitated discussion led by a rural sociologist, in which the research team asked participants to discuss the history of the region with particular emphasis on the impact on native vegetation. Points discussed during this session were recorded on a board so that participants could ensure the facilitator had correctly interpreted their perspective. We also had hard-copy maps, aerial photographs and charts of historical rainfall to help prompt discussion during the workshops.

The workshops were run over a single afternoon, but a considerable amount of work went into each one. Each workshop involved eight to 12 local landholders and NRM agency staff. Participants were initially contacted by letter, with a follow-up phone call from an NRM manager about one month before the workshop. To ensure key events in each region's history were discussed, the researchers had to be familiar with the broad history of the region. To improve our regional knowledge, we took part in a half-day field trip through the study area with a local guide who was familiar with the natural, social and agricultural history.

Following the workshop, a succinct report (less than six pages) was sent to participants within six weeks seeking their corrections and additional information. The report was a synthesis of the key points that emerged from the workshop, combined with relevant information from other sources such as historical wool prices and rainfall records. After receiving feedback from workshop participants, the research team revised and published the landscape history summaries in a user-friendly style for distribution in hard copy and electronic forms (Race *et al.* 2009a, b, c). It should be stressed that activities associated with vegetation change identified in the summaries may not be 'drivers' of change, but represented a preliminary synthesis to guide our subsequent in-depth field assessments and landholder interviews.

As part of the wider Landscape Logic effort, Bayesian network models (Chapter 18, this volume) were developed to identify and explore the importance of factors influencing the type of vegetation actions being implemented in the case study areas, and the influence of these vegetation actions on vegetation condition (given the landscape context). Outcomes from the landscape history workshops were valuable in refining the influence or conceptual diagrams which represented our understanding of the cause-and-effect relationships between drivers (e.g. land management, climate), processes (e.g. recruitment of indigenous vegetation species) and outcomes (change in vegetation condition).

RESULTS: A PERIOD OF DYNAMIC AND VARIABLE CHANGE

This section provides a synthesis of the drivers of land use change across our three study areas. The various influences on the extent of native vegetation on farmland in the study areas were identified as multiple and complex, with uncertain interactions and implications (see Fig. 11.3 for an example of how we recorded the multi-dimensional change within a single case study area). Some of the drivers of change had an immediate impact on the extent of native vegetation (e.g. conversion from livestock grazing to cropping) while others were incremental (e.g. native species favoured by decline in use of fertiliser). Also, some drivers had a direct impact

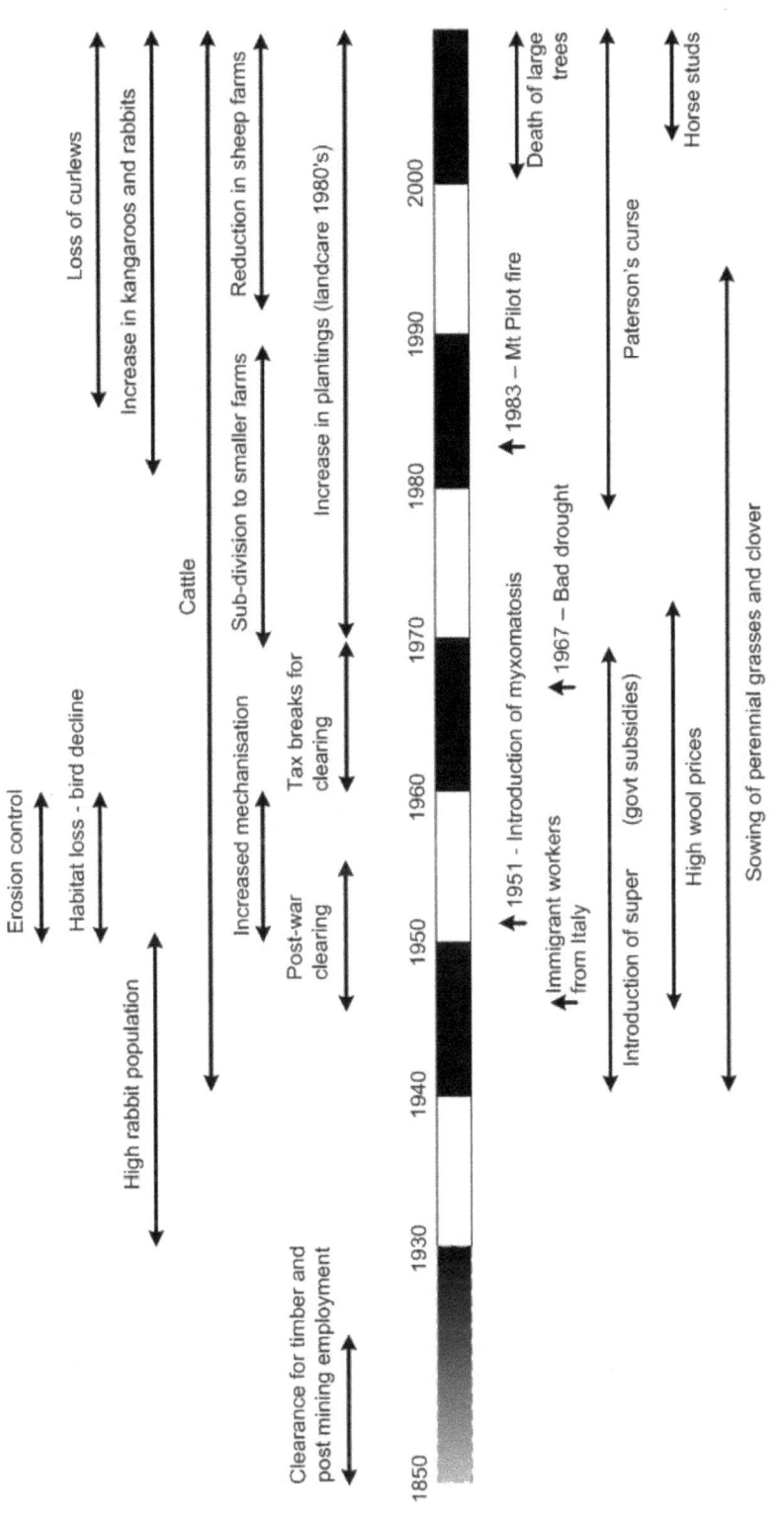

Figure 11.3: Timeline of key activities, events and changes in the Chiltern–Springhurst study area, 1930–2006.

(e.g. revegetation or sowing seeds of native vegetation) and others indirect (e.g. control of rabbits, change in demographics). Furthermore, some influences were highly visible (e.g. clearing of native vegetation for town settlements) while others were only significant in aggregate (e.g. drought plus increases in pests). We found that local knowledge added a new level of understanding to such complexities.

CHANGES IN FARMING

In broad terms, the magnitude and type of changes in agriculture in our study areas reflect the national trend, with a steady increase in farmland, mostly for livestock farming, until its peak in the mid 1970s. The 1950s were characterised by the widespread promotion and establishment of introduced pasture species, mainly phalaris (*Phalaris tuberosa*), subclover (*Trifolium subterraneum*) and lucerne (*Medicago sativa*), together with increased applications of fertiliser (superphosphate and molybdenum, subsidised in the 1960s), which led to a notable increase in the productivity and profitability of farming. Returns from wool production in particular were buoyant between the mid 1950s and early 1970s, although prices fluctuated from year to year. These changes led to the decline of native grasses in production lands.

While wool production was the dominant farming enterprise in the region (in terms of area of land), areas of flatter and more fertile farmland were used for cropping and cattle (beef and dairying). The steep hills that had been cleared of native vegetation were aerially sown with phalaris in an attempt to prevent soil erosion and the spread of weeds. The high wool prices during the mid 1950s led to many farmers investing in farm infrastructure (e.g. increase in drainage works) and equipment (e.g. purchase of tractors). The availability of tractors meant that farmers could cultivate much greater areas for both pasture improvement (e.g. spreading superphosphate) and fodder conservation (e.g. hay production). Larger farm machinery made tree removal easier and more desirable, as paddock trees tended to impede the movement of larger machinery. Stubble burning also contributed to a noticeable decline in paddock trees, as trees that burned were often removed soon after.

Since the 1980s there has been a decline in terms of trade for most small to medium-sized wool producers in Victoria. Since the early 1990s there has been a noticeable increase in the number of small high-value rural properties sold to 'lifestylers' or 'hobby farmers', who are often willing to pay more for rural land than its perceived value for productive agriculture alone. The past decade has also seen increasing costs for many inputs (e.g. fertilisers, fuel) with decreasing farm-gate returns.

For many farmers, the poor returns from conventional livestock enterprises (wool, meat) and continued dry seasons have resulted in destocking of much of the region's more marginal farmland and the subdividing of properties for sale to lifestylers. Consequently, there has been an increase in the regeneration of native vegetation in destocked areas. Some workshop participants noted that the combination of declining use of superphosphate, reduced cropping and lower livestock numbers has led to an increase in native tree regeneration and recolonisation by native grasses since the mid 1990s. However, workshop participants also suggested that, while there may have been an increase in the area of native vegetation, there has been a loss in the quality of native vegetation (decline in biodiversity) over this period. The main indications of this decline were increasing weed cover, decreasing habitat for birds and other native animals, and the continued loss of large paddock trees throughout the study areas. It was also reported that there is little regeneration of understorey vegetation on the plains, with soil erosion on localised sites.

Other industries also contributed to clearing of native vegetation during the 1960s–90s through the demand for power poles by utility companies, realignment and expansion of major roads and gravel pits and quarries for road construction.

Government policies and programs that were reported to have had an impact on the extent of native vegetation (decrease and increase) included:

- requirement by government and banks (holding mortgages) for landholders with government-issued leases to convert bush to permanent farmland from the 1880s to 1930s;
- income tax deductions for expenses related to clearing native vegetation from farm-land (ending in the 1970s);
- establishment of soldier settlement properties after World War II;
- subsidies of superphosphate during the 1960s;
- sowing phalaris and planting willows and poplars for erosion control from the 1950s–80s;
- commencement of the state/national Landcare program in the late 1980s, raising awareness of native vegetation and provision of plants and fencing materials;
- the introduction in 1989 of state legislation restricting the clearance of native vegeta-tion;
- promotion of 'whole farm planning', including promoting the benefits of retaining remnant native vegetation.

A FINER SCALE: A CHANGING COMMUNITY

During the Depression, many unemployed people came from Melbourne to Chiltern to work in the region's timber industry (harvesting, milling, transporting). The influx of migrants following World War II led to many being employed in the region as farm labourers, with some later establishing small farming properties. However, the region's farming communities began to decline noticeably after the 1970s.

The Victorian Department of Soil Conservation (DSC) had been active in the region since the 1950s, focusing on working with farmers to control soil erosion. DSC staff offered farmers subsidised materials (fencing, grass seed) and on-ground works (rock structures) to control soil erosion. Until the 1980s, poplars and willows were still being recommended to control riverbank and gully erosion.

Workshop participants reported that in the 1960s schools began to promote environmen-tal awareness among students, which was supported in the 1970s by local projects to revegetate along creeks and by larger-scale native plantings (e.g. the Albury-Wodonga Development Corporation's plantings around Leneva, Barnawartha and Baranduda in north-east Victoria). Several Farm Tree and later Landcare groups were formed in the 1980s. These groups fostered awareness and interest in planting native species (not necessarily local) on farms for livestock shade and shelter, for native wildlife habitat and for erosion control. By the 1990s, this interest had evolved into widespread interest in planting and protecting local native species (including understorey species), and more recently interest in local wildlife. Following the increased awareness in native vegetation management created by Farm Tree and Landcare groups, and the influx of lifestylers, there was considerable interest in joining the Victorian Department of Sustainability and Environment's Land for Wildlife network during the 1990s. More recently, there has been interest in the conservation covenants promoted by Trust for Nature.

CONCLUSIONS

MESHING SCIENCE WITH A BOTTOM-UP APPROACH TO UNDERSTAND LANDSCAPE HISTORY

Our experience highlights a range of benefits for researchers when they actively engage the local community in the process to document and analyse the history of a local landscape. These include:

- efficient assembling of data and information from which to acquire a shared knowledge about the history of the local landscape;
- better comparisons between regions, and placing these regions within the context of a wider geographic understanding of landscape dynamics;
- understanding and evaluating conflicting uses and interpretations of data/information ('evidence');
- improved awareness within local communities of elements of continuity and change in their local landscape;
- a greater understanding of the history of a region, and avoiding judgements based exclusively on the views of 'outsiders';
- improved understanding of the way that landholder experience, circumstance and capacity influence land use over time;
- a stronger relationship between researchers and local communities.

The combination of the regional workshops and the interdisciplinary research team proved an invaluable platform from which to direct further research. The workshops provided important, locally referenced knowledge of each study area, much of which was either undocumented or would have required intensive research to acquire. While we found our approach to be a relatively efficient means of building a rich picture of the history of a rural landscape, a key ingredient was the researchers' willingness to engage in a cross-disciplinary conversation and be receptive to the knowledge of many non-researchers (e.g. landholders, staff of NRM organisations, agribusiness specialists) who actively manage the local landscape.

The approach described in this chapter is an attempt to bridge the social and natural sciences, and efficiently build a shared epistemology between research disciplines and between researchers and the local community. However, it is an approach that should not be underestimated in terms of planning, preparation (people and materials) and implementation. Our experience showed that considerable effort needs to be given to:

- ensure adequate resourcing and time for an effective participatory process;
- generate the background information and materials (e.g. adequate mining of existing data);
- engage a range of experienced people (landholders, agency staff);
- cross-reference scientific and local knowledge;
- involve people with a range of expertise in the data analysis and interpretation.

By listing the steps involved in the approach we used for this research, we are attempting to provide some transparency to the process that was effective for our purpose and to encourage other researchers to consider using a similar process for integrated participatory research.

ACKNOWLEDGEMENTS

The authors express their sincere gratitude to the many landholders, NRM professionals and other researchers, particularly David Duncan, who contributed to the generation of the collective knowledge reflected in this chapter. The authors also thank Imogen Fullagar and an anonymous reviewer for their valuable comments on an earlier draft of this chapter.

REFERENCES

Alessa L, Kliskey A and Brown G (2007) Social-ecological hotspots mapping: a spatial approach for identifying coupled social-ecological space. *Landscape and Urban Planning* **85**(1), 27–39.

Anadon JD, Gimenez A, Ballestar R and Perz I (2009) Evaluation of local ecological knowledge as a method for collecting extensive data on animal abundance. *Conservation Biology* **23**(3), 617–625.

Aslin HJ and Brown VA (2004) *Towards Whole of Community Engagement: A Practical Toolkit*. Murray-Darling Basin Commission, Canberra.

Australian National Audit Office (ANAO) (2008) 'Regional delivery model for the Natural Heritage Trust and National Action Plan for Salinity and Water Quality'. ANAO Report no. 21. Australian Government, Canberra.

Barr N, Wilkinson R and Karunaratne K (2005) 'Understanding rural Victoria'. Dept of Primary Industries, Melbourne.

Braimoh AK (2009) Agricultural land-use change during economic reforms in Ghana. *Land Use Policy* **26**(3), 763–771.

Brook BW, Sodhi NS and Bradshaw CJA (2008) Synergies among extinction drivers under global change. *Trends in Ecology and Evolution* **23**(8), 453–460.

Chambers R (1983) *Rural Development: Putting the Last First*. Longman, London.

Chambers R (1997) *Whose Reality Counts? Putting the First Last*. Intermediate Technology Publications, London.

Cork SJ and Delaney K (2005) 'Thinking about the futures of Australia's landscapes'. Technical Report. Land and Water Australia, Canberra.

Crosthwaite J, Callaghan J, Farmar-Bowers Q, Hollier C and Straker A (Eds) (2004) 'Land use changes, their drivers and impact on native biodiversity – driver research phase one: overview report'. Dept of Sustainability and Environment, Melbourne.

Dept of Sustainability and Environment (DSE) (2009) *Securing Our Natural Future: A White Paper for Land and Biodiversity at a Time of Climate Change*. Dept of Sustainability and Environment, Melbourne.

Duncan D and Kyle G (2010) 'Management of native vegetation on private land in northern Victoria: the magnitude, motivation and resourcing of revegetation and restoration'. Technical Report no. 23. Landscape Logic, Hobart.

Duncan DH and Wintle BA (2008) Towards adaptive management of native vegetation in regional landscapes. In *Landscape Analysis and Visualisation: Spatial Models for Natural Resource Management and Planning*. (Eds C Pettit, W Cartwright, I Bishop, K Lowell, D Pullar and D Duncan) pp. 159–182. Springer-Verlag, Berlin.

Duncan DH, Kyle G and Race D (2010) Combining facilitated dialogue and spatial data analysis to compile landscape history. *Environmental Conservation* **37**(4), 432–441.

Martin P and Werren K (2009) 'Discussion paper: an industry plan for the Victoria environment?' Report prepared for the Victorian Dept of Sustainability and Environment, Melbourne.

Mottet A, Ladet S, Coqué N and Gibon A (2006) Agricultural land-use change and its drivers in mountain landscapes: a case study in the Pyrenees. *Agriculture, Ecosystems and Environment* **114**(2–4), 296–310.

Race D, Duncan DH, Kyle G and Merritt W (2009a) 'Landscape history and vegetation change in the Chiltern–Springhurst region of Victoria'. Landscape Logic, Hobart. <www.landscapelogic.org.au/publications/Brochures/Chiltern_Springhurst_brochure.pdf>.

Race D, Duncan DH, Kyle G, Merritt W and Park G (2009b) 'Landscape history and vegetation change in the Muckleford region of Victoria'. Landscape Logic, Hobart. <www.landscapelogic.org.au/publications/Brochures/Muckleford_brochure.pdf>.

Race D, Duncan DH, Kyle G and Merritt W (2009c) 'Landscape history and vegetation change in the Violet Town–Longwood region of Victoria'. Landscape Logic, Hobart. <www.landscapelogic.org.au/publications/Brochures/Violet_Town–Longford_brochure.pdf>.

Seabrook L, McAlpine C and Fensham R (2006) Cattle, crops and clearing: regional drivers of landscape change in the Brigalow Belt, Queensland, Australia, 1840–2004. *Landscape and Urban Planning* **78**(4), 373–385.

12

Development of a state-and-transition model to guide investment in woodland vegetation condition

Libby Rumpff, David Duncan, Peter Vesk and
Brendan Wintle

SUMMARY

Reversing the decline in the extent and condition of native vegetation is a national priority in Australia, with substantial investments in native vegetation restoration and management being made across the country. However, considerable uncertainty surrounds the determinants of success and the most cost-effective strategies to achieve restoration goals. Targeted monitoring is exceedingly rare in natural resource management (NRM) and consequently there is no reliable way to justify government investment on the basis of measured environmental benefits. This is a critical shortcoming that is frequently exposed as NRM agencies report on their progress towards reaching native vegetation targets.

Process models can represent beliefs about the dynamics of an ecological system and how the ecosystem responds to management. Our team's aim was to develop a process model that could be used to represent, inform, learn about and report on the ecological impact of investment in native vegetation management. We briefly describe the conceptual and quantitative models already used by our stakeholders in the context of guiding investment decisions and outcome reporting. The shortcomings of those approaches are highlighted. We also discuss the early stages of our own model development and exploration, culminating in the decision to pursue a quantitative state-and-transition model (STM) framework, implemented as a Bayesian network. We are optimistic that the STM can be utilised to support adaptive management of native woodland vegetation in south-eastern Australia.

INTRODUCTION

The long history of widespread clearing, utilisation, overutilisation and conversion of Australian ecosystems has extinguished species and threatened the habitat of those that remain and the ecological functions they support. Reversing that long-term decline in native vegetation extent and quality to achieve a net gain has become a priority for investment in the environment (DNRE 2002). However, despite decades of investment, resource management agencies have found it difficult to demonstrate how their investment of public funds has contributed to positive ecological change (ANAO 1998, 2008; DSE 2008b).

For the most part, ecologists and those charged with managing and implementing management have failed to convert the rhetorical commitment to monitoring into an active element of the NRM culture (Lindenmayer and Likens 2010). This is certainly true of native vegetation management, where little structured learning has emerged from the many successes and failures (Rumpff *et al.* 2011). The management of natural resources such as native vegetation, habitat and biodiversity are ecologically complex and the desired outcomes are expected to be highly variable in space and time. In Australia, additional complexity arises from the fact that regional native vegetation management agencies are charged with reporting on changes in vegetation condition arising from their investment on lands for which they have no direct management responsibility or control (Duncan and Wintle 2008). There are many investment programs in native vegetation and many government and non-government delivery agents (Crowley 2001; Duncan and Kyle 2010), further adding to the complexity in information aggregation and interpretation. On the other hand, only two main types of funded active management exist: the exclusion of grazing and disturbance from remnant vegetation; and revegetation, the reintroduction of propagules and seedlings into formerly cleared land. Nonetheless, at present it is difficult to predict the likely outcomes of a specific investment at a given site, let alone the cumulative impact of many such investments over a region or catchment of interest (DSE 2008b).

Process models are an inherently useful tool that can be used to characterise and, in some instances, overcome uncertainties such as those described above for native vegetation management. At the very least, a collaborative approach to process model development helps to tease apart semantic ambiguities from genuine epistemic uncertainty (Burgman 2005). A process model is a synthesis of beliefs and knowledge about ecological system dynamics and the response of the system to management interventions. It is used to explore knowledge gaps and uncertainties, guide investment decisions and support reporting requirements. Quantitative models of ecosystem response help resolve uncertainties about management as they can be iteratively updated using monitoring data (Rumpff *et al.* 2011). As such, quantitative process models are a core component of an adaptive management framework, as they enable us to evaluate the extent to which management actions contribute to achieving management objectives (Nichols and Williams 2006; Duncan and Wintle 2008; Rumpff *et al.* 2011).

In this chapter we describe the roles and limitations of existing models used to guide investment decisions in native vegetation management in Victoria. We then present our attempts to develop a site-based quantitative process model to provide the basis for a targeted monitoring strategy for a Victorian catchment management authority (CMA). The aim was to design a model that would fit within the decision context of the Goulburn Broken CMA and so could be used to represent, inform, learn about and report on the ecological impact of investment in native vegetation management. Our model development so far has culminated in a quantitative state-and-transition model, implemented as a Bayesian network (Rumpff *et al.* 2011). We are optimistic that the STM we present can be utilised to support adaptive management of native woodland vegetation in south-eastern Australia.

EXISTING NATIVE VEGETATION CHANGE MODELS USED BY STATE AND REGIONAL NRM BODIES

The models that inform the resource management and investment decision-making of NRM agencies are not often made explicit, if they are recognised as 'models' at all (Rumpff *et al.* 2011; Chapter 2, this volume). However, the Victorian Department of Sustainability and Environment and a number of CMAs have developed a series of conceptual and quantitative models to guide investment and estimate change in the extent and condition of native vegetation.

The general theory about the effect of government investment in native vegetation management is given by Figure 12.1. Under many relevant public investment programs, private land owners can receive financial support to change management (e.g. remove stock, signified by a new fence; refrain from cutting or removing timber) in the expectation this will improve the condition of the relevant site (Fig. 12.1, trajectory A). Funding is allocated on the assumption that either the average pattern of utilisation is degrading the condition of the land (Fig. 12.1, trajectory B) or, even if the current management is not degrading, funding will reduce the risk of a later change to a more degrading use (Fig. 12.1, trajectory C). The intention of this kind of conceptual investment model may be sound, but as an accountable model it is difficult to decompose and interrogate the underlying assumptions. For example, what kind of improvement is anticipated, how will it be achieved and over what time-frame?

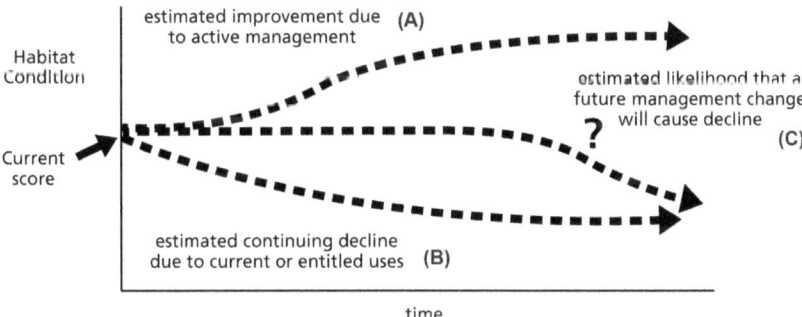

Figure 12.1: The change in habitat condition expected over time with and without management intervention (DSE 2008b, p. 13). 'Current score' is a measure of habitat condition, which is calculated using the Habitat Hectares approach (Parkes *et al.* 2003; DSE 2004).

A nominal model of site improvement that has been used in a number of reporting contexts suggested that the quality of a native vegetation site under conservation management agreements increases by 10% over 10 years (Brunt and McLennan 2006; VCMC 2007; DSE 2008a). Though arbitrary, these assertions are at least explicit and consequently they invite new data to refine the assumption. But what might this 10% increase in quality look like? Presumably, gains in quality differ in amount and type depending on the age and quality of the vegetation being managed and on a host of other environmental and management history variables. In Victoria, as in most other states, investment in quality at a given site is conceived as improving vegetation structure (diversity and cover of life-forms) and composition (diversity and cover of native species within life-forms). The Habitat Hectares metric is widely employed in Victoria to make assessments of vegetation condition. The method records large trees, tree canopy cover, understorey life-form cover and diversity, amount of fallen logs, weediness, organic litter and plant species recruitment in categorical classes (Parkes *et al.* 2003). Functionality is assessed by scoring the size and position of the site in relation to the surrounding landscape (i.e. landscape context) (DSE 2006). The desired endpoint is sites which have the structural and compositional characteristics approaching long-undisturbed reference sites. Using these reference sites as a guide, benchmark levels are estimated for each attribute and candidate sites for investment are compared against benchmark conditions (Parkes *et al.* 2003). 'Naturalness' is a common goal in vegetation management and benchmarks are used as a measure of this goal (Gibbons *et al.* 2008). However, both their definition and use is contentious (Hunter 1996; Oliver *et al.* 2002; McCarthy *et al.* 2004).

To calculate a biodiversity benefit score from conservation management proposals or evaluate offset proposals, a spreadsheet calculator model (Gain Calculator) was developed (Fig. 12.2; DSE 2006, 2008a). The Gain Calculator and its companion document (DSE 2006, 2008a) provide an explicit logic for change within individual attributes of a native vegetation condition score. It includes the type and area of a vegetation site, the current score of that site described in Habitat Hectares currency, the current controls and management being applied to that site, the proposed new controls and the predicted gain in quality of that site under those circumstances. It is among the most detailed vegetation change models in use in NRM. However, several problems inhibit the model from supporting learning about how vegetation condition changes over time. First, the Gain Calculator model deals only with the influences on native vegetation attributes due to the prescribed or proscribed management actions. It does not include important influences on vegetation change influenced by broader landscape or climatic factors, nor the land use history or socio-economic drivers that may contribute to or constrain native vegetation change. Second, by necessity the model is generic because it is required to be implemented in all vegetation forms across the state. For example, there is only one set of calculations for all wooded vegetation types and another for unwooded vegetation types. Last, a key factor that limits the ability of this model to support learning is the opacity of the underlying assessment metric, the Habitat Hectares method (Parkes *et al.* 2003; DSE 2004). The Habitat Hectares method is inherently unsuitable for monitoring change over time as it is based on roving visual assessments and broad categorical cover classes that ensure a high degree of subjectivity during the assessment process (Duncan and Vesk, unpub. data). Habitat Hectares lacks sensitivity to confidently detect or apportion change to different management regimes in the context of background environmental and random site variation. This is unsurprising, as Habitat Hectares was primarily developed to serve a distinct function as a conservation planning tool to aid decisions about the allocation

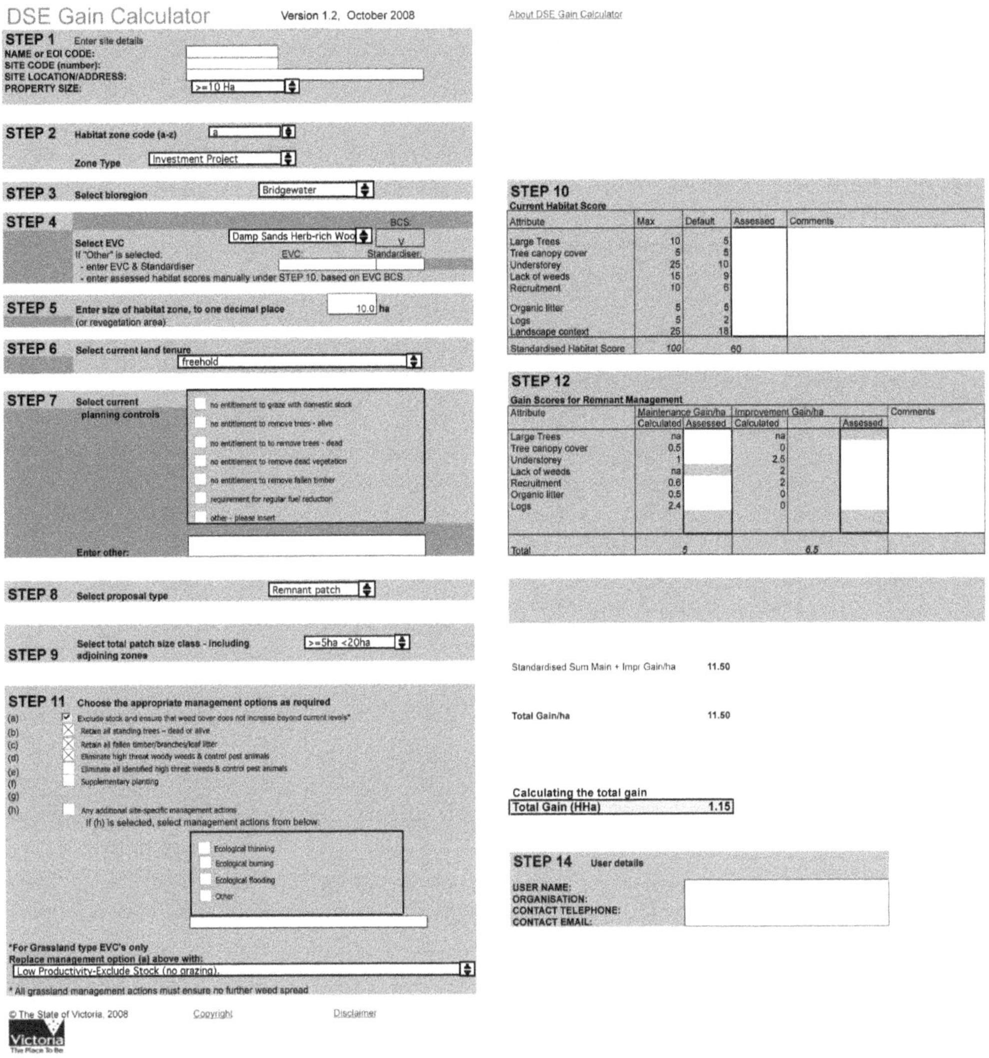

DSE Gain Calculator   Version 1.2, October 2008   About DSE Gain Calculator

STEP 1   Enter site details
NAME or EOI CODE:
SITE CODE (number):
SITE LOCATION/ADDRESS:
PROPERTY SIZE: >=10 Ha

STEP 2   Habitat zone code (a-z)   a
         Zone Type   Investment Project

STEP 3   Select bioregion   Bridgewater

STEP 4
         Select EVC   Damp Sands Herb-rich Woo   BCS: v.
         If "Other" is selected:
         - enter EVC & Standardiser   EVC: Standardiser:
         - enter assessed habitat scores manually under STEP 10, based on EVC BCS

STEP 5   Enter size of habitat zone, to one decimal place   10.0 ha
         (or revegetation area)

STEP 6   Select current land tenure
         freehold

STEP 7   Select current
         planning controls
         ☐ no entitlement to graze with domestic stock
         ☐ no entitlement to remove trees - alive
         ☐ no entitlement to to remove trees - dead
         ☐ no entitlement to remove dead vegetation
         ☐ no entitlement to remove fallen timber
         ☐ requirement for regular fuel reduction
         ☐ other - please insert

         Enter other:

STEP 8   Select proposal type   Remnant patch

STEP 9   Select total patch size class - including   >=5ha <20ha
         adjoining zones

STEP 11 Choose the appropriate management options as required
(a) ☑ Exclude stock and ensure that weed cover does not increase beyond current levels*
(b) ☒ Retain all standing trees – dead or alive
(c) ☒ Retain all fallen timber/branches/leaf litter
(d) ☒ Eliminate high threat woody weeds & control pest animals
(e) ☐ Eliminate all identified high threat weeds & control pest animals
(f) ☐ Supplementary planting
(g)
(h) ☐ Any additional site-specific management actions
        If (h) is selected, select management actions from below:
         ☐ Ecological thinning
         ☐ Ecological burning
         ☐ Ecological flooding
         ☐ Other

*For Grassland type EVC's only
Replace management option (a) above with:
Low Productivity-Exclude Stock (no grazing)
* All grassland management actions must ensure no further weed spread

© The State of Victoria, 2008   Copyright   Disclaimer

Victoria
This Place To Be

STEP 10
Current Habitat Score

Attribute	Max	Default	Assessed	Comments
Large Trees	10	5		
Tree canopy cover	5	5		
Understorey	25	10		
Lack of weeds	15	9		
Recruitment	10	6		
Organic litter	5	5		
Logs	5	2		
Landscape context	25	18		
Standardised Habitat Score	100	60		

STEP 12
Gain Scores for Remnant Management

Attribute	Maintenance Gain/ha Calculated	Assessed	Improvement Gain/ha Calculated	Assessed	Comments
Large Trees	na		na		
Tree canopy cover	0.5		0		
Understorey	1		2.5		
Lack of weeds	na		2		
Recruitment	0.6		2		
Organic litter	0.5		0		
Logs	2.4		0		
Total	5		6.5		

Standardised Sum Main + Impr Gain/ha   11.50

Total Gain/ha   11.50

Calculating the total gain
Total Gain (HHa)   1.15

STEP 14   User details
USER NAME:
ORGANISATION:
CONTACT TELEPHONE:
CONTACT EMAIL:

Figure12.2: Annotated screenshot of the DSE native vegetation gain calculator (Ticehurst and Duncan 2008). Version 1.2 of the calculator model is available from www.dse.vic.gov.au (DSE 2008a).

of resources and was purposefully broad so that botanical expertise was not required for its use (Parkes *et al.* 2003).

A NEW MANAGEMENT MODEL OF SITE CHANGE IN NATIVE VEGETATION

Our pursuit of a model of native vegetation change better suited to encapsulating and learning about ecological changes in response to management was undertaken with a number of principles in mind. First, the model should allow for distinct outcomes for different vegetation types, rather than a generic state- or region-wide picture. Second, the model should incorporate the influence of site management, but also ecological processes, climate and land use legacies (i.e. a process model, see Rumpff *et al.* 2011). Last, the model should deal in raw continuous data on vegetation structure and composition. As a principle of development, we decided to be consistent with the habitat attributes described in the Habitat Hectares framework to aid the use and adoption of such a model by NRM agency staff.

To this end, we carried out a pilot field study of a field method sensitive enough to detect changes in structure and composition over the time-scale of investment programs and associated reporting requirements (Rumpff *et al.* 2009). The information currency of the model determines whether it is possible to detect meaningful changes over time and whether the model can effectively be updated with new data.

The early phase of model-building through influence diagrams inevitably led to considerable complexity and poor spatio-temporal resolution (Fig. 12.3). In common with the Gain Calculator model, this conceptual model recognised that the initial condition (at the time of assessment) and the applied management should be key drivers of the observed change in condition. It decomposed change in some Habitat Hectare attributes in order to capture existing knowledge about differential sensitivity to management interventions. For example, understorey life-form was redefined into finer divisions of shrub, herb and graminoid elements to capture the variable effects of vertebrate herbivore grazing on different life-form groups. The model also attempted to capture the basic environmental conditions that would give rise to the ecological vegetation type under consideration. It incorporated the influence of former land use regimes, recognising that the capacity to change may be influenced by legacies of former land use type and intensity.

Despite the apparent complexity of this model, there was a relatively simple underlying structure. Environmental drivers influence vegetation type and the site history and land use history determine the initial state for each of the vegetation components (attributes). These factors, plus any applied management interventions, influence the change observed in the vegetation attributes over time. However, early attempts at parameterising this model made it abundantly clear that it would be an arduous task to translate this influence diagram structure into a workable quantitative model. For example, the initial state (for each vegetation component) was implicit only, as the trajectory of vegetation change was captured in a deterministic relationship with past land use. As a result, updating the model would be difficult and we would not be able to capture and resolve many of the uncertainties surrounding the response of system dynamics to management. However, the model provided an important foundation for the desired quantitative process model. We recognised that the model implicitly carried the hallmarks of an existing and increasingly popular modelling approach for native vegetation management: state-and-transition models (STMs).

LOGIC BEHIND DEVELOPMENT OF THE STM

At this stage our efforts became more focused on developing a site-based model that would fit within the decision context of a Victorian CMA. We collaborated with the Goulburn Broken CMA to achieve this. Within its regional catchment strategy, the Goulburn Broken CMA has a specified target for improving vegetation condition but it was uncertain how to measure progress toward that objective. Its goal was to resolve the uncertainties surrounding the current

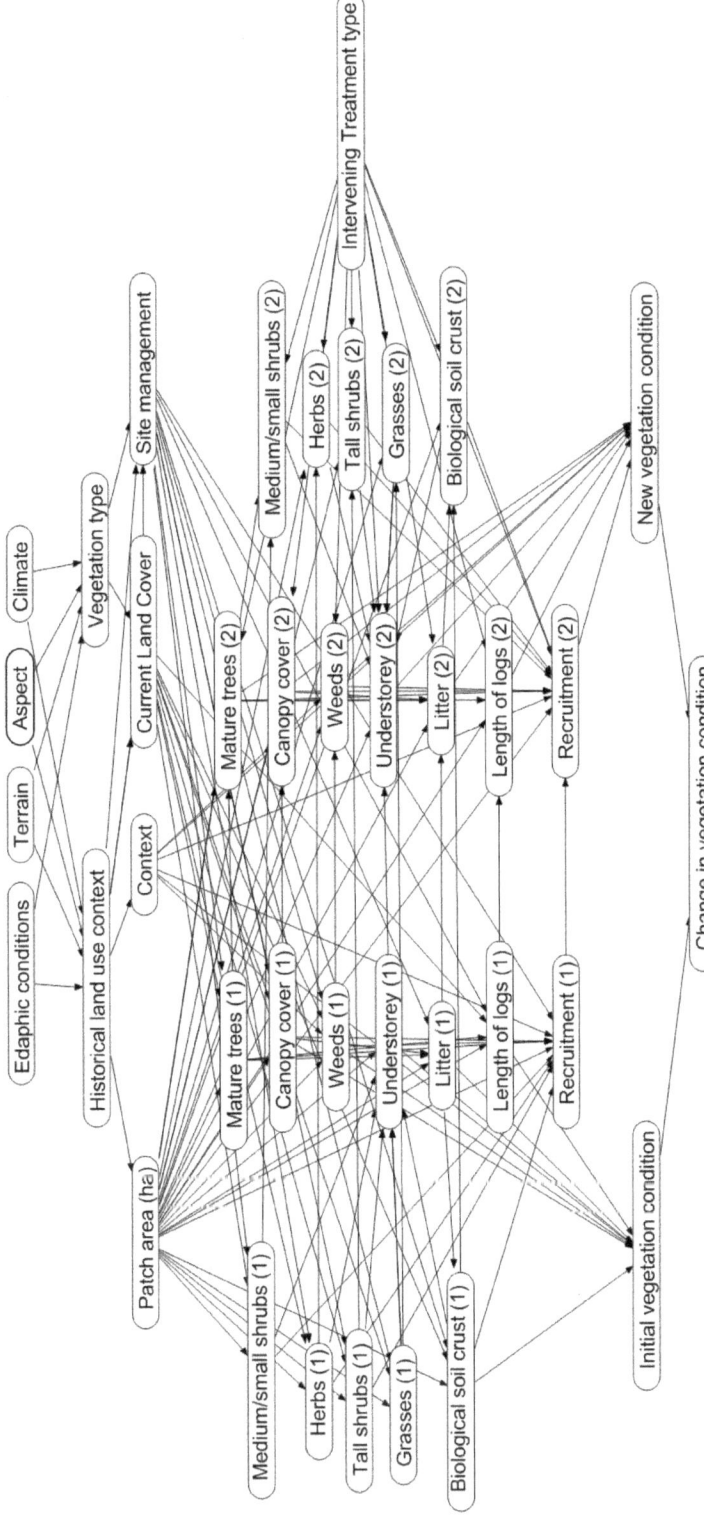

Figure 12.3: Early developmental stage of a model of native vegetation change designed to detect ecological changes in response to management. The model includes vegetation type, vegetation attributes measured over two time periods (1, 2), measures of vegetation condition (Initial, New, Change), site management, management intervention (Intervening treatment), ecological processes, climatic and land use legacy factors.

assumptions about the links between management actions and achieved outcomes. The Goulburn Broken CMA had expressed a desire to set up an adaptive management framework (GBCMA 2003), so our aim was to develop a quantitative model that could be utilised within a targeted monitoring strategy. Our plan for the next stage of model development was to retain the main elements presented in the previous model (Fig. 12.3) and develop a quantitative process model utilising an STM framework. We adopted a Bayesian network modelling framework (Cain 2001; Marcot *et al.* 2006; McCann *et al.* 2006), which had recently been used to structure an STM for the management of rangelands (Bashari *et al.* 2009).

STMs are models that describe different states of vegetation condition that may occur and the possible transitions between these states (Westoby *et al.* 1989). The increasing adoption of STMs in native vegetation management signalled a move away from the foster paradigm, whereby an initial management intervention (e.g. fencing and in-planting) will theoretically result in a single linear successional trajectory towards a 'climax' or reference vegetation community (Palmer *et al.* 1997; Rumpff *et al.* 2011). STMs are favoured because of recognition that there are different states of vegetation condition in the landscape that have arisen from different land use histories, multiple non-linear pathways of change between states, threshold behaviour and possible barriers to restoration (Stringham *et al.* 2003; Suding *et al.* 2004; Hobbs 2007; Bestelmeyer *et al.* 2009; Rumpff *et al.* 2011).

The STM concept is valued because it reflects that the reference state may not be achievable with management intervention, nor is it always the desired endpoint. The logic of this modelling approach has been recognised by the Australian government, which currently utilises a woodland STM to help managers classify sites according to states and provides guidance on the management intervention required to achieve a specified outcome (Zammit *et al.* 2010). This is a typical use of STMs in a conservation planning context.

STMs are commonly represented as conceptual models, providing a framework for describing overall changes in vegetation condition that is easy to understand and communicate (Yates and Hobbs 1997b; Prober *et al.* 2002; Thackway and Lesslie 2006; McIntyre and Lavorel 2007; Zammit *et al.* 2010). This is an important point as, while the current metric used for reporting does allow numerical accounting of change, the units of change are somewhat abstract and thus difficult to conceptualise. Also, similar scores can arise from different kinds of sites. For example, while a densely wooded thicket may change very slowly without management intervention, an open site of a similar score may change rapidly under the same circumstances. Thus, the decontextualised change in score does not shed light on likely causes. Many CMAs have specific objectives about the degree to which vegetation condition should increase across a catchment but are uncertain about how they should define and measure their progress toward that objective. Developing descriptive 'states' as a quantitative unit of measurement for vegetation condition, in consultation with CMA staff, is a fundamentally useful undertaking.

There were several reasons why it was important to retain the individual vegetation attributes within the model. We recognised that the structure of the previous model (Fig. 12.3) had a key advantage over most STMs in that it tried to capture how multiple individual attributes of vegetation might change in response to management. One of the primary limitations of the current form of STMs is that states are generally described qualitatively. As such, they are typically too coarse to detect meaningful changes in individual vegetation attributes, they cannot adequately represent uncertainty in the dynamics of those attributes and they cannot be updated with new data (Rumpff *et al.* 2011). Management actions are often targeted at individual attributes (e.g. weed cover, weed control), so a quantitative understanding of how those attributes respond to management at a site scale is crucial. Second, the reporting demands of CMAs are likely to continue in a similar fashion, so the attributes included within the model should fit the decision and reporting context for our targeted end users. Last, a quantitative model which defined states of vegetation condition in relation to vegetation attributes would enable us to measure progress towards a transition to a new state, or qualify apparent inertia or confinement within a given state (Rumpff *et al.* 2011). This is an important consideration as most CMAs are required to report on progress on an annual basis but actual transitions can occur over much longer time-frames.

In summary, our decision was to model vegetation condition in two ways: as individual vegetation attributes and as overall 'states' of vegetation condition. This enabled a novel tool for guiding management decisions, assessing and learning about the response of a system to management (to improve decision-making) and facilitating a logical way to report on changes in vegetation condition. We felt our site-based model would aid decision-making on two scales: consideration of which management interventions are required at the site scale and consideration of the suite of sites (in different states) to invest in at the catchment scale.

DEVELOPMENT OF THE STM

There were several key steps involved in restructuring and parameterising the model, which are outlined in detail in Rumpff *et al.* (2011) and summarised here. These steps were:

1 create a conceptual STM;
2 propose states according to individual vegetation attributes;
3 identify and add the driving environmental and land use factors and the current and potential future management interventions to the conceptual model;
4 parameterise the model as a Bayesian network.

DEFINING STATES AND TRANSITIONS

The first stage was to develop the conceptual STM, which was initiated in a workshop with a team of ecologists from the University of Melbourne and the Department of Sustainability and Environment. It was decided that woodland environments would be used as a target system for the model. Woodlands were once widespread across the Goulburn Broken catchment, but like the rest of Australia they have been subject to a long history of clearance and degradation (Yates and Hobbs 1997a). As a result these ecosystems are often the focus of restoration efforts across the catchment. We identified the different states of woodland condition we believe occur in the landscape, possible transitions between the different states and the main management interventions or environmental conditions required to trigger those transitions (Fig. 12.4). This conceptual model was then discussed with NRM agency staff, landholders and other ecologists.

DEFINING STATES ACCORDING TO INDIVIDUAL VEGETATION ATTRIBUTES

The next step was to quantify the states according to their structural and compositional attributes (Table 12.1). This required each state to be defined according to ranges in values for each individual vegetation attribute (hereafter called state variables). At present these are defined by expert opinion, though we are verifying those states in the field with the aid of CMA staff and other land managers. Within the model it is possible to record the state (or transition

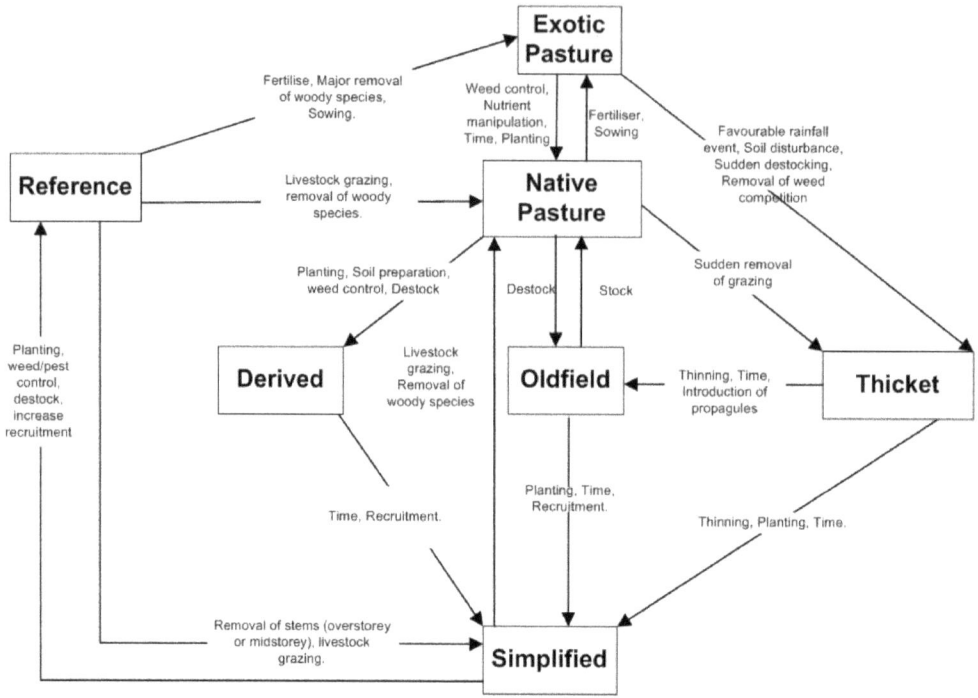

Figure 12.4: The conceptual STM, depicting the controlling factors required to trigger a transition between states (adapted from Rumpff *et al.* 2011).

probability) based on the values specified for each of the state variables. However, there is often large overlap between states, and not all state variables will effect a transition to another state. The variables which do cause a transition are modelled, while the other variables may be monitored for reporting or surveillance purposes. For example, the Native Pasture and Exotic Pasture states are structurally and compositionally similar. It is likely that the composition of the understorey distinguishes the states, with weed cover, native understorey richness and native understorey cover as the primary distinguishing variables. Additionally, we may discover some systems are undergoing slow but incremental changes, rather than crossing thresholds and transitioning to new states (Bestelmeyer 2009). In this case, the model structure will require revision. However, this does not matter if we use continuous state variables to define and update the model; we retain the ability to detect changes in vegetation condition.

IDENTIFYING FACTORS AND ADDING TO THE STM

The next stage in developing the conceptual model was to specify the role of management interventions, controlling environmental and land use variables, and time on the native vegetation attributes. Management interventions currently employed by the Goulburn Broken CMA were incorporated, as were interventions not widely used but with potential for future implementation (e.g. nutrient manipulation, kangaroo-proof fencing). Environmental and land use factors were labelled as independent factors or process variables. Independent factors are those that can influence system dynamics at the site scale but that cannot be modified by management interventions (e.g. rainfall, time). Process variables also control system dynamics at the site scale but can be manipulated by management interventions (e.g. soil compaction, seed

Table 12.1: State variables, management interventions, independent factors and process variables incorporated into the STM

State variables	Management interventions	Independent factors	Process variables
Mature stem density	Grazing exclusion	Time	P enrichment
Immature stem density	Pest control	Time since grazing	N enrichment
Native understorey richness	Soil treatment	Probability rainfall	Soil compaction
Native understorey cover	Soil disturbance	Land use	Overstorey seed availability
Native midstorey richness	Direct seeding	Time since cropping	Grazing pressure
Native midstorey cover	Tube stock	Carrying capacity	
Weed cover	Applied thinning	Removal of grazing or logging or mining	
Recruitment	Weed control	Natural regeneration zone	
	Livestock density	Sheep camp	
	Stocking rate	Sown	

Source: Rumpff *et al.* (2011).

availability). Process variables directly influence the state variables and generally provide the link in the model between the independent factors and the management interventions (Fig. 12.5). Time was incorporated as a continuous variable, but as the STM was to be implemented as a Bayesian network 'time' was discretised into states. These states were relevant to the time-frames in which we might expect to see changes in the different state variables, and accommodated the reporting frequency of the Goulburn Broken CMA.

PARAMETERISING THE STM AS A BAYESIAN NETWORK

The STM was implemented as a Bayesian network. There were many perceived benefits to this approach. First, we felt retaining the graphical depiction of the model would make it easier for the end users to understand and communicate. Second, Bayesian networks are particularly useful in an adaptive management strategy because they can represent prior beliefs about the response of the system to management, and this belief can be iteratively updated as new monitoring data is collected. Last, the model can be parameterised using a combination of quantitative and qualitative information sources (Marcot *et al.* 2006; McCann *et al.* 2006; Nyberg *et al.* 2006; Rumpff *et al.* 2011). Within this model we used a combination of information from the literature, and expert opinion. This makes it a particularly appealing tool given the poor availability of information in native vegetation management. We can take advantage of the available data to produce a prior model to provide direction for decisions about management, and update the model as appropriate data are collected.

THE NEXT STAGE

Through the process described above we have produced a process STM that represents current knowledge and belief about the dynamics of woodland ecosystems within the Goulburn Broken catchment, and how they do or might respond to management. We have attempted to tailor this model so that it is suitable for an adaptive management strategy. Monitoring data can and should be used to iteratively update the state of knowledge represented within the model, and

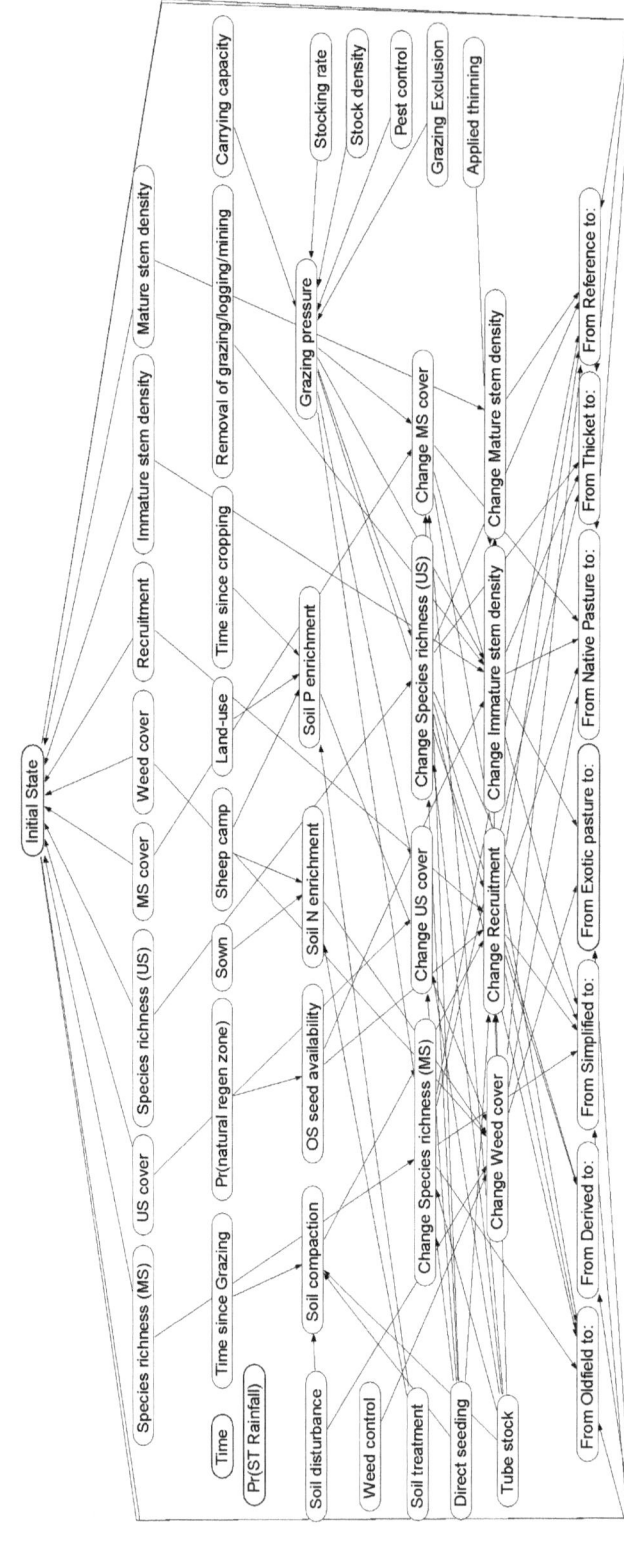

Figure 12.5: The full conceptual STM, depicting the initial state and transition nodes (e.g. 'From Oldfield to'), the state variables (e.g. 'Species richness'), management interventions (e.g. 'Soil disturbance'), independent factors (e.g. 'Time since grazing') and process variables (e.g. 'Grazing pressure') (adapted from Rumpff *et al.* 2011).

improved decisions can be made as outcomes from management actions and other events become better understood (Williams *et al.* 2009). We have already been able to demonstrate 'learning' by incorporating data from a sample of investment sites into the STM and examining the changes in the posterior probability of transition between states (Rumpff *et al.* 2011). We have also investigated some important considerations for the rate of learning, such as incomplete datasets, the number of monitoring samples and model confidence (Rumpff *et al.* 2011).

As a part of the STM development process, we explored and represented the uncertainties of ecologists and CMA staff about the response of woodland ecosystems to management. For example, there is considerable uncertainty around the rate of habitat development in revegetated sites (Vesk *et al.* 2008) and the circumstances under which release from grazing pressure might result in thickets of woody vegetation (Lunt *et al.* 2010). In both cases, what additional management is required, if any, is also uncertain.

The next steps for this model will involve resolving some of the uncertainties exposed throughout this process. We began with the aim of developing a tool to help resolve uncertainties regarding the efficacy of management actions but uncovered important uncertainties surrounding many of the parameter estimates, state boundaries and rates of transition between states. Additionally, we are uncertain whether certain states are discrete entities or simply represent slow changes that we have been so far unable to observe (Bestelmeyer *et al.* 2009). Some of these uncertainties may be resolved through the collection and analysis of field data (e.g. the state boundaries). For others, resolution may require us to explore and reproduce different opinions between experts as competing models. As an important note, this project is a part of an evidence-based approach to management and our model is driven by the decision context of the Goulburn Broken CMA. As such, our aim is to focus on investigating and resolving the critical uncertainties in the model (Duncan and Wintle 2008; Martin *et al.* 2009; Rumpff *et al.* 2011). There will doubtless be many sources of uncertainty but the most critical are the uncertainties that, if resolved, can result in a changed management decision (Rumpff *et al.* 2011).

The STM model presented in this chapter provides an example of a tighter ecological process model that can be used to guide investment and reporting by state and regional NRM agencies for native vegetation management. However, it represents only one broad and widespread vegetation type. The task of developing a set of models which can be applied to the full range of ecosystems in a region or state is undeniably complicated. In the next phase of development, we believe the focus should be on verification and demonstration of the STM's performance in its native vegetation investment context and in reporting for woodland vegetation change. Second, the model structure and behaviour can begin to be tested with perhaps one additional ecosystem type. We are confident that the model has a serviceable balance of complexity and simplicity, if these qualities are carefully managed as we work toward broader applicability.

ACKNOWLEDGEMENTS

This project was funded by the Applied Environmental Decision Analysis and Landscape Logic CERF hubs, the Goulburn Broken CMA and the University of Melbourne. Thanks to workshop attendees: Matt White, Steve Sinclair, Garreth Kyle, Graeme Newell, Wendy Merrit, Chris Jones, Vanessa Keogh and Tim Barlow. For assistance with Bayesian networks and NETICA software we would like to thank Terry Walshe and Yung En Chee. Brendan Wintle was supported by ARC Fellowship DP0774288. This chapter was improved by the comments of Ted Lefroy, James Todd and an anonymous reviewer.

REFERENCES

Australian National Audit Office (ANAO) (1998) 'Preliminary inquiries into the Natural Heritage Trust'. Australian National Audit Office, Canberra.

Australian National Audit Office (ANAO) (2008) 'Regional delivery model for the Natural Heritage Trust and the National Action Plan for Salinity and Water Quality'. Australian National Audit Office, Canberra.

Bashari H, Smith C and Bosch OJH (2009) Developing decision support tools for rangeland management by combining state and transition models and Bayesian belief networks. *Agricultural Systems* **99**, 22–34.

Bestelmeyer BT, Tugel AJ, Peacock GL, Robinett DG, Shaver PL, Brown JR, Herrick JE, Sanchez H and Havstad KM (2009) State-and-transition models for heterogeneous landscapes: a strategy for development and application. *Rangeland Ecology and Management* **62**, 1–15.

Brunt K and McLennan R (2006) 'Biodiversity Monitoring Action Plan'. Goulburn Broken Catchment Management Authority, Shepparton, Victoria.

Burgman M (2005) *Risks and Decisions for Conservation and Environmental Management.* Cambridge University Press, Cambridge.

Cain J (2001) 'Planning improvements in natural resources management: guidelines for using Bayesian networks to support the planning and management of development programmes in the water sector and beyond'. Centre for Ecology and Hydrology (Crowmarsh and Gifford), Oxon.

Crowley K (2001) Effective environmental federalism? Australia's Natural Heritage Trust. *Journal of Environmental Policy and Planning* **3**, 255–272.

Dept of Natural Resources and Environment (DNRE) (2002) 'Victoria's native vegetation management: a framework for action'. Dept of Sustainability and Environment, Melbourne.

Dept of Sustainability and Environment (DSE) (2004) *Vegetation Quality Assessment Manual: Guidelines for Applying the Habitat Hectares Scoring Method. Version 1.3.* Dept of Sustainability and Environment, Melbourne.

Dept of Sustainability and Environment (DSE) (2006) 'Native vegetation gain approach: technical basis for calculating gains through improved native vegetation management and revegetation'. Dept of Sustainability and Environment, Melbourne.

Dept of Sustainability and Environment (DSE) (2008a) DSE Gain Calculator. Dept of Sustainability and Environment, Melbourne.

Dept of Sustainability and Environment (DSE) (2008b) 'Native vegetation net gain accounting first approximation report'. Dept of Sustainability and Environment, Melbourne.

Duncan D and Kyle G (2010) 'Management of native vegetation on private land in northern Victoria: the magnitude, motivation and resourcing of revegetation and restoration'. Technical Report no. 23. Landscape Logic, Hobart.

Duncan DH and Wintle BA (2008) Towards adaptive management of native vegetation in regional landscapes. In *Landscape Analysis and Visualisation: Spatial Models for Natural Resource Management and Planning.* (Eds C Pettit, W Cartwright, I Bishop, K Lowell, D Pullar and D Duncan) pp. 159–181. Springer-Verlag, Berlin.

Gibbons P, Briggs SV, Ayers DA, Doyle S, Seddon J, McElhinny C, Jones N, Sims R and Doody JS (2008) Rapidly quantifying reference conditions in modified landscapes. *Biological Conservation* **141**, 2483–2493.

Goulburn Broken CMA (2003) Regional catchment strategy. Goulburn Broken Catchment Management Authority, Shepparton, Victoria.

Hobbs RJ (2007) Setting effective and realistic restoration goals: key directions for research. *Restoration Ecology* **15**, 354–357.

Hunter M (1996) Benchmarks for managing ecosystems: are human activities natural? *Conservation Biology* **10**, 695–697.

Lindenmayer DB and Likens GE (2010) The science and application of ecological monitoring. *Biological Conservation* **143**, 1317–1328.

Lunt ID, Winsemius LM, McDonald SP, Morgan JW and Dehaan RL (2010) How widespread is woody plant encroachment in temperate Australia? Changes in woody vegetation cover in lowland woodland and coastal ecosystems in Victoria from 1989 to 2005. *Journal of Biogeography* **37**, 722–732.

Marcot BG, Steventon JD, Sutherland GD and McCann RK (2006) Guidelines for developing and updating Bayesian belief networks applied to ecological modeling and conservation. *Canadian Journal of Forest Research* **36**, 3063–3074.

Martin J, Runge MC, Nichols JD, Lubow BC and Kendall WL (2009) Structured decision making as a conceptual framework to identify thresholds for conservation and management. *Ecological Applications* **19**, 1079–1090.

McCann RK, Marcot BG and Ellis R (2006) Bayesian belief networks: applications in ecology and natural resource management. *Canadian Journal of Forest Research* **36**, 3053–3062.

McCarthy MA, Parris KM, van der Ree R, McDonnell MJ, Burgman MA, Williams NSG, McLean N, Harper MJ, Meyer R, Hahs A and Coates T (2004) The habitat hectares approach to vegetation assessment: an evaluation and suggestions for improvement. *Ecological Management and Restoration* **5**, 24–27.

McIntyre S and Lavorel S (2007) A conceptual model of land use effects on the structure and function of herbaceous vegetation. *Agriculture, Ecosystems and Environment* **119**, 11–21.

Nichols JD and Williams BK (2006) Monitoring for conservation. *Trends in Ecology and Evolution* **21**, 669–673.

Nyberg JB, Marcot BG and Sulyma R (2006) Using Bayesian belief networks in adaptive management. *Canadian Journal of Forest Research* **36**, 3104–3116.

Oliver I, Smith PL, Lunt I and Parkes D (2002) Pre-1750 vegetation, naturalness and vegetation condition: what are the implications for biodiversity conservation? *Ecological Management and Restoration* **3**, 176–178.

Palmer MA, Ambrose RF and Poff NL (1997) Ecological theory and community restoration ecology. *Restoration Ecology* **5**, 291–300.

Parkes D, Newell G and Cheal D (2003) Assessing the quality of native vegetation: the 'habitat hectares' approach. *Ecological Management and Restoration* **4**, S29–S38.

Prober SM, Thiele KR and Lunt ID (2002) Identifying ecological barriers to restoration in temperate grassy woodlands: soil changes associated with different degradation states. *Australian Journal of Botany* **50**, 699–712.

Rumpff L, Duncan D and Tolsma A (2009) 'Detecting short-term change in the structure and composition of dry grassy forest'. Technical Report no. 10. Landscape Logic, Hobart.

Rumpff L, Duncan DH, Vesk PA, Keith DA and Wintle B (2011) State-and-transition modelling for adaptive management of native woodlands. *Biological Conservation* **144**(4), 1175–1178.

Standish RJ, Cramer VA and Yates CJ (2009) A revised state-and-transition model for the restoration of woodlands in Western Australia. In *New Models for Ecosystem Dynamics and Restoration*. (Eds RJ Hobbs and KN Suding) pp. 189–205. Island Press, Washington.

Stringham TK, Krueger WC and Shaver PL (2003) State and transition modeling: an ecological process approach. *Journal of Range Management* **56**, 106–113.

Suding KN, Gross KL and Houseman GR (2004) Alternative states and positive feedbacks in restoration ecology. *Trends in Ecology and Evolution* **19**, 46–53.

Thackway R and Lesslie R (2006) Reporting vegetation condition using the Vegetation Assets, States and Transitions (VAST) framework. *Ecological Management and Restoration* **7**, S53–S62.

Ticehurst JL and Duncan DH (2008) Vegetation Gain Bayesian Network: explanatory documentation. In *iCAM Technical Report 2008/02*. Integrated Catchment Assessment and Management Centre, Fenner School of Environment and Society, Australian National University, Canberra.

VCMC (2007) 'Catchment Condition Report 2007'. Victorian Catchment Management Council, Melbourne.

Vesk PA, Nolan R, Thomson JR, Dorrough JW and Mac Nally R (2008) Time lags in provision of habitat resources through revegetation. *Biological Conservation* **14**, 174–186.

Westoby M, Walker B and Noy-Meir I (1989) Opportunistic management for rangelands not at equilibrium. *Journal of Range Management* **42**, 266–274.

Williams BK, Szaro RC and Shapiro CD (2009) *Adaptive Management: The US Department of the Interior Technical Guide*. (Ed. Adaptive Management Working Group). US Dept of the Interior, Washington.

Yates CJ and Hobbs RJ (1997a) Temperate eucalypt woodlands: a review of their status, processes threatening their persistence and techniques for restoration. *Australian Journal of Botany* **45**, 949–973.

Yates CJ and Hobbs RJ (1997b) Woodland restoration in the Western Australian wheatbelt: a conceptual framework using a state and transition model. *Restoration Ecology* **5**, 28–35.

Zammit C, Attwood A and Burns E (2010) Using markets for woodland conservation on private land: lessons from the policy–research interface. In *Temperate Woodland Conservation and Management*. (Eds DL Lindenmayer, A Bennett and RJ Hobbs) pp. 297–308. CSIRO Publishing, Melbourne.

Patch Data Viewer: a tool for planning investment in vegetation extent and condition from patch to regional scales

Tony Norton and Michael Lacey

SUMMARY

The management of native vegetation is an important aspect of natural resource management (NRM) in Australia, and a focal area for Landscape Logic. Patch Data Viewer (PDV) is a GIS-based decision support tool that was designed as part of Landscape Logic to allow NRM groups and other natural resource managers to examine and map the distribution of vegetation communities and patches of native vegetation from the level of individual properties to landscapes and whole catchments. We developed the tool initially for Tasmania. It can be easily modified for use in any region of the world and by any organisation that has access to spatial information on native vegetation and wishes to use that knowledge in decision-making. The application grew out of a need to characterise the distribution of remnant native vegetation across landscapes and to provide desktop ways to search identifiable characteristics of individual remnant vegetation patches. The use of PDV allows the characteristics of individual vegetation patches to be examined at a property, landscape and regional level in ways that support management decisions. For example, the unique location, vegetation community composition, size, perimeter length and distance to the nearest neighbouring patch can be determined and mapped. Information about the designated use of land both around and supporting the patch of native vegetation can be tabulated and used to consider management options such as revegetation and habitat restoration. In addition to examining vegetation extent, PDV enables information on the condition of native vegetation to be linked to each patch. Other Landscape Logic teams developed new remote-sensing technologies, spatial modelling techniques and field-based assessment protocols to enable the capture of vegetation condition data at a regional level. Incorporating associated data into PDV provides important additional information to support decision-making.

INTRODUCTION

Improving the conservation and management of Australia's native vegetation is a national priority (Australian Government 2009a). The NRM, government and research partners of Landscape Logic identified native vegetation as a major focus for landscape research as part of their research programs in Victoria and Tasmania. As a consequence, a number of complementary research activities were undertaken to improve the science underpinning the management of native vegetation ecosystems and the conservation of biodiversity dependent on these ecosystems for habitat.

New spatial science techniques for characterising the structure of native vegetation and its configuration in the landscape using remotely sensed datasets and spatial technologies were developed by Landscape Logic teams (Chapter 14, this volume), as were new quantitative methods to calculate the spatial metrics of vegetation and land cover (Lechner *et al.* 2008; Norton 2009a; Sheffield *et al.* 2009; Michaels *et al.* 2010). In addition, two national technical workshops were held to promote integrative approaches to the assessment and monitoring of native vegetation in Australia (Williams 2007; Norton and Michaels 2010). These workshops considered new and efficient technical means to track changes in the extent and condition of native vegetation over time at multiple scales, and indicated priorities for future research and management to enhance the increase in biodiversity habitat resulting from improved vegetation management (Norton and Michaels 2010).

This chapter describes the development of a novel computer-based decision support tool for analysing and visualising vegetation data from patch to regional scale (Norton 2009a, b; Lacey and Norton 2010). This tool was developed using Tasmania as a case study, but can be readily modified for use in any management region in the world. In this chapter, we introduce our tool, Patch Data Viewer, discuss its development and illustrate its applications for informing the conservation and management of native vegetation ecosystems.

PATCH DATA VIEWER

Patch Data Viewer (PDV) was designed to allow NRM groups and other resource managers and decision-makers to examine and map the distribution of vegetation communities and patches of native vegetation from individual property to landscape and whole-catchment scales. The application grew out of a clear need by land owners and managers to have better and faster access to scientific information about the status and significance of vegetation on their land and under their stewardship. Prior to the development of PDV, no free and simple-to-use analysis and mapping tool was available to support the management of native vegetation.

The core of PDV is a computationally efficient search engine for managing, accessing, analysing and displaying complex spatial data. The computer tool employs open-source GIS software (freely available licence) that can be operated by people who do not have dedicated GIS or mapping experience. Mapped displays of native vegetation at different scales are supported by detailed tabulated information presented in a series of associated tables or forms. The forms within the database contain information about various aspects of native vegetation and were constructed using Microsoft Access 2000 and later versions. This software package is required to use PDV. The PDV application was written by Dr Michael Lacey. It is less than 200 MB in size and can be run from a data stick on desktop and laptop computers.

PDV is supported by a range of digital, spatial data layers chosen purposefully for the target region and the primary applications of the decision support tool. These data layers may include the digital boundaries of the region, a digital elevation model, regional settlements and infra-structure such as cities, towns and roads, land cover and land ownership (e.g. cadastre) (Norton 2009a). Land cover data may include satellite images and aerial photographs of the target region. These layers complement those showing the distribution of native vegetation patches and communities. The volume of supporting data will vary depending on the size of the target region and the number and resolution of spatial data layers. In the case of Tasmania (~7.5 million ha in area), the associated spatial data required a storage capacity of approximately 3.5 GB.

The primary purpose of PDV was to provide a ready means of identifying, mapping and reviewing information about remnant patches of native vegetation in a target landscape or region. This was done by first creating a digital layer to represent the spatial extent of patches of native vegetation using the Fragstats package developed by Dr Kevin McGarigal and colleagues (McGarigal and Marks 1994). Every patch of native vegetation identified for the target region was uniquely numbered and a number of new metrics (e.g. perimeter length of patch, distance to nearest patch neighbour) were calculated for each patch. The patch number-ing scheme provides an index by which information from many existing spatial datasets related to vegetation distribution in the target region can be easily accessed, cross-referenced and analysed. For example, the unique location, vegetation community composition, habitat condi-tion, size, perimeter length and distance to the nearest neighbouring patch can be quickly determined and mapped using this system. The patch layer can be linked to a cadastre. Thus, patch data within the database can be searched using the unique number of the patch or its associated property code, the latter a key attribute for identifying managers of patches on private lands. Alternatively, visual searches can be made by using the interactive mapping tool within PDV. In addition, information about the designated use of land both around and supporting the patch of native vegetation can be tabulated and used to consider management options such as revegetation and habitat restoration.

Tabulated data used in PDV was sorted by catchment, first to enable catchment-based analyses and second to split the processing task of viewing the datasets into manageable parts. The tool can be readily modified to support other data presentation formats including, for example, terrestrial and marine bioregions, biodiversity hotspots, indigenous protected areas, local government areas and special management areas with vegetation covenants.

PDV uses a visual menu to help users explore the different spatial analysis options and map native vegetation at multiple scales (Fig. 13.1). The map viewer enables vegetation data to

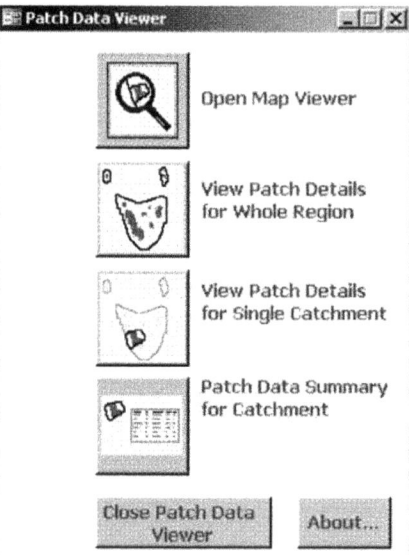

Figure 13.1: Initial window in Patch Data Viewer showing options for decision support analysis and mapping.

be viewed by catchment over the whole of the target region, supports the selection of information layers for each catchment in the target region including the patch reference layer, property boundaries, vegetation communities and Landsat imagery, and provides reference layers including (in the case of Tasmania) the coastline, catchment boundaries, catchment names and names of major urban centres.

By clicking the map of the targeted region (Fig. 13.2) it is possible to select individual native vegetation patches, use the patch reference number to view data related to that patch and shown in associated forms, identify individual properties and identify individual vegetation community polygons. Catchment-wide distribution of individual native vegetation communities can be viewed by selecting from a list on the 'Catchments Data Summary' form. Individual properties associated with a native vegetation patch can be selected from a list on the 'Native Vegetation Patch Data _Single Catchment' form. It is possible to open other layers such as orthophotos by dragging them onto the map viewer window.

The latest version of PDV is V4.2; a user guide is available in hard copy and electronic form for the software tool (Lacey and Norton 2010). Users are required to register to access the tool. This can be done by contacting the Tasmanian Department of Primary Industries, Parks, Water and Environment or the authors.

TASMANIA: A CASE STUDY TO ILLUSTRATE THE FUNCTIONALITY OF PDV

Tasmania comprises a land area of over 68 000 km^2 that includes a main island and over 240 small islands (ABS 2006) (Fig. 13.3). Its natural resources are diverse (CCNRM 2009; NRM North 2009; NRM South 2009; Tasmanian Planning Commission 2009). The state has 158 mapped native vegetation communities and supports a diverse range of land uses that span

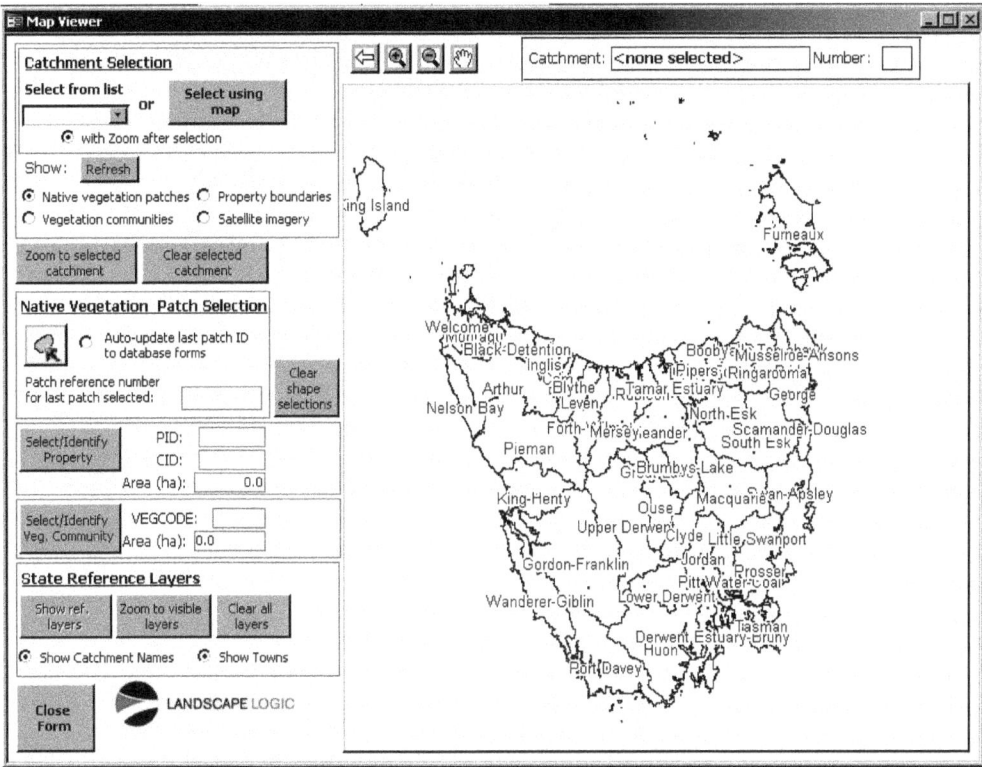

Figure 13.2: Open window in Map Viewer. Function buttons are shown at left, and at top of map for panning across image and zooming.

nine bioregions and 48 water catchments. A number of terrestrial and marine ecosystems and ecological communities are endemic and many provide important ecosystem services that underpin agriculture, nature-based tourism and other key aspects of the state's economy (Tasmanian Planning Commission 2009; Williams *et al.* 2010). Tasmania is promoted as the 'natural state' and the 'island of inspiration' owing to its large and relatively unspoiled natural environment that includes globally significant areas of national park and World Heritage (Tourism Tasmania 2010).

The native vegetation data used for the case study were sourced from the TASVEG coverage developed by the Tasmanian Department of Primary Industries, Parks, Water and Environment. The native vegetation patch layer was created using the Fragstats package (McGarigal and Marks 1994) and patches of vegetation were sampled originally with GIS (ArcMap V9.3, ESRI) using a 25 m grid spacing to derive metrics (i.e. if the cell centre of each vegetation polygon was >25 m apart they were considered to be separate patches). The vegetation community composition of each patch was determined by intersecting the patch layer with the TASVEG coverage (Lacey and Norton 2010).

STATUS OF NATIVE VEGETATION

The development of PDV allowed the derivation of a range of new spatial data on native vegetation to help assess the status of these communities in Tasmania. The mapped extent of Tasmania's native vegetation is approximately 5.06 million ha, or 74% of the land area. This is

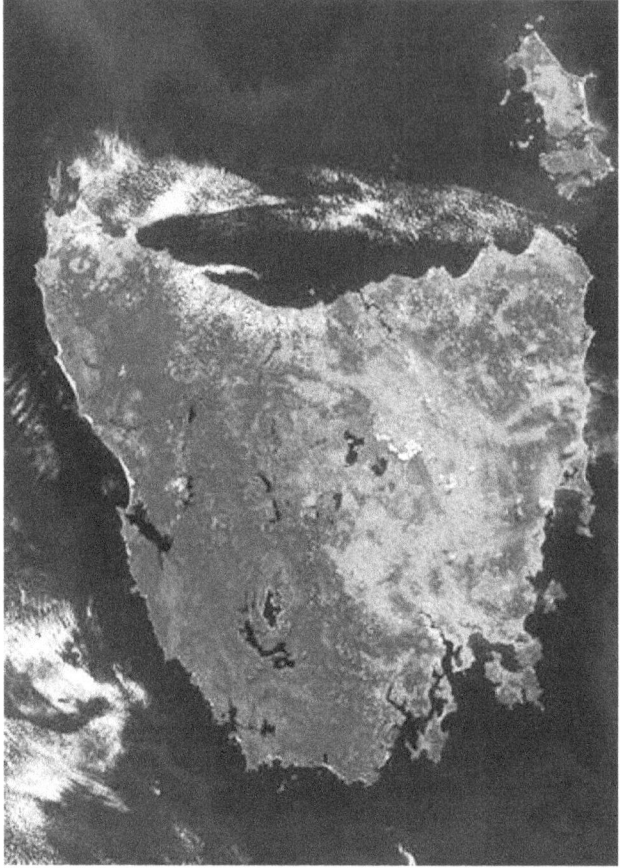

Figure 13.3: Landsat™ image of Tasmania indicating areas that retain native vegetation and those that have been modified (e.g. Midlands) to support agriculture. Image source: Geoscience Australia 2009. Note: King Island is not shown.

relatively high in comparison to many regions of the Australian continent. The extent of native vegetation varies across bioregions from a low of around 36% (Tasmanian Northern Midlands) to a high of 94% (Tasmanian West) (Table 13.1). The proportional contribution of Tasmanian West to the state's total native vegetation extent is very large. The Tasmanian South East bioregion also makes a proportional contribution to the extent of native vegetation in the state. Together, these two bioregions support around 50% of the total extent of native vegetation in Tasmania.

Figure 13.4 summarises the degree of modification of native vegetation across Tasmanian bioregions using the approach of McIntyre and Hobbs (1999). Intact ecosystems are based on the extent of native vegetation exceeding 90% cover by area, a threshold that is commonly used worldwide to evaluate landscape-level modification and the implications for ecosystem degradation (Millennium Ecosystem Assessment 2005).

The Tasmanian West bioregion and the Tasmanian Central Highlands bioregion were identified as intact in this analysis whereas the King, Tasmanian Northern Slopes and Tasmanian Northern Midlands bioregions were fragmented. The degree of native vegetation modification within bioregions is masked at this level of analysis, but is not random. The extent of

Table 13.1: Extent of mapped native vegetation in Tasmania's nine bioregions

Bioregion	Total area (ha)	NV per region (ha)	% NV per region
Ben Lomond	657 040	459 228	69.8
Flinders	492 313	308 091	62.5
King	425 085	251 164	59.0
Tasmanian Central Highlands	767 330	664 834	86.6
Tasmanian Northern Midlands	415 121	152 257	36.6
Tasmanian Northern Slopes	622 696	343 990	55.2
Tasmanian South East	1 105 306	739 152	66.8
Tasmanian Southern Ranges	781 177	666 265	85.2
Tasmanian West	1 563 293	1 472 687	94.2
Total	**6 829 365**	**5 057 671**	**74.0**

NV = native vegetation.

remaining native vegetation is closely related to European settlement patterns and the clearing and modification of specific Tasmanian landscapes to permit agricultural production (Tasmanian Planning Commission 2009).

Based on these analyses using the criteria of McIntyre and Hobbs (1999), the Tasmanian landscape can be described as medium variegated (see also Michaels *et al.* 2010). The state

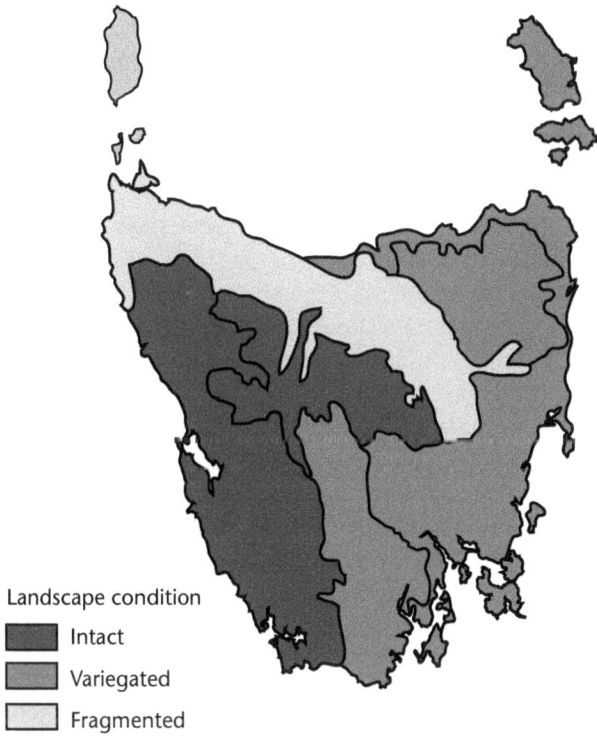

Landscape condition
- Intact
- Variegated
- Fragmented

Figure 13.4: Landscape modification of Tasmanian bioregions based on native vegetation cover.

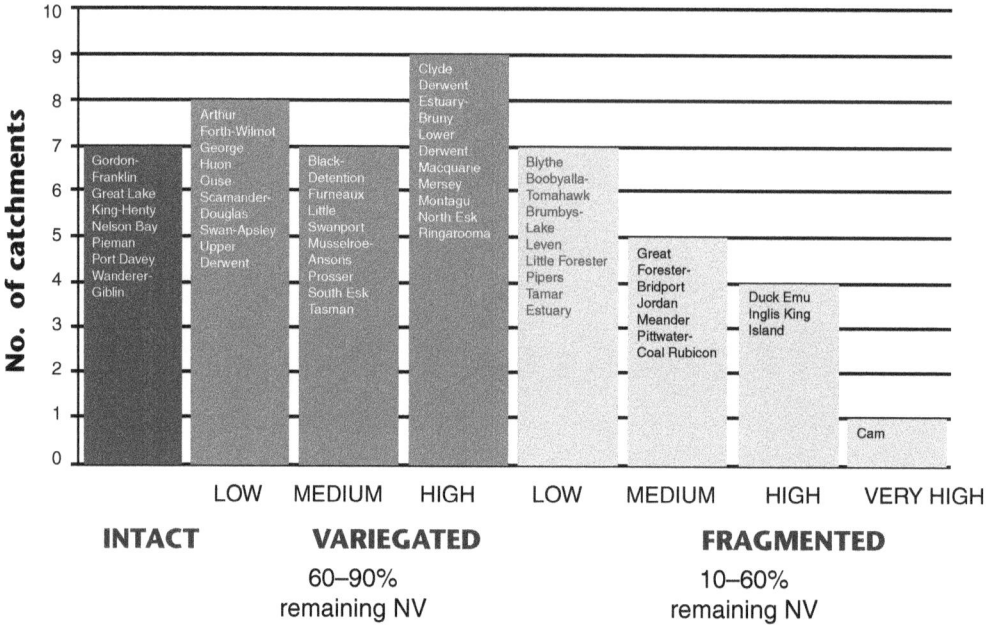

Figure 13.5: Degree of modification of native vegetation within the 48 water catchments in Tasmania based on the methods of McIntyre and Hobbs (1999), as derived from data within Patch Data Viewer.

retains 76% cover of native vegetation, by area. Some catchments, such as the Pieman and Port Davey in western Tasmania, are intact (retain >90% by area of native vegetation), some are variegated (retain 60–90% native vegetation cover), and some which occur in agricultural landscapes are fragmented (retain 10–60% cover), such as the Cam, Rubicon and Pittwater-Coal. This intact, variegated and fragmented distribution of Tasmanian regions is shown in Figure 13.5.

The creation of a vegetation patch layer and database using Fragstats indicated that Tasmania supports over 33 000 patches of native vegetation. For convenience, these patches were classified in terms of size: <10 ha in area, 10–50 ha in area and >50 ha (Fig. 13.6). Analysis of the size, frequency and connectivity of these patches of native vegetation across bioregions demonstrates the relationship between the modification of native vegetation and land use (Michaels *et al.* 2010). For example, less than 1% of the total number of Tasmania's native vegetation patches are found in the Tasmanian West bioregion compared to over 50% in the King, Tasmanian Northern Slopes and Tasmanian Northern Midlands bioregions combined, where a proportion of the state's intensive and extensive agriculture is located.

The intensity of modification of native vegetation at a catchment level across Tasmania can be illustrated by an examination of the frequency of occurrence of native vegetation patches in a catchment (Fig. 13.7). Catchments with a high patch density (high frequency of occurrence of patches per unit area) of remnant native vegetation invariably support a range of agricultural land uses (including plantation forestry). It is in these landscapes and environments where past and ongoing threatening processes have increased the likelihood of population decline and extinction (Tasmanian Planning Commission 2009; Williams *et al.* 2010).

Forty-six threatened native vegetation communities are recognised under Tasmanian government legislation (Harris and Kitchener 2005). These communities include eucalypt

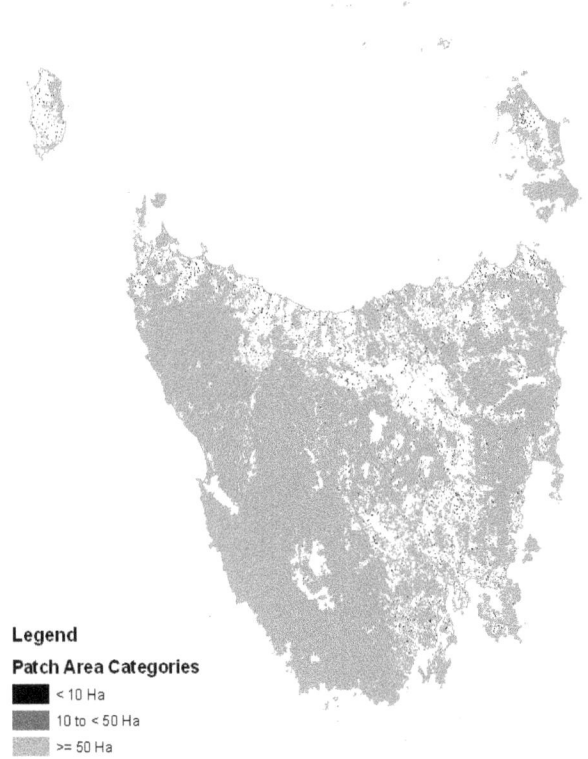

Figure 13.6: Distribution of over 33 000 patches of native vegetation in Tasmania, based on a Fragstats analysis of the TASVEG coverage (V2.0).

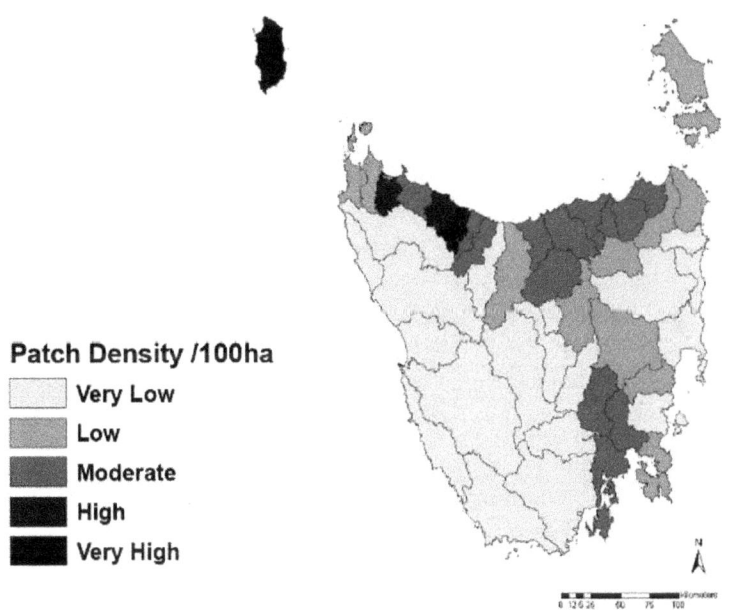

Figure 13.7: Estimated density of patches of native vegetation per 100 ha for each catchment in Tasmania.

Table 13.2: Profile of threatened vegetation communities in Tasmanian bioregions

Bioregion	Total area (ha)	TVC per region (ha)	% TVC per region
Ben Lomond	657 040	10 783	1.6
Flinders	492 313	18 515	3.7
King	425 085	15 468	3.6
Tasmanian Central Highlands	767 330	84 745	11.0
Tasmanian Northern Midlands	415 121	28 172	6.7
Tasmanian Northern Slopes	622 696	20 526	3.3
Tasmanian South East	1 105 306	103 727	9.3
Tasmanian Southern Ranges	781 177	20 746	2.6
Tasmanian West	1 563 293	12 513	0.8
Total	**6 829 365**	**315 199**	**4.6**

TVC = threatened vegetation communities.

forest and woodland, rainforest and related scrub, highland treeless vegetation, sedgeland, wetland and saltmarsh. By area, the largest extent of threatened vegetation communities occurs in the Tasmanian Central Highlands (~11% of region) and Tasmanian South East bioregions (9.38%) (Table 13.2), a total area in excess of 188 000 ha. A number of threatened vegetation communities were found in the Tasmanian Northern Midlands bioregion; this observation was consistent with the degree of modification of native vegetation recorded in this bioregion (Norton 2009a, 2010). Lowland native grassland communities in Tasmania were recently listed as nationally threatened communities under the *Environmental Protection and Biodiversity Conservation Act 1999* (Australian Government 2009b). The distribution of these communities spans bioregions in central and eastern Tasmania (Fig. 13.8).

DECISION SUPPORT APPLICATIONS

PDV can be used to perform a range of functions that support investment and other management decisions, such as:

- searching for all the patches containing a particular native vegetation community within a management region (e.g. NRM North). This type of analysis may assist in prioritising investment in native vegetation at a regional scale;
- identifying the different native vegetation communities that occur in a selected catchment. This type of analysis may facilitate identifying and mapping the major assets in a catchment or region. An example of asset options that could be viewed for a region is given in Figure 13.9;
- locating all the mapped small patches of remnant vegetation occurring between two large isolated patches that contain the same particular native vegetation community. This type of information may be important for managers where connectivity is critical to habitat or species conservation (Fig. 13.10).

DISCUSSION

All remaining patches of vegetation are important for biodiversity conservation. Large patches of native vegetation are important as they can support more species and larger populations of

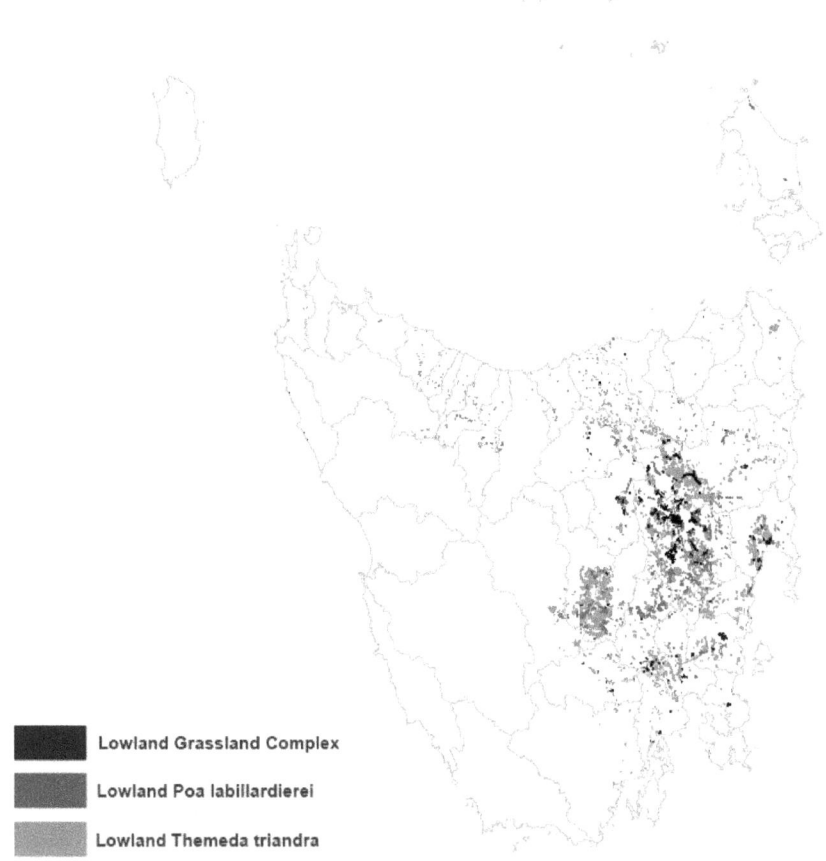

Figure 13.8: Distribution of lowland native grassland ecosystems in Tasmania that are listed as nationally threatened ecological communities (Australian Government 2009). Base data theLIST © State of Tasmania.

species. Small patches may support different species to large patches, and can provide important habitat for rare and threatened plant species.

Linking fragments by planting corridors of native vegetation has been proposed as part of the solution to the problem of habitat fragmentation (Michaels *et al.* 2008; Norton 2009a). The example from the Meander catchment (Fig. 13.10) illustrates the decision support input that PDV can provide for quantifying options for the revegetation of highly modified landscapes where improving connectivity for biodiversity conservation is important. The technique can provide objective information to support cost–benefit analyses of different management options directed to vegetation enhancement at a landscape level (Michaels *et al.* 2010).

Climate change may disproportionately affect terrestrial ecosystems that are already modified and degraded as a result of intensive land use. For example, the native vegetation of the Tasmanian Northern Midlands bioregion and Cam catchment in north-western Tasmania has been cleared, highly modified and fragmented. Patches of native vegetation in these areas provide important habitat for a diverse range of species (Williams *et al.* 2010). Over 5900 patches of remnant native vegetation <10 ha in size and 1895 patches 10–50 ha in size are located in the Tasmanian Northern Midlands bioregion (Williams *et al.* 2010). Although the

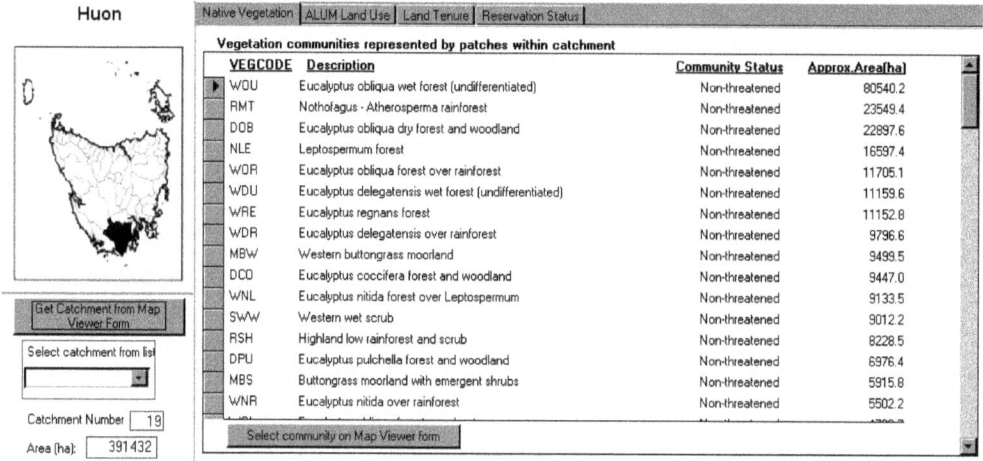

VEGCODE	Description	Community Status	Approx.Area(ha)
WOU	Eucalyptus obliqua wet forest (undifferentiated)	Non-threatened	80540.2
RMT	Nothofagus - Atherosperma rainforest	Non-threatened	23549.4
DOB	Eucalyptus obliqua dry forest and woodland	Non-threatened	22897.6
NLE	Leptospermum forest	Non-threatened	16597.4
WOR	Eucalyptus obliqua forest over rainforest	Non-threatened	11705.1
WDU	Eucalyptus delegatensis wet forest (undifferentiated)	Non-threatened	11159.6
WRE	Eucalyptus regnans forest	Non-threatened	11152.8
WDR	Eucalyptus delegatensis over rainforest	Non-threatened	9796.6
MBW	Western buttongrass moorland	Non-threatened	9499.5
DCO	Eucalyptus coccifera forest and woodland	Non-threatened	9447.0
WNL	Eucalyptus nitida forest over Leptospermum	Non-threatened	9133.5
SWW	Western wet scrub	Non-threatened	9012.2
RSH	Highland low rainforest and scrub	Non-threatened	8228.5
DPU	Eucalyptus pulchella forest and woodland	Non-threatened	6976.4
MBS	Buttongrass moorland with emergent shrubs	Non-threatened	5915.8
WNR	Eucalyptus nitida over rainforest	Non-threatened	5502.2

Catchment Number 19
Area (ha): 391432

Figure 13.9: Patch data summary for the Huon catchment in southern Tasmania showing the range, status and area of some of the native vegetation communities within the selected catchment. The catchment number and area for the Huon is provided by PDV (bottom left).

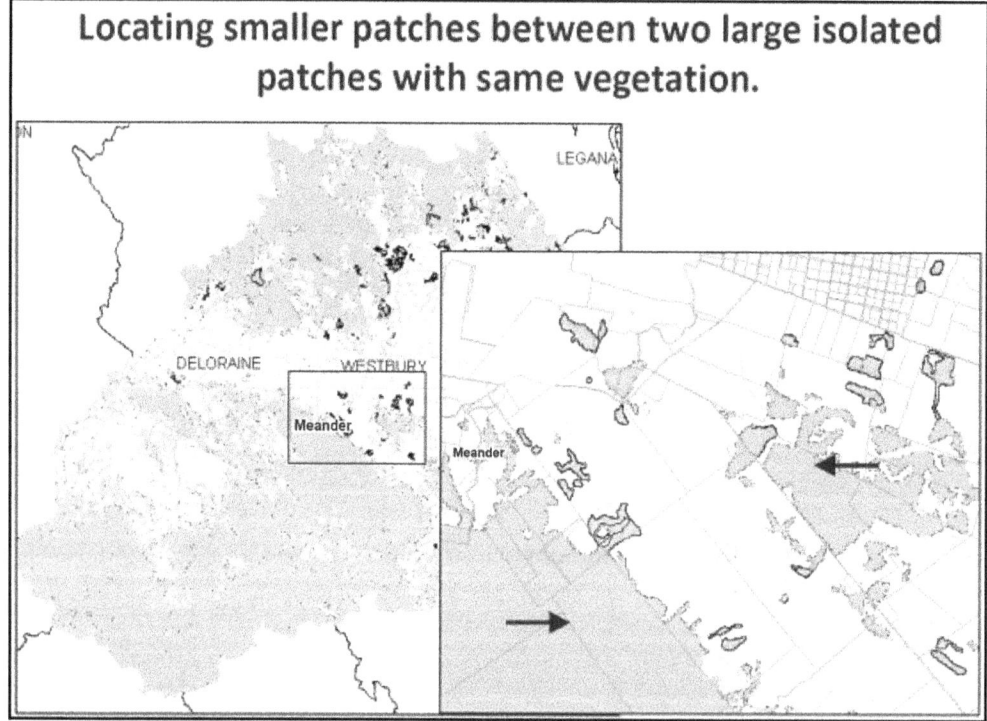

Locating smaller patches between two large isolated patches with same vegetation.

Figure 13.10: PDV map showing the location of smaller patches of the same native vegetation community between two large isolated patches of vegetation in the Meander catchment in northern Tasmania. PDV allows the unique identification number and metrics of each patch to be displayed in tabular form to complement the mapping.

range in elevation in the bioregion is from sea level to >500 m above sea level, most of the patches are concentrated in the range 100–300 m above sea level.

A defining feature of the state's native vegetation is the large tract of natural and near-natural ecosystems found in western Tasmania (Williams *et al.* 2010). These native vegetation ecosystems exceed 2 million ha in total area and span much of the west coast, including seven contiguous catchments. These ecosystems are recognised as of global significance (Tasmanian Planning Commission 2009). They include the South West Tasmanian World Heritage area and the Tarkine Wilderness that is the subject of a proposal for formal conservation reservation (Williams *et al.* 2010).

The active modification of native vegetation for agriculture in Tasmania has turned large continuous patches of native vegetation into numerous small patches that may be spatially isolated. PDV can be used to help understand the nature of changes in native vegetation resulting from landscape modification, and to explore options for vegetation restoration and enhancing connectivity between patches. Limiting the loss and modification of remnant patches of native vegetation in these landscapes appears essential for mitigating further declines in biodiversity, and we hope that tools such as PDV will help to improve the decision support available to landscape managers.

Other Landscape Logic researchers (Chapter 14, this volume) developed new remote-sensing technologies, spatial modelling techniques and field-based assessment protocols to enable the capture of vegetation condition data at a regional level. A major benefit of PDV is that it allows datasets from these specific areas of progress to be viewed through a common user-friendly interface, and in a way that supports integrated and better-informed decision-making.

CONCLUSIONS

The management of native vegetation is an important aspect of NRM in Australia and, as a consequence, Landscape Logic chose to invest a significant proportion of its effort into improving the decision support available for vegetation management. Patch Data Viewer grew out of a need by planners, managers and decision-makers to be able to map and quantify the distribution of remnant native vegetation across landscapes and to provide readily accessible ways to search for important information on the status, condition, ownership and management of vegetation. The new tool can be used for planning investment in vegetation extent and condition from a single patch to landscape and regional scales. The development of the PDV has been well received by natural resource managers and other users and stakeholders in Tasmania. Since the tool can be modified for use by any of the 56 NRM regions in Australia, there is a significant opportunity for its much wider application across the continent's diverse range of landscapes and native vegetation communities.

ACKNOWLEDGEMENTS

The research reported here was conducted as part of a coordinated spatial modelling project within the CERF Landscape Logic National Research Hub. Dr Karyl Michaels, Professor Jann Williams and James Shaddick made important contributions to the research on native vegetation assessment and monitoring in Tasmania. Professor Simon Jones of RMIT University co-led the project with important support from local colleagues Dr Gang-Jun Liu, Dr Elizabeth Farmer, Dr Karin Reinke, Grant Dickins, Kathryn Sheffield, Alex Lechner and Naoko Miura. We are grateful for the intellectual contributions from them all.

REFERENCES

ABS (2006) 'Statistics on the Australian coastline'. Australian Bureau of Statistics, Canberra.

Australian Government (2009a) *Caring for Our Country Business Plan 2010–2011.* Australian Governmen, Canberra.

Australian Government (2009b) Lowland native grasslands of Tasmania. A nationally threatened ecological community. *Environment Protection and Biodiversity Conservation Act 1999* – Policy Statement 3.18. Canberra.

CCNRM (2009) *Cradle Coast NRM Regional NRM Strategy.* <www.nrmtas.org/regions/cradle/regional_strategy.shtml> (accessed 11 February 2010).

GeoScience Australia (2009) Landsat™ image of Tasmania. <http://www.ga.gov.au/earth-observation/satellites-and-sensors/landsat.html>.

Harris S and Kitchener A (2005) *From Forest to Fjaeldmark: Descriptions of Tasmania's Vegetation.* Dept of Primary Industries, Water and Environment, Hobart.

Lacey MJ and Norton TW (2010) 'Patch Data Viewer: user guide. A decision support tool for the spatial analysis of native vegetation'. Unpublished Technical Report. Landscape Logic, Hobart.

Lechner AM, Jones SD and Bekessy SA (2008) A study on the impact of scale-dependent factors on the classification of land-cover maps. In *Quality Aspects in Spatial Data Mining.* (Eds A Stein, J Shi and B Wietske) pp. 745–754. CRC Press, Boca Raton.

McGarigal K and Marks BJ (1994) Fragstats: spatial pattern analysis program for quantifying landscape structure. Forest Science Dept, Oregon State University, Corvallis, Oregon.

McIntyre S and Hobbs RJ (1999) A framework for conceptualizing human impacts on landscapes and its relevance to management and research. *Conservation Biology* 13, 1282–1292.

Michaels K, Lacey M, Norton TW and Williams JE (2008) Vegetation futures for Tasmania. <www.greeningaustralia.org.au/uploads/Veg%20Futures%20Proceedings/VF_W_Michaels.pdf>.

Michaels K, Norton TW, Lacey MJ and Williams JE (2010) Spatial analysis of Tasmania's native vegetation cover and potential implications for biodiversity conservation. *Ecological Management and Restoration* 11, 193–199.

Millennium Ecosystem Assessment (2005) *Ecosystems and Human Well-Being: Synthesis.* Island Press, Washington.

Norton TW (2009a) Monitoring for biodiversity: making the right connections. In *Proceedings of the Australian Academy of Science 2009 Fenner Conference on the Environment.* 11–12 March 2009, Canberra. <www.landscapelogic.org.au/Fenner/Norton.pdf>.

Norton TW (2009b) *Cradle Coast Regional Land Use Planning Initiative: Mapping Agricultural Land in the Cradle Coast Region of Tasmania.* Cradle Coast Authority.

Norton TW (2010) *Cradle Coast Regional Land Use Planning Initiative, Phase 2: Mapping Constraints Imposed on Agricultural Land by Existing Rural Development.* Tasmanian Institute of Agricultural Research, University of Tasmania.

Norton TW and Michaels K (2010) 'Third Australian National Workshop on Native Vegetation Assessment'. Technical Report no. 21. Landscape Logic, Hobart.

NRM North (2009) 'NRM North Annual Report 08–09'. NRM North, Launceston.

NRM South (2009) NRM South Regional NRM Strategy. <www.nrmtas.org/regions/south/regional strategy.shtml> (accessed 7 February 2010).

Sheffield K, Jones S, Ferwerda JG, Gibbons P and Zerger A (2009) Linking biological survey data to remote sensing datasets. In *Innovations in Remote Sensing and Photogrammetry Lecture Notes in Geoinformation and Cartography.* (Eds S Jones and K Reinke) pp. 51–64. Springer, Berlin.

Tasmanian Planning Commission (2009) *Tasmanian State of the Environment Report.* Dept of Justice, Hobart.

Tourism Tasmania (2010) *Highlights of Tourism in Tasmania.* Tasmanian Government, Hobart.

Williams JE (2007) 'Second Australian National Workshop on Native Vegetation Assessment'. Technical Report no. 2. Landscape Logic, Hobart.

Williams JE, Dickins G, Norton TW and Jones S (2010) 'Spatial analysis of priorities for NRM investment in Tasmania and integrated information and data management systems: a comparative assessment of Tasmania's mapped NRM assets'. Research report by RMIT University to Cradle Coast NRM, Tasmania.

14

Measuring the components of vegetation condition using remote sensing

Simon Jones, Alex Lechner, Kathryn Sheffield,

Naoko Miura, Elizabeth Farmer, Karin Reinke and

Tony Norton

SUMMARY

This chapter describes work undertaken within the Landscape Logic project that extends existing knowledge of remote sensing and how it can contribute to the measuring and mapping of vegetation condition. Its focus is providing tools and techniques for monitoring and reporting changes in vegetation and its components. Understanding the relationship between remotely sensed and field-based measures of different vegetation biophysical characteristics underlies the development of any such new tool or technology, as does understanding issues of spatial data uncertainty and the implications of this for mapping vegetation and condition.

We use three different case studies to explore the development of these important foci. Our first case study considers the utility of vegetation indices and texture analyses for mapping vegetation condition through the use of operational satellite-based multi-spectral remotely sensed data. Our second case study describes the potential role of new active remote sensing technologies, such as airborne laser scanning (LiDAR, Light Detection and Ranging), in assessing the structural components of vegetation condition. Our third case study illustrates how theoretical research using simulation modelling can aid the quantification of spatial uncertainty in remote sensing. Through these case studies, this chapter assesses the contemporary strengths and limitations of remote sensing for vegetation condition assessment.

INTRODUCTION

Vegetation condition assessment is an important part of regional landscape management and was a major research theme of Landscape Logic (Williams 2007; Miura and Jones 2010; Norton and Michaels 2010; Chapter 12, this volume). Monitoring and reporting changes in vegetation condition is critical to assessing the return on public investment programs for landscape management. Vegetation attributes, used as indicators or as composite measures of vegetation condition, typically have strong associations with a range of taxa or ecological functions and are often used to assess the impact of policy and management objectives (Reinke and Jones 2006). Currently, vegetation condition in Australia is often reported relative to benchmarks, tracking vegetation structural, compositional and functional attributes as well as the spatial extent of vegetation communities. For the purpose of the work presented here, vegetation condition will be used to broadly describe an *absolute* measure of one or more biophysical parameters that are typically used as a contributing parameter or indicator to an overall estimate of vegetation condition. Quantifying the primary individual biophysical parameters enables greater flexibility in accommodating future changes in the way in which vegetation condition is derived and/or reported.

There are presently three broad approaches to assessing vegetation condition: on-ground site assessment, spatial modelling and remote sensing (Gibbons *et al.* 2006). Remote sensing is a technology with great potential to deliver timely vegetation condition information at landscape and regional scales, as it offers standardised sampling units, synoptic data capture and regular repeat acquisition. Through this chapter we introduce recent developments in remote sensing technologies and processing methods, and how they apply to the assessment of vegetation condition.

Current methods for assessing vegetation condition involve a laborious process with many logistically expensive point/plot-based measurements. Field-based measurements are usually conducted on a relatively small percentage of the total area within a landscape (total <0.5%), not more than annually, and may not represent a whole landscape's condition accurately due to the spatial heterogeneity inherent in natural systems. Remote sensing data provide a means by which vegetation assessment can be conducted at temporal and spatial sampling intensities that are adequate to identify changes in vegetation condition over time with a high degree of confidence. While the development of remote sensing methods to measure vegetation condition is still in its infancy, this technology is demonstrating a powerful capability to support landscape management. For example, over the past decade numerous new earth-observing multi-spectral sensors with high spatial resolution (<10 m pixel size) (Table 14.1) have been launched, along with new airborne remote sensing imaging technologies such as LiDAR and hyperspectral sensors that can aid in the assessment of vegetation condition.

Techniques that link remote observations with actual biophysical parameters used in the assessment of vegetation condition are being researched. There are two main foci of such research. The first is on developing new classification methods or utilising new remote sensing

Table 14.1: Spatial and spectral details of SPOT-5, IKONOS and other high spatial resolution (GSD <10 m) multi-spectral satellite observing systems used for vegetation condition assessments

	Spatial resolution (m) Ground sample distance	Spectral sensitivity (band passes in nm)	Radiometric resolution	Swath width (km)	Launch date/ local overpass time/orbit height
Ikonos	3.2/4 (up to 26° off-nadir)	4: blue (445–0.516), green (506–595), red (632–698), near IR (757–853)	11 bits	11.3	1999/10.30 am/681 km
	1	Panchromatic (455–853)			
SPOT 5	10 (nadir)	4: green (500–590), red (610–680), near IR (780–890), shortwave IR (1580–1750)	8 bits	60	2002/10.30 am/822 km
	SWIR: 20 (nadir)				
	5	Panchromatic (480–710)			
Quickbird	2.44 m GSD at nadir	4: blue (450–520), green (520–600), red (630–690), near IR (760–900)	11 bits	16.5	2001/10.30 am solar time/450 km
	0.61	Panchromatic (450–900)			
GeoEye 1	1.65	4: blue (450–510), green (510–580), red (655–690), near IR (780–920)	11 bits	15.2	2008/10.30 am/684 km
	0.41	Panchromatic (450–800)			
WorldView 2	1.8 m GSD at nadir, 2.4 m GSD at 20° off-nadir	8: coastal (400–450), blue (450–510), green (510–580), yellow (585–625), red (630–690), red edge (705–745), near IR 1 (770–895), near IR 2 (860–1040)	11 bits	16.4	2009/10.30 am/770 km
	0.46	Panchromatic (400–900 or 450-800)			
Rapideye	6.5 m GSD at nadir	5: blue (440–510), green (520-590), red (630–690), red edge (690–730), near IR (760–880)	12 bits	25	2009/11.00 am

technology with the goal of deriving land cover maps describing vegetation condition for a particular area. The second focus investigates the uncertainty in classification methods by quantifying the factors affecting classification accuracy. In this chapter, we use three case studies to give examples of new research in these two areas. Our first case study involves the use of operational multi-spectral remotely sensed data from two high spatial resolution sensors (Ikonos and SPOT 5; Table 14.1) for the assessment of vegetation condition. It demonstrates the utility of vegetation indices and texture analyses for deriving vegetation condition. Our second case study describes the potential role of new active remote sensing technologies in assessing the structural components of vegetation condition. In particular, the use of airborne laser scanning (LiDAR, Light Detection and Ranging) for characterising the ecological structure of a forested landscape is examined, and we describe a forest structural characterisation scheme derived from LiDAR point clouds. The third and final section of this chapter illustrates how theoretical research using simulation modelling can aid in the quantification of spatial uncertainty in remote sensing. It examines issues surrounding the accurate mapping of small (often fragmented) and linear (often ecologically significant) vegetation patches. Our results demonstrate that the spatial resolution of the raster grid should be many times finer than the vegetation patches of interest, to ensure their accurate extraction.

RELATIONSHIP BETWEEN REMOTELY SENSED INDICES AND FIELD-BASED MEASURES

This section presents an overview of vegetation condition assessments using operational multi-spectral satellite remote sensing platforms, namely SPOT 5 and Ikonos (Table 14.1) and explores their utility in estimating field-based biophysical measures of vegetation. Remote sensing-based vegetation measures range from highly detailed fine-scale assessments, to regional and global applications (Thomlinson *et al.* 1999; Defries *et al.* 2000; Armston *et al.* 2004; Johansen *et al.* 2007). Satellite remotely sensed data provide an efficient method to measure stands of vegetation in a timely manner, particularly over larger tracts of vegetation (Coops and Culvenor 2000; Zawadzki *et al.* 2005). The spectral response of vegetation is characterised by lower reflectance in the visible portion of the electromagnetic spectrum and high reflectance of near-infrared wavelengths. Key factors that affect the spectral reflectance characteristics of vegetation include leaf size and orientation, physical structure of the plant, species distribution, vegetation density and the influence of other land covers (Bannari *et al.* 1995; Armitage *et al.* 2000; Nagendra 2001). The temporal influence on vegetation spectral response is also a key consideration due to vegetation phenology (and consequently spectral reflectance) and associated plant growth stages (Nagendra 2001; McCoy 2005).

A common approach to the characterisation of vegetation using multi-spectral remotely sensed data is the use of vegetation indices. Many different vegetation indices have been developed to provide information on a range of vegetation characteristics such as vegetation cover, leaf density or leaf water content. The Normalized Difference Vegetation Index (NDVI) is widely used to derive estimates of vegetation cover (Bannari *et al.* 1995; Defries *et al.* 2000; North 2002; Carreiras *et al.* 2006). Other routinely used vegetation indices include the Soil Adjusted Vegetation Index (SAVI) and the Enhanced Vegetation Index (EVI) (Huete 1988; Bannari *et al.* 1995; Nagler *et al.* 2001; Huete *et al.* 2002).

In this study, a number of different types of remotely sensed variables were calculated from imagery obtained from SPOT 5 and Ikonos to examine any relationship between satellite observations and different vegetation biophysical parameters collected in the field. Full details of the method employed in this study can be found in Sheffield *et al.* (2009). Variables investigated included mean spectral reflectance values, vegetation indices and image texture measures. Mean spectral reflectance values and vegetation indices were found to be strongly correlated with vegetation condition attributes. Of the suite of vegetation indices calculated for this study, the simple ratio, SAVI and NDVI had the highest correlations with stem density (adjusted r^2 values ranging from 0.60 to 0.70, $p < 0.05$). However, adjusted r^2 values for overall (or composite) site vegetation condition were 0.47 (SPOT 5 data) and 0.51 (Ikonos data) at $p < 0.05$ respectively. This relationship occurs because a number of vegetation attributes were not strongly correlated ($r < 0.40$) with any remotely sensed variables. These attributes included grass cover, bare ground, exposed rock, understorey and exotic species cover, organic litter cover and hollow-bearing trees.

We found texture metrics (e.g. measures of contrast, correlation, entropy and homogeneity) were strongly correlated with a number of vegetation attributes. Texture measures derived using near-infrared reflectance were more strongly correlated with vegetation attributes than texture measures derived from other sensor bands. The suite of texture measures identified differed for each sensor. Texture analysis makes use of the fine-scale patterns of vegetation that are present in imagery derived from high spatial resolution sensors. While measures of image texture are not widely used in this type of application, our analysis highlighted the potential contribution of texture measures for the recovery of vegetation structural attributes from remotely sensed data. A comprehensive presentation of these results is given in Sheffield (2009) and Sheffield *et al.* (2009).

ADVANCES IN REMOTE SENSING TECHNOLOGIES: UTILISING LIDAR FOR VEGETATION CONDITION MAPPING

Forest structure – the architectural arrangement of plant material – has received less attention than species composition in terms of description and/or classification, yet diagnostically structure is considered just as important in characterising a forest as its composition (Florence 1996; Spies 1998; Stone and Porter 1998). This analogy was used by Noss (1990), who suggested that vegetation condition, when assessed in the context of biodiversity, should be considered in terms of structure, composition and function. Spies (1998) described the essential attributes of forest structure as structural type, size, shape and spatial distribution (vertical and horizontal) of components and examined their roles and importance in the functioning and diversity of ecosystems. For example, foliage layering or vertical foliage distribution is a component of forest structure that plays important roles in absorption of solar radiation, the microclimate of the forest and providing wildlife habitat. Many authors have noted the association between biodiversity and measures of the variety and/or complexity of arrangement of structural components within an ecosystem (Mac Nally *et al.* 2001; Sullivan *et al.* 2001). Most forest stand structure descriptors are based on measures easily obtainable from the ground level (e.g. diameter at breast height (DBH), stem density or ground-assessed canopy cover). Currently, appraisal or scoring methods for structural complexity require a laborious process that involves site visits and many logistically expensive point-based measurements. An automated or semi-automated method would be ideal.

Airborne laser scanning (LiDAR) is recognised as a powerful tool for forest structure characterisation. Numerous papers have documented its utility for the estimation of forest attributes in forestry (Lefsky *et al.* 1999; Means *et al.* 1999; Nilsson 1996). Næsset (1997) showed the potential of LiDAR to estimate fractional cover. Similar methods utilising the point density of LiDAR returns to estimate fractional cover were presented in other studies (Coops *et al.* 2007; Hopkinson and Chasmer 2007, 2009; Morsdorf *et al.* 2006; Solberg *et al.* 2006) and showed promising results. While the majority of previous LiDAR research focuses on its application for forestry, the purpose of our case study is to report on a protocol for characterising the ecological structure of a dry eucalypt forest landscape using LiDAR data alone for the assessment of vegetation condition. LiDAR data were acquired over the study area using a RIEGL LMS-Q560 sensor. The full forest characterisation scheme is presented in Miura and Jones (2010).

The forest structural characterisation scheme is derived from both the range, i.e. the distance information from the LiDAR system (how long it takes for a pulse of energy to travel from the sensor to the object and return to the sensor) and the nature of the return type. Four return types are distinguished:

Table 14.2: Forest characterisation scheme (FCS)

Category	Description	LiDar return ratio
1 OG	Opening above the ground	Ground Type 1
2 OL	Opening above low vegetation	Low veg Types 1 and 2
3 VL	Presence of understorey vegetation	Low veg total (Types 1, 2, 3 and 4)
4 CC	Canopy cover	Medium veg Types 1 and 2 and High veg Types 1 and 2
5 OM	Opening above medium vegetation	Medium veg Types 1 and 2
6 VM	Presence of mid-storey vegetation	Medium veg total (Types 1, 2, 3 and 4)
7 VH	Presence of high trees	High veg total (Types 1, 2, 3 and 4)
8 DH	Vertically dense canopy of high trees	High veg Types 3 and 4

- Type 1 are singular returns – only one return was recorded from each emitted pulse of energy;
- Type 2 are first of many returns – part of the pulse of incident energy has interacted with a plant facet and been reflected back to the sensor but much of the energy has continued through the tree, interacting with other structural elements along its path;
- Type 3 are intermediate returns – the subsequent interactions of the pulse described in Type 2;
- Type 4 are the last of many returns – the last pulse returned to the sensor from an incident pulse.

From these, the forest condition score (FCS) is derived (Table 14.2) and validated against field-derived equivalents of these metrics.

Of the eight proposed FCS categories, all showed good association with field-based metrics. For example:

- LiDAR-derived CC and field-measured CC assessments all displayed significant positive correlations;
- LiDAR-derived VL (Category 3, presence of understorey vegetation) was a good predictor of field-recorded leaf area for vegetation <1 m in height (r^2=0.82, p < 0.05);
- LiDAR-derived CC (Category 4, canopy cover) was significantly correlated with two ground-based measures of CC (photosynthetic/photosynthetic and non-photosynthetic), with r^2=0.86 and 0.84 respectively;
- LiDAR-derived VH (Category 7, presence of high trees) and LiDAR-derived DH (Category 8, vertically dense canopy of high trees) showed strong correlations with field-derived sum of DBH weighted canopy depth in High veg with r^2=0.73 and 0.71 respectively in both medium vegetation (1–5 m) and high vegetation (>5 m);
- LiDAR-derived OM (Category 5, opening above medium vegetation) was strongly correlated with field-derived opening above Medium veg with r^2=0.73.

We validated these results using a second LiDAR dataset that had been acquired over the same area (Miura and Jones 2009).

The proposed FCS method has the ability to characterise some elements of the ecological structure of a dry eucalypt forest landscape. Regression analysis showed that LiDAR-derived variables were good predictors of field-recorded variables across a different range of forest structural types. The proposed scheme demonstrated the potential of different laser pulse return properties from a full waveform LiDAR to provide information on the complexity of habitat structure in an efficient and cost-effective manner. In terms of Spies' (1998) proposed components of forest structure, the FCS effectively reports on all elements of component 1 (foliage) as well as elements of component 4 (tree boles), the volume of fallen trees (component 7) and potentially the biomass element of component 8 (shrub, herb and moss layers). The FCS can be adapted for use in any designed 3D ecological characterisation scheme. It is anticipated that the FCS may have wide applicability in characterising forest structure over a range of scales from patch to landscape, although this clearly requires further investigation.

SPATIAL DATA UNCERTAINTY AND THE IMPLICATIONS FOR VEGETATION MAPPING

Thematic maps derived from remote sensing imagery are used to characterise landscape structure and composition and relate these to landscape processes or landscape change (Metzger 2008). Management of native vegetation requires careful attention to how the spatial element of the vegetation is described and measured (Williams 2004; Michaels *et al.* 2010). Providing quality spatial data in map form or as an area statement results in the spatial element of the variable (whether vegetation condition or one of the underlying parameters making up condition) being important. Indeed, any time we look to quantify the amount of a vegetation attribute in a landscape the spatial element becomes explicit and the resultant measures are inevitably influenced by the source of the data (e.g. satellite system), as well as the processing and analysis methods used to extract the variable of interest. While remote sensing offers many opportunities for mapping vegetation biophysical and condition parameters, it is important to consider issues of spatial uncertainty. This section explores issues of spatial uncertainty resulting from information derived from simulated remote sensing sources and classification activities within the context of remnant vegetation patches. Simulated rather than actual data were used due to the large volume of examples needed to support the statistical analysis in this problem.

Features such as small remnant and linear vegetation patches have ecological value that is often proportionally greater than their areal extent. The presence or absence of these features change landscape pattern-related properties such as connectivity and fragmentation. Accordingly, the logic and process which supports the mapping of these patches is critical. Small or linear vegetation patches are ecologically significant and can be found as roadside vegetation, hedgerows, scattered trees, riparian areas and greenways or can be purposely built to facilitate connectivity (Bennett 1990; Gergel *et al.* 2007; Hilty *et al.* 2006; Manning *et al.* 2006). However, due to the relatively narrow width of corridors they may be underrepresented in the landscape when mapped using remote sensing (Vogt *et al.* 2007) or traditional field-based mapping. To effectively use a mapping tool, map users need to know (for that tool) the smallest discernible feature at any given spatial resolution, and the accuracy at which these features are mapped. It has been suggested that high spatial resolution imagery of 0.25–10 m is required for mapping in complex or urban areas (Jensen and Cowen 1999). Congalton *et al.* (2002) suggested that sensors with finer spatial resolutions, such as Ikonos with 4 m multi-spectral sensors, will be more appropriate for features with smaller areas such as riparian vegetation. Research on the appropriate spatial resolution for mapping small and linear objects is often based on qualitative examinations. Quantification with a simulation model describing the theoretical limitations of remote sensing due to pixel size was presented in Lechner *et al.* (2009); a summary of that study now follows.

Extraction probability (of a feature such as a tree or vegetation community) and classification accuracy are a function of the size, shape and random position of a feature with respect to a remote sensing systems array's grid. Additionally, they are a function of both its spectral characteristics and those of the surrounding objects. This, our third case study, simulates imagery in order to model the subpixel location of features with respect to the image grid; it tests the effect of grid position, contrast and feature size and shape in isolation. Since a feature's position is random relative to the position of the imaging sensors, grid small features may be lost when they make up only a portion of a cell or are found at the intersection of several cells (Cunningham 2006). Problems of this type are particularly common in highly fragmented environments such as urban and peri-urban areas. For example, Australian roadside vegetation is typically 2–4 m wide, while high spatial resolution satellites such as Quickbird and SPOT XS have a multi-spectral spatial resolution of 2.4 m and 10 m respectively (Table 14.1).

Detectability is scene- and sensor-specific (Adams and Gillespie 2006) and decreases with increasing spectral similarity between target and surrounding objects and the sensors' sensitivity (Forshaw *et al.* 1983). Classification will be affected by other factors such as image registration, view angle, radiometric calibration, image acquisition time and sensor characteristics such as spatial and radiometric resolution and bandwidth (Cracknell 1998; Townshend *et al.* 1991). Thus, classification accuracy and extraction probability in this study are the result of the geometric properties of the grid alone, representing the highest achievable accuracy for remote sensing where the above factors are ignored. The results of this study allow:

- determination of the effect of the position of the raster grid in relationship to small and linear landscape features on classification;
- calculation of the appropriate spatial resolution required to extract features of various degrees of elongation and area;
- examination of the effect of differing spectral contributions of the object and its surrounding classification.

Using a statistical simulation model, we tested the effect of patch size and shape, classification threshold and grid location on the classification of small and linear features using remote sensing data with a statistical simulation model. Our model considered rectangles of a variety of lengths, widths and total areas with different classification thresholds and orientations to simulate the mapping of small and linear patches. In the simulation, each feature was represented by a high spatial resolution raster and was resampled to a low resolution raster representing the remote sensing raster grid in order to simulate its subpixel patch location. Approximately 225 combinations of length:width ratios and 255 areas were tested for three classification thresholds. For each combination of length, width and area the computer simulation repeatedly systematically shifted the grid vertically and horizontally with respect to patch location. As well as changes in the vertical and horizontal position of the feature, we simulated different orientations of the feature (0–45°). For each length:width combination, there were 51 005 combinations of x, y positions and orientations tested. For each grid position, mean accuracy and extraction probability were calculated by comparing the difference between the area classified as patch versus matrix in the grid and the subgrid. Differences in the area of each class were calculated through the high spatial resolution subgrid using a range of pixel weightings to test for differences in classification thresholds. Accuracy was measured with errors of commission and omission as well as mapping accuracy. Using the simulation model described above, our study investigated seven aspects of mapping small and linear patches. These focused on the effects of patch area, elongation and position on classification accuracy and extraction probability. A full account of the results is available in Lechner *et al.* (2009). Our results from four measures are reported in this chapter:

- effect of area on accuracy and extraction probability;
- effect of elongation on accuracy and extraction probability;
- effect of feature orientation on accuracy and extraction probability;
- overview of effect of length:width ratio, area and classification threshold on accuracy and extraction probability.

This study found that larger patches were always extracted and, once the area of a square was greater than 2 pixels, the probability of extraction was 100% for 0.5 classification threshold (Fig. 14.1). For classification thresholds of 0.75 and 0.25, the area of a square feature needed to be at least 1 and 3 pixels respectively for a 100% probability of extraction. As the patch size increased, the mean accuracy also increased. For patches with an area of 0–2 pixels and a classification threshold of 0.5, the initial increase in accuracy was steep – up to around 2 pixels area. The effect of increasing elongation was a decrease in the probability of extraction and mean accuracy. A patch of a single pixel in area had a probability of extraction of around 64% when it was a square, however, once the length:width ratio was greater than 6 it could no longer be extracted for 0.5 classification threshold. If the classification threshold favoured patches the effect of elongation was less dominant, although it resulted in greater errors of commission. Smaller and more elongated patches had larger errors of omission than commission. However, as they became larger and more compact these errors balanced out, as misidentification occurred around the edge of patches. The interior pixels of a patch were always classified correctly, as they were in the majority regardless of their position. Mean accuracy was lower when patches had an orientation similar to the grid. Increasing elongation resulted in greater differences in mean accuracy between orientations.

This case study highlights several issues in using the pixel model to describe geographic phenomena. It focuses on understanding misclassification resulting from patch size approaching the scale of the sensor's spatial resolution. Small and/or elongated patches have a reduced probability of extraction, a reduced mapping accuracy and an increased variability in accuracy due to the effects of grid position. To extract those patches accurately, the grid spatial resolution should be many times finer. For example, a square patch needed an area of at least 11

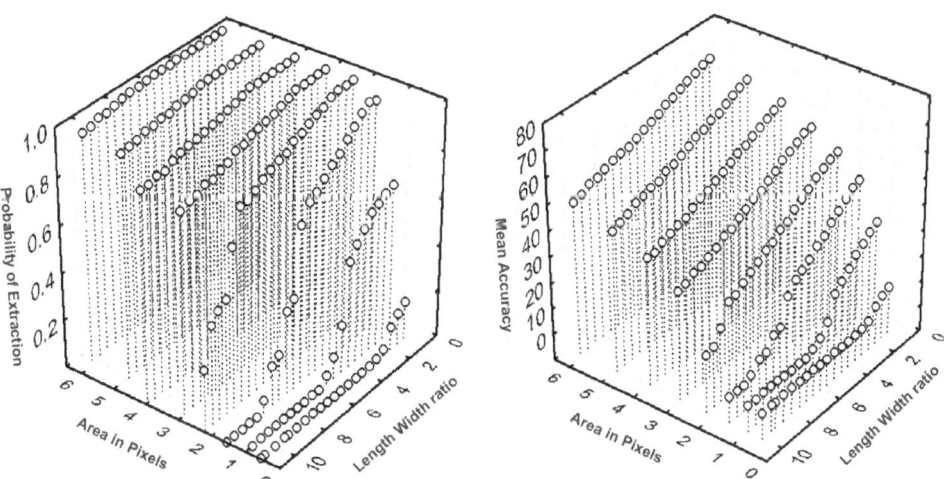

Figure 14.1: Output of simulation model describing the probability of extraction versus length:width ratio and area, and mean accuracy versus length:width ratio and area.

pixels to achieve a mean accuracy of 75%, while a linear patch with a width:length ratio of 4 needed an area of 12.3 pixels.

CONCLUSIONS

This chapter presented a series of case studies to illustrate the opportunities and limits of remote sensing in the mapping of native vegetation. The first case study assessed the utility of operational satellite-based sensors for reporting site-scale estimates of vegetation condition parameters. We found that vegetation indices and texture metrics exhibited significant correlations with a number of ground-assessed vegetation condition attributes. The second case study presented a useful method for using LiDAR remote sensing technology to characterise the three-dimensional structural components of vegetation condition. The third and final case study investigated the impact that remote sensing resolution, heterogeneity and classification methods can have when mapping and reporting on native vegetation.

Operational remote sensing is indispensible for NRM, allowing the rapid detection of change in vegetation condition as a result of human disturbances ranging from habitat fragmentation to introduced species and climate change. New developments in remote sensing technology and classification methods can yield increased attribution accuracy and the ability to measure a greater number of vegetation condition metrics. While remote sensing shows great promise, it is important that significant effort is devoted both to developing new technologies and attribution methods, and understanding the uncertainties that may result from using these technologies and methods. Remote sensing systems have the potential to provide total and repeat coverage information about vegetation condition at the site, regional and landscape scales and perhaps, in the future, at a comprehensive global scale.

REFERENCES

Adams JB and Gillespie AR (2006) *Remote Sensing of Landscapes with Spectral Images: A Physical Modeling Approach*. Cambridge University Press, New York.

Armitage RP, Weaver RE and Kent M (2000) Remote sensing of semi-natural upland vegetation: the relationship between species composition and spectral response. In *Vegetation Mapping: From Patch to Planet*. (Eds R Alexander and AC Millington) pp. 83–102. John Wiley and Sons, UK.

Armston JD, Danaher TJ and Collett LJ (2004) A regression approach to mapping woody foliage projective cover in Queensland with Landsat data. In *Proceedings of the 12th Australasian Remote Sensing and Photogrammetry Conference*, Fremantle, Western Australia.

Bannari A, Morin D, Bonn F and Huete AR (1995) A review of vegetation indices. *Remote Sensing Reviews* **13**, 95–120.

Bennett AF (1990) Habitat corridors and the conservation of small mammals in a fragmented forest environment. *Landscape Ecology* **4**, 109–122.

Carreiras JMB, Pereira JMC and Pereira JS (2006) Estimation of tree canopy cover in evergreen oak woodlands using remote sensing. *Forest Ecology and Management* **223**, 45–53.

Congalton RG, Birch K, Jones R and Schriever J (2002) Evaluating remotely sensed techniques for mapping riparian vegetation. *Computers and Electronics in Agriculture* **37**, 113–126.

Coops NC and Culvenor DS (2000) Utilizing local variance of simulated high spatial resolution imagery to predict spatial pattern of forest stands. *Remote Sensing of Environment* **71**, 248–260.

Coops N, Hilker T, Wulder M, St-Onge B, Newnham G, Siggins A and Trofymow J (2007) Estimating canopy structure of Douglas-fir forest stands from discrete return LiDAR. *Trees – Structure and Function* **21**, 295–310.

Cracknell AP (1998) Synergy in remote sensing: what is in a pixel? *International Journal of Remote Sensing* **19**, 2025–2047.

Cunningham MA (2006) Accuracy assessment of digitized and classified land cover data for wildlife habitat. *Landscape and Urban Planning* **78**, 217–228.

Defries RS, Hansen M, Townshend JRG, Janetos AC and Loveland TR (2000) A new global 1-km dataset of percentage tree cover derived from remote sensing. *Global Change Biology* **6**, 247–254.

Florence RG (1996) *Ecology and Silviculture of Eucalypt Forests*. CSIRO Publishing, Melbourne.

Forshaw MRB, Haskell A, Miller PF, Stanley DJ and Townshend JRG (1983) Spatial resolution of remotely sensed imagery: a review paper. *International Journal of Remote Sensing* **4**, 497–520.

Gergel S, Stange Y, Coops N, Johansen K and Kirby K (2007) What is the value of a good map? An example using high spatial resolution imagery to aid riparian restoration. *Ecosystems* **10**, 688–702.

Gibbons P, Zerger A, Jones S and Ryan P (2006) Mapping vegetation condition in the context of biodiversity conservation (Editorial). *Ecological Management and Restoration* **7**(S1), 1–2.

Hilty JA, Lidicker WZ Jr and Merenlender AM (2006) *Corridor Ecology: The Science and Practice of Linking Landscapes for Biodiversity Conservation*. Island Press, Washington.

Hopkinson C and Chasmer L (2007) Using discrete laser pulse return intensity to model canopy transmittance. *Photogrammetric Journal of Finland* **20**(2), 16–26.

Hopkinson C and Chasmer L (2009) Testing LiDAR models of fractional cover across multiple forest ecozones. *Remote Sensing of Environment* **113**, 275–288.

Huete AR (1988) A soil-adjusted vegetation index (SAVI). *Remote Sensing of Environment* **25**, 295–309.

Huete AR, Didan K, Miura R, Rodriguez EP, Gao X and Ferreira LG (2002) Overview of the radiometric and biophysical performance of the MODIS vegetation indices. *Remote Sensing of Environment* **83**, 195–213.

Jensen JR and Cowen DC (1999) Remote sensing of urban/suburban infrastructure and socio-economic attributes. *Photogrammetric Engineering and Remote Sensing* **65**, 611–622.

Johansen K, Coops NC, Gergel SE and Stange Y (2007) Application of high spatial resolution satellite imagery for riparian and forest ecosystem classification. *Remote Sensing of Environment* **110**, 29–44.

Lechner A, Stein A, Jones S and Ferwerda J (2009) Remote sensing of small and linear features: quantifying the effects of patch size and length, grid position and detectability on land cover mapping. *Remote Sensing of Environment* **113**(10), 2194–2204.

Lefsky MA, Cohen WB, Acker SA, Parker GG, Spies TA and Harding D (1999) LiDAR remote sensing of the canopy structure and biophysical properties of Douglas fir–western hemlock forests. *Remote Sensing of Environment* **70**, 339–361.

Mac Nally R, Parkinson A, Horrocks G, Conole L and Tzaros C (2001) Relationships between terrestrial vertebrate diversity, abundance and availability of coarse woody debris on south-eastern Australian floodplains. *Biological Conservation* **99**, 191–205.

Manning AD, Fischer J and Lindenmayer DB (2006) Scattered trees are keystone structures: implications for conservation. *Biological Conservation* **132**, 311–321.

McCoy RM (2005) *Field Methods in Remote Sensing*. Guilford Press, New York.

Means JE, Acker SA, Harding DJ, Blair JB, Lefsky MA, Cohen WB, Harmon ME and McKee WA (1999) Use of large-footprint scanning airborne LiDAR to estimate forest stand characteristics in the Western Cascades of Oregon. *Remote Sensing of Environment* **67**, 298–308.

Metzger J (2008) Landscape ecology: perspectives based on the 2007 IALE world congress. *Landscape Ecology* **23**, 501–504.

Michaels K, Norton TW, Lacey M and Williams JE (2010) The contribution of small patches of native vegetation to biodiversity conservation in Tasmanian agricultural landscapes. *Ecological Management and Restoration* **11**, 193–199.

Miura N and Jones SD (2009) Testing the performance of a forest characterization scheme using multiple dataset comparison. *Silvilaser 2009 Conference*. 14–16 October, College Station, Texas.

Miura N and Jones SD (2010) Characterizing forest ecological structure using airborne laser scanning. *Remote Sensing of Environment* **114**(5), 1069–1076.

Morsdorf F, Kötz B, Meier E, Itten KI and Allgöwer B (2006) Estimation of LAI and fractional cover from small footprint airborne laser scanning data based on gap fraction. *Remote Sensing of Environment* **104**, 50–61.

Næsset E (1997) Estimating timber volume of forest stands using airborne laser scanner data. *Remote Sensing of Environment* **61**, 246–253.

Nagendra H (2001) Using remote sensing to assess biodiversity. *International Journal of Remote Sensing* **22**, 2377–2400.

Nagler PL, Glenn EP and Huete AR (2001) Assessment of spectral vegetation indices for riparian vegetation in the Colorado River delta, Mexico. *Journal of Arid Environments* **49**, 91–110.

Nilsson M (1996) Estimation of tree heights and stand volume using an airborne LiDAR system. *Remote Sensing of Environment* **56**, 1–7.

North PRJ (2002) Estimation of fAPAR, LAI, and vegetation fractional cover from ATSR-2 imagery. *Remote Sensing of Environment* **80**, 114–121.

Norton TW and Michaels K (2010) 'Third National Workshop on Native Vegetation Assessment'. Technical Report no. 21. Landscape Logic, Hobart.

Noss RF (1990) Indicators for monitoring biodiversity: a hierarchical approach. *Conservation Biology* **4**, 355–364.

Reinke R and Jones SD (2006) Issues arising from the integration of regional scale remotely sensed data with site based assessments of native vegetation condition. *Ecological Management and Restoration* **7**(S1), 18–23.

Sheffield K (2009) Multi-spectral remote sensing of native vegetation condition. PhD thesis. RMIT University, Melbourne.

Sheffield K, Jones S, Ferwerda JG, Gibbons P and Zerger A (2009) Linking biological survey data to remote sensing datasets. In *Innovations in Remote Sensing and Photogrammetry/Lecture Notes in Geoinformation and Cartography*. (Eds S Jones and K Reinke) pp. 51–64. Springer, Berlin.

Solberg S, Næsset E, Hanssen KH and Christiansen E (2006) Mapping defoliation during a severe insect attack on Scots pine using airborne laser scanning. *Remote Sensing of Environment* **102**, 364–376.

Spies TA (1998) Forest structure: a key to the ecosystem. *Northwest Science* **72**, 35–39.

Stone JN and Porter JL (1998) What is forest stand structure and how to measure it? *Northwest Science* **72**, 25–26.

Sullivan TP, Sullivan DS and Lindgren PMF (2001) Stand structure and small mammals in young lodgepole pine forest: 10-year results after thinning. *Ecological Applications* **11**, 1151–1173.

Thomlinson JR, Bolstad PV and Cohen WB (1999) Coordinating methodologies for scaling landcover classification from site-specific to global: steps toward validating global map products. *Remote Sensing of Environment* **70**, 16–28.

Townshend JRG, Justice C, Li W, Gurney C and McManus J (1991) Global land cover classification by remote sensing: present capabilities and future possibilities. *Remote Sensing of Environment* **35**, 243–255.

Vogt P, Riitters K, Estreguil C, Kozak J, Wade T and Wickham J (2007) Mapping spatial patterns with morphological image processing. *Landscape Ecology* **22**, 171–177.

Williams JE (2004) Metrics for assessing the biodiversity values of farming systems and agricultural landscapes. *Pacific Conservation Biology* **10**, 145–163.

Williams JE (2007) 'Futures for native vegetation condition research in Tasmania and Victoria'. Workshop summary. Landscape Logic, Hobart.

Zawadzki J, Cieszewski CJ, Zasada M and Lowe RC (2005) Applying geo-statistics for investigations of forest ecosystems using remote sensing imagery. *Silva Fennica* **39**, 599–618.

15

The role of social norms in natural resource management

Wendy Minato, Allan Curtis and Catherine Allan

SUMMARY

This chapter examines the influence of social norms on the management of native vegetation in a small rural community in north-eastern Victoria. Examining change from a social norms perspective was a novel way to explore the influence of government intervention and the inter-action between farming and non-farming landholders in a particular setting. From this study we have gained some understanding of the relationship between social norms and natural resource management (NRM) in a changing rural community. Well-established farming norms (such as keeping the land 'tidy' and weed-free) are influencing newcomers to the area. New norms and practices (e.g. tree planting) have emerged as a result of government invest-ment in Landcare. Although the community is changing, there does not generally seem to have been a decrease in the level of social trust and neighbourliness between new residents and old. This is a positive finding, given that the stock of social capital in rural communities is likely to be a key factor in determining the success or failure of future investment in community-based NRM. We suggest that better understanding of social norms in different contexts will lead to more effective delivery of NRM policy and program interventions.

INTRODUCTION

This study used qualitative methods to examine the relative influence of NRM investment and demographic change on the management of native vegetation on private land (Minato *et al.*

2010). The extent of native vegetation on rural properties in north-eastern Victoria is thought to have increased in recent decades (see Chapters 10 and 11, this volume) but there has been little research exploring the social factors that might have contributed to the change.

Key assumptions underpinning the research included the following: that recent improvements in native vegetation extent and quality (Chapter 12, this volume) can be partly attributed to a range of policy instruments aimed at protecting and enhancing native vegetation, that increasing numbers of lifestyle land owners and a decrease in the number of full-time farmers are having an impact on the way native vegetation is managed on private land and that this trend is likely to continue, with implications for future NRM investment in amenity landscapes.

We review the social norms literature then present and analyse selected findings from the study from a social norms perspective. We then consider the degree to which changing community structure is influencing NRM, and the implications of our research for future policy interventions in rural areas.

SOCIAL NORMS

Social norms are shared cultural understandings of how to behave in common social situations. They are informal rules, implicit in the day-to-day running of a society and we learn these rules by observing others and through a long process of trial and error, repeating the strategies that appear to have successful outcomes (Bicchieri 2006; Wallace 2009). As such, the norms people learn will vary from one culture to another, and within cultures there will be variations across families and groups, and with exposure to norms that exist in various situations (Ostrom 2000). There are norms associated with all aspects of human behaviour, encouraging people to act in ways which are consistent with other people's expectations. Sunstein (1995) observed that norms reduce personal freedom by dictating what individuals should and shouldn't do, but also give us the freedom to socialise in ways that would not be possible without the unspoken understandings that norms rely upon.

Two main types of norms can be identified. The first type are norms that have a moral imperative associated with them, in that people behave in a certain way because they perceive that they 'ought' to, irrespective of their beliefs and the outcome for them personally. Most commonly referred to as 'injunctive norms', this type of obligation involves social attitudes of approval and disapproval and some form of external sanctioning as a means of enforcement. This type of norm most often applies in situations where there may be a conflict between what an individual might prefer to do and what is best for the group or society at large.

The second type of norm is commonly referred to as a 'descriptive norm' (Cialdini et al. 1991; Bicchieri 2006). These norms provide 'social proof' (Axelrod 1986) that a certain behaviour in a given situation is an acceptable one, even if we are not sure why. These norms generate social expectations without the same level of moral obligation associated with injunctive norms (Hechter and Opp 2001). People conform to these norms because they want to fit in, or because it is easier to imitate others' behaviour in situations where they might be unsure of what to do (White et al. 2009). There is also an information-processing advantage associated with these norms in that they allow individuals, without much conscious deliberation, to make quick decisions about how to behave.

Although norms are widely accepted, in practice they tend to be acted out as 'locally constituted phenomena' in specific situations (Fine 2001). This is because deciding how to behave in a particular situation depends on an individual's interpretation of what is happening in a particular context. This will be based on past experience, knowledge about other actors involved and potential outcomes in that particular setting. In any given circumstance there may be numerous conflicting norms requiring an individual to choose from a range of options. A combination of situational factors, personal preferences and immediate contingencies determine what action shall be taken (Fine 2001; Bicchieri 2006).

NORMS AND NRM

In Australia, the need to address complex resource management problems has led to the development and application of a variety of policy instruments. In particular, there has been substantial investment in community-based NRM. It was recognised that existing voluntary local groups were proving a useful vehicle for change, and this led to the development of the federally funded national Landcare program. It was hoped that Landcare groups would engage rural landholders and facilitate community participation in activities designed to raise awareness, knowledge and management skills and increase adoption of more sustainable land management practices (Curtis and De Lacy 1996). Initially, the focus was on education and activities such as field days and farm walks. National Heritage Trust funding later supported a range of on-ground works including fencing, tree planting, seed collecting and pest and weed control projects (Curtis and De Lacy 1996).

Given the social success of Landcare and the long history of Landcare groups in north-eastern Victoria and within the case study area, the activities of the local Landcare group were an obvious focus for this research.

THE CASE STUDY

The Indigo Valley subcatchment covers an estimated 35 500 ha within the Indigo Shire local government area of north-eastern Victoria, Australia (Fig. 15.1). The valley is a rural landscape within commuting distance of Albury–Wodonga, one of the largest inland centres (population 87 500) in Australia (ABS 2010). Fieldwork and the examination of available maps indicated that the valley has a mix of around 160 small farms (<400 ha) and lifestyle blocks (40 ha). When first settled in the mid 1800s the valley would have been a small farming community with strong social and family networks. This is changing as the last generation of full-time farmers leave the land and their properties are passed on, or subdivided and sold, to a socially diverse cohort of new owners, often from urban backgrounds.

North-eastern Victoria has been the focus of many government programs aimed at improving the quality and extent of native vegetation. Indigo Valley in particular has been targeted for various land, water and vegetation initiatives. From the late 1980s the valley was a high-priority area for dryland salinity management, with a focus on reducing groundwater discharge through an increase in perennial vegetation in the landscape. Another project aimed to establish more native vegetation, firewood plantations and perennial pastures in the valley. Native vegetation protection is the current major focus for funding of on-ground works and there are payments available for the conservation of native pastures or remnant vegetation patches within designated areas. The need to protect recognised assets (agricultural land, soil structure, water quality) threatened by salinity is ongoing and funding is available for fencing and revegetation projects.

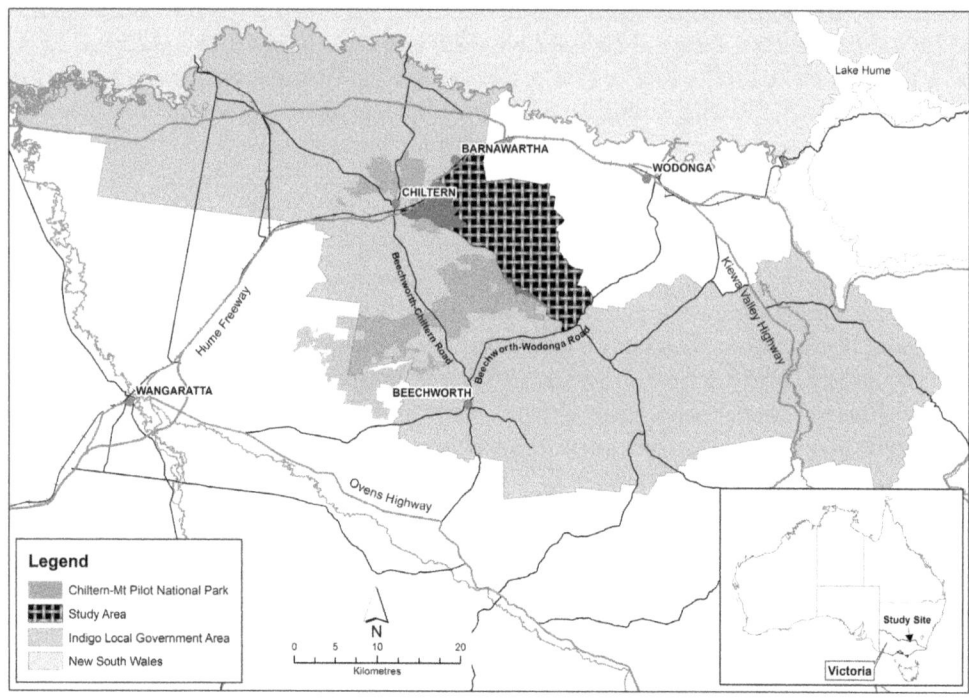

Figure 15.1: The case study area within Indigo Valley Shire, north-eastern Victoria.

A number of purposeful sampling techniques were used to identify a mix of farming and lifestyle property owners who were willing to participate in the research. Purposeful sampling was also used to identify key informants known to have many years of experience in NRM; these individuals would provide an alternative perspective on government programs and change in native vegetation within the case study area.

Over a period of about eight months during 2008–09, 36 landholders were interviewed; interviews were semi-structured and loosely based on an interview guide listing topics such as involvement in government investment programs, level of regard for native vegetation, future plans for the land, past and present land use and the nature of the local community.

Seven key informant interviews were also conducted during this period. The interviewees represented agencies such as Trust for Nature, the Department of Sustainability and Environment, the Department of Primary Industries and the North-East Catchment Management Authority.

With prior consent, all of the interviews were taped and transcribed. Inductive data analysis was carried out concurrently with data collection. NVivo software was used to organise and make sense of the data. Participants were allocated a unique code which has been used for the interview excerpts quoted in this chapter. The LH prefix refers to landholders, GA to agency staff.

FINDINGS

The following section is presented in two parts. Part 1 uses selected findings from the interview data to illustrate some key norms and the strategies used to establish and enforce them. In

Part 2 the results are interpreted with reference to the literature on social norms, focusing on the role of Landcare in establishing and enforcing local norms and social interactions between the farming community and newer lifestyle residents.

PART 1: THE EVIDENCE

As with many Landcare groups in agricultural areas, the initial focus of the Indigo Valley Landcare group (established 1988) was on planting trees to address erosion and salinity problems. Committee members deliberately set out to make tree planting an annual event. They raised awareness of the importance of planting trees via the newsletter and, more importantly, made it easy to follow through. They pushed the simple message that 'it is important to plant trees' and said 'here's how you can do it' (advice on which trees to plant and how to order). They made the trees affordable by subsidising their purchase and providing tree guards at no cost. A local nursery supplied the trees and landholders would pick them up, usually after rain. The group successfully established a new pattern (norm) of behaviour that resulted in the planting of thousands of trees in the valley. This is a good example of a changing practice norm which has arisen out of a need to address an environmental problem.

> *We were all in the habit, from Landcare, of every year, you'd just buy another hundred trees and put in another plantation somewhere because it was easy (LH36).*

Planting trees is a very visual activity and Indigo Valley locals seemed generally well informed about who was planting what and where. Larger plantings such as those involving the assistance of the Indigo Valley Incentive Project were invariably commented upon by neighbouring land owners and there is visual evidence throughout the valley of fenceline, break of slope and gully plantings that are easy to spot from the road. One of the key informants drew our attention to the fact that planting in prominent places can be a deliberate strategy to advertise a group's activities. From our perspective, this strategy can also be considered a mechanism for establishing a new descriptive norm encouraging landholders to plant trees.

In addition to its tree planting activities, the Indigo Valley Landcare group was involved in weed control. The group had almost eradicated blackberries throughout the valley and there was an expectation that every land owner should 'do their bit' so as to not undermine the efforts of others. One family was perceived to be negligent in this respect, consistently failing to eliminate weeds on their property, and the committee discussed ways of tackling the problem:

> *I remember us having this big chat about how we were going to broach the fact they had blackberries everywhere (LH36).*

The committee chose a group member who was not closely connected to the family to ring and proffer help, on the grounds that it was acknowledged to be a large and difficult job. The group set up a roster of about 25 people to spray on a regular basis and the spray cart was supplied without charge; all the landholder needed to do was pay for the spray. We see this as evidence of a community group using external sanctioning to enforce an injunctive norm relating to weed control, thereby achieving an outcome that was perceived to be of benefit to the local community.

The problem of weed control was a recurring theme throughout the interviews and was not confined to discussions about Landcare activities. The majority view was that weed control was a duty and that 'looking after your weeds' was something everyone was expected to do.

Anecdotally, there were stories of farmers taking matters into their own hands if they thought they had no other option:

> So next day he came down with his spraying equipment and sprayed it all out. He said I didn't want to have farmland each side of this huge seed bank of heliotrope. He told him how serious it was and the chap felt a bit sort of guilty about it and said it probably came with feed I bought for feeding my horses. So he was right on top of the man who was totally ignorant of the seriousness of this little soft-looking weed (LH9).

Although there were some exceptions, the majority of newer non-farmer property owners in Indigo Valley acknowledged that they had a responsibility to carry out weed and pest control. These newer landholders were conscious that if they didn't, the local farmers would be 'on to them'; this might take the form of gossip, direct communication or, as in the example above, direct action to enforce the norm.

> Well I can tell you now, the three farmers around here will come up and if they see a flower in the middle of your paddock, they'll come up and say to you 'go and get that damn thing out of there now' (LH23).

Newer owners looked to older residents for guidance:

> I want to be as good as my neighbours are ... the neighbours who I look up to and respect for what they've done ... I think a lot of work, good work, has been done and I'd hate to be the person that undoes it (LH33).

Another lifestyle owner spoke of weed control in a manner that conveyed a moral imperative to 'do the right thing'.

> ... you've got an obligation to look after your weeds (LH18).

Another quite deliberate strategy devised by the local Landcare group involved choosing a number of older farmers ('eight or so') to serve as mentors, and linking them to new people coming into the valley. According to one Landcare committee member, the idea was to make the older farmers feel good about themselves and ensure that local knowledge was passed on to new residents.

With the help of a paid coordinator, the Landcare group was involved in a number of more ambitious projects to address dryland salinity. To be successful, these required group consensus about what needed to be done and the participation of landholders along the entire length of the valley. One project involved mapping the break of slope with the intention of fencing the entire length to better manage up-slope vegetation and reduce salinity outbreaks below. We interviewed the Landcare coordinator who worked with the group at the time. This person reflected, almost 20 years later, on the outcomes. On first reflection, the project was deemed a failure:

> ... there was still a denial process going on with a lot of people in the valley ... you know you could see the salt in their paddock but they were just in denial ... the big

plan to do these things on their property with them putting up half the money just wasn't going to happen ... I don't think the total community was ready then for it. And if you want something on a scale that big to work, to show something 10 or 15 years down the track, you've got to have total commitment (GA4).

However, further reflection led to a slightly different assessment of the outcomes:

The success of that project as I see it would have been that there was a change of thinking, community recognition of salt and the problem that did exist in the valley, that it was catchment wide and not just on anyone's particular property. I think most people now would put their hand up to that and if you started again now you'd probably have a better success or better chance of making things work (GA4).

The former coordinator was not the only person to reflect on changing attitudes to native vegetation as a result of Landcare activities:

I think there's been quite a bit of change on a social level you'd have to say ... I think if you just tried to protect remnants without that phase having been done there probably would have been less achieved (GA7).

Indigo Valley has a number of families and individuals who hold considerable status within the valley. Some of the changes in attitudes and behaviour within the farming population, and willingness to be involved in Landcare activities, can be partially attributed to their influence. Many interviewees, including newer residents in the valley and key informants, mentioned the pivotal role of these people. These families and individuals were all long-term residents, part of the farming community and actively involved in local groups such as the Country Fire Authority and Indigo Valley Landcare.

To return to the theme of tree planting, one individual in particular became a staunch advocate for revegetation and was able to persuade others to plant native trees. This farmer was well known locally and commanded a high level of local esteem, as evidenced by the following comment:

[X] was great because he went over lots of different types of people, people that respected him, and it worked well and people were keen to be involved (LH36).

He persuaded not only farmers but newer property owners to follow his example:

I had a very good neighbour down the road down there, he's gone now, he's passed away, and he said if I didn't donate 10% of my land to trees, in 20 years time I wouldn't have the other 90% anyway, because that's what salinity was going to do ... he's donated 10% of his place, over 1000 acres ... So I did the same thing (LH23).

PART 2: INTERPRETATION FROM A SOCIAL NORMS PERSPECTIVE

Although the conditions for the emergence of norms are not always clear (Hechter and Opp 2001), social norms are thought to emerge in a group setting if members believe that certain norms are instrumental for the attainment of their goals (Opp 2001). Indigo Valley Landcare deliberately set out to make tree planting an annual habit, thereby establishing a new descriptive norm of behaviour which resulted in the planting of thousands of trees. The more that landholders planted trees, the more likely it was that this activity would provide 'social proof' that planting trees was what everyone did. Tree planting is a visual activity; visibility increases the salience of a particular behaviour and the likelihood that others will follow suit (Cialdini *et al.* 1991; Foss 1983).

The norms around weed control are examples of injunctive norms. Indigo Valley Landcare's approach to weed control, from a norms perspective, can be seen as a strategy aimed at minimising negative externalities associated with not practising weed control. Not making every effort to eradicate weeds on your property violates a strongly held norm within the farming community stipulating that every landholder has a responsibility to undertake this task. Weed control is a duty and no one questions the right of individuals to sanction others who fail to conform to this norm. The example given of a farmer trespassing on another land owner's property to remove a perceived threat to his livelihood was not judged to be improper because the norm condones such behaviour.

The group strategy to deal with a violation of the norm pertaining to weed control is a good example of how an existing norm can be enforced. The group was able to decide on an action and enlist enough volunteers to do the necessary work. In agreeing to have the spraying done, we could also infer that the property owners recognised the legitimacy of the group's actions, although they may have felt some embarrassment at being singled out for special help, particularly because other property owners in the area had complied with the expectation that you 'look after your weeds', making violations of the norm more costly. In this case there was a noticeable amount of judgemental gossip about the family by other landholders in the valley (Merry 1984).

New knowledge can be a catalyst for norm change (Sunstein 1995; Ellickson 2001). With the support of government funding and a paid coordinator, the Landcare group disseminated new information about the causes of land degradation and the benefits of planting and protecting native vegetation. It is known that people tend to interpret information in a way that sustains their existing beliefs. However, if contrary evidence is convincing, people cannot interpret it away (McAdams 1997). The local Landcare group raised awareness of the reasons for planting trees so that there was, and is, an environmental imperative associated with the action.

DISCUSSION

Community social networks and local organisations are important for community cohesion and knowledge transfer, and both individuals and groups can act as 'surrogate extension officers' and 'models for better conservation practices' (Harrington *et al.* 2006). Indigo Valley has had some key individuals supporting Landcare and making new residents feel welcome. Even though the valley lacks a township, the local school and Country Fire Authority are well supported and act as social glue, binding people together. Many family networks still exist, a small community church runs a weekend service in the lower valley and weekend newspaper delivery is carried out by local volunteers on a roster basis. Because they choose to live in a rural area, the majority of newer landholders are keen to become involved in the local community. They have integrated well, forging new social capital as an outcome of their interaction with longer-term residents.

The extent to which changing community structure has and will influence NRM in rural areas is an important issue for land managers. A key question is whether newcomers are influencing the older residents, or vice versa. This research indicates that well-established existing norms are likely to hold sway. We might expect that in a community undergoing demographic change, the existing local norms would slowly diffuse as newcomers develop social ties with longer-term residents. The Landcare group accelerated this process by setting up mentors for new residents. Although we doubt that this was an explicit goal, the welcoming and mentoring of new residents, in combination with their desire to fit in, was an effective strategy for developing trust and reciprocal obligations and for ensuring that local codes of behaviour were adopted.

We do not mean to imply that newer residents are not influencing land management practices. The key difference between lifestyle owners and farmers is that the former are not as dependent (or solely dependent) on the land for income. Without the need to have land in production, non-farming land owners are more inclined to plant trees and retain native vegetation for aesthetic reasons. Their land management behaviour is less influenced by tradition and many are thirsty for knowledge about how to manage the land. They are generally receptive to advice from NRM practitioners and are less suspicious of government programs aimed at increasing biodiversity on private land.

The persuasive power of cultural expectations should not be underestimated. Norms create expectations about how people should act but they need to be maintained by constant interaction and the knowledge that others *are* behaving as expected. This research demonstrates the value of investing in community-based NRM organisations such as Landcare, and future NRM policy should capitalise on the ability of groups to achieve social change. This would facilitate the emergence and maintenance of norms that will predispose individuals to cooperate, value and conserve native vegetation.

Support in the form of increasing government aid or incentives can destroy social capital by making individuals less dependent on one another (Coleman 1990). Norms that promote cooperation and dictate behaviour in particular contexts are known to be less costly and more effective than material rewards. Financial incentives are known to 'crowd out motivation' by reducing the intrinsic motivation to engage in certain behaviours (Dwyer *et al.* 1993; Stern 1999). This is backed up by comments by key informants, two of whom independently observed that government incentives have created a hand-out mentality among a proportion of landholders, who are reluctant to adopt new practices unless they are paid to do so.

Regulation can work against the emergence of trust and voluntary cooperative behaviour, which are known to decline as state intervention increases (Kollock 1994). An example from this study was the way in which the Landcare group acted in lieu of the law to ensure that weed control was carried out in the upper valley. Policies that target social norms are likely to be more effective in the long term as state enforcement will be unnecessary (normative controls will be used) and the strength of the community will be maintained (Horne 2000). Norms that motivate individuals to behave in a certain way are more likely to result in long-term commitment to desired outcomes, such as biodiversity conservation.

Because norms are often based on beliefs about relevant facts they can change, and often quite quickly, in response to new information (Sunstein 1995). There is evidence that once the value of established habits is questioned, there is room for other factors to have an influence on behaviour (Dahlstrand and Biel 1997). New information about land degradation and land management practices was partly the catalyst for change in Indigo Valley, but it was effective because it was disseminated via a group and taken up by local people who had the power to influence others. This supports the view that information, in combination with other factors, can be an effective catalyst for change (Stern 1999).

Key players in a community have the connections and influence to initiate (or resist) change. Involving these people in any new venture is important because they legitimise the activity or behaviour and draw in other people (McDermott 2000). The social costs (negative sanctions, loss of respect) of changing to a new practice can outweigh other benefits (Richards *et al.* 2005), so the uptake or acceptance of a new practice by a person of local standing can make all the difference. NRM agency personnel already target 'low-cost landholders' when they want to roll out a new incentive because they are known to be receptive and have local influence. The findings from this study validate this strategy; rather than being seen as taking the path of least resistance, the targeting of key people in communities can be seen as crucial if new norms of behaviour are to be established.

Perceived normative support for an activity can increase the level of support – visibility is important (Foss 1983; Strahilevitz 2003). A straightforward way to promote the establishment of new norms of land management behaviour is simply to make more visible the activities or behaviours that will produce desired outcomes. This strategy will also increase the likelihood that noncompliance will warrant social disapproval.

If conditions change so that compliance with a norm becomes excessively costly, individuals will look for alternative options that may result in further changes to norms (Horne 2001). Therefore, if compliance with particular norms is considered desirable then steps should be taken to ensure that compliance does not become too onerous. This would require, in the case of tree planting for instance, that trees are readily available and cheap to purchase.

CONCLUSIONS

As rural areas such as Indigo Valley become increasingly heterogeneous it is unlikely that the values and concerns of the farming population will remain the dominant influence. Although the community is changing, there does not seem to have been a decrease in the level of social trust and neighbourliness between new residents and old, and as long as the mix of farming and lifestyle owners changes slowly this will probably continue to be the case. This is a positive finding, given that the stock of social capital in rural communities is likely to be a key factor in determining the success or failure of future investment in community-based NRM.

Key messages from our research are that supporting community groups, targeting community gatekeepers and increasing the visibility of desirable behaviours may be effective ways to achieve practice change. Reducing the social costs of altering land management practices by establishing positive norms in the community will foster change. Other forms of government intervention may not only be less necessary, but also less desirable as they may undermine the formation of social capital necessary for norm establishment and enforcement. Program evaluations should include some consideration of the extent to which positive social norms have been established or reinforced.

Social norms guide behaviour no matter what the situation or context. New norms will naturally evolve as social contexts and specific practices change over time. Interventions should aim to work with naturally occurring social processes, not against them. Future research on social norms in different contexts may lead to the development of more effective policy instruments aimed at long-term social change rather than at short-lived measurable outcomes.

REFERENCES

Australian Bureau of Statistics (ABS) (2010) 'Estimated resident population'. Annually updated population figures. Australian Bureau of Statistics, Canberra. <http://profile.id.com.au/Default.aspx?id=264&pg=210&prn=1>.

Axelrod R (1986) An evolutionary approach to norms. *American Political Science Review* **80**, 1095–1111.

Bicchieri C (2006) *The Grammar of Society: The Nature and Dynamics of Social Norms.* Cambridge University Press, New York.

Cialdini R, Kallgren C and Reno R (1991) A focus theory of normative conduct: a theoretical refinement and re-evaluation of the role of norms in human behaviour. In *Advances in Experimental Social Psychology. Vol. 24.* (Ed. MP Zanna) pp. 201–234. Academic Press, San Diego.

Coleman JS (1990) *Foundations of Social Theory.* Harvard University Press/Belknap Press, Cambridge.

Curtis A and De Lacy T (1996) Landcare in Australia: does it make a difference? *Journal of Environmental Management* **46**, 119–137.

Dahlstrand U and Biel A (1997) Pro-environmental habits: propensity levels in behavioural change. *Journal of Applied Social Psychology* **27**, 588–601.

Dwyer W, Leeming F, Cobern M, Porter B and Jackson M (1993) Critical review of behavioral interventions to preserve the environment: research since 1980. *Environment and Behavior* **25**, 275–321.

Ellickson R (2001) The evolution of social norms: a perspective from the legal academy. In *Social Norms.* (Eds M Hechter and K-D Opp) pp. 35–75. Russell Sage Foundation, New York.

Fine G (2001) Enacting norms: mushrooming and the culture of expectations and explanations. In *Social Norms.* (Eds M Hechter and K-D Opp) pp. 35–75. Russell Sage Foundation, New York.

Foss RD (1983) Community norms and blood donation. *Journal of Applied Social Psychology* **13**, 281–290.

Harrington C, Curtis A and Black R (2008) Locating communities in natural resource management. *Journal of Environmental Policy and Planning* **10**, 199–215.

Hechter M and Opp K-D (Eds) (2001) *Social Norms.* Russell Sage Foundation, New York.

Horne C (2000) Community and the state: the relationship between normative and legal controls. *European Sociological Review* **16**, 225–243.

Horne C (2001) Sociological perspectives on the emergence of social norms. In *Social Norms.* (Eds M Hechter and K-D Opp) pp. 35–75. Russell Sage Foundation, New York.

Kollock P (1994) The emergence of exchange structures: an experimental study of uncertainty, commitment, and trust. *American Journal of Sociology* **100**, 313 345.

McAdams RH (1997) The origin, development, and regulation of norms. *Michigan Law Review* **96**, 338–433.

McDermott R (2000) *Knowing in Community: 10 Critical Success Factors in Building Communities of Practice.* IHRIM journal, 19–26 March 2000. <http://www.knowledgeboard.com/download/3465/Knowing-in-Community.pdf> (accessed 15 April 2011).

Merry S (1984) Rethinking gossip and scandal. In *Toward a General Theory of Social Control. Vol. 1.* (Ed. D Black) pp. 271–302. Academic Press, New York.

Minato W, Curtis A and Allan C (2010) Social norms and natural resource management in a changing rural community. *Journal of Environmental Policy and Planning* **12**, 381–402.

Opp KD (2001) Social networks and the emergence of protest norms. In *Social Norms.* (Eds M Hechter and K-D Opp) pp. 234–273. Russell Sage Foundation, New York.

Ostrom E (2000) Collective action and the evolution of social norms. *Journal of Economic Perspectives* **14**, 137–158.

Richards C, Lawrence G and Kelly N (2005) Beef production and the environment: is it really 'hard to be green when you are in the red'? *Rural Society* **15**, 192–209.

Stern P (1999) Information, incentives, and pro-environmental consumer behaviour. *Journal of Consumer Policy* **22**, 461–478.

Strahilevitz LJ (2003) Social norms from close-knit groups to loose-knit groups. *University of Chicago Law Review* **70**, 359–372.

Sunstein C (1995) Social norms and social rules. *Chicago Working Papers in Law and Economics (Second Series) No. 36.* Law School, University of Chicago.

Wallace JD (2009) *Norms and Practices.* Cornell University Press, Ithaca.

White KM, Smith JR, Terry, DJ, Greenslade JH and McKimmie BM (2009) Social influence in the theory of planned behaviour: the role of descriptive, injunctive, and in-group norms. *British Journal of Social Psychology* **48**, 135–158.

Understanding rural landholders' responses to climate change

Maureen Rogers, Allan Curtis and Nicki Mazur

SUMMARY

Global climate modelling has identified south-eastern Australia as a 'hotspot' for more frequent climatic extremes. Rural landholders may be vulnerable to the risks this presents if they are not considering climate change in management decisions. Australia's rural landholders have long been accustomed to responding to climatic variability. However, some of these adaptations may lead to downstream impacts. It is therefore important to understand what landholders are doing and the extent to which climate change is influencing their decisions. This chapter draws on a case study in two Victorian districts to address these knowledge gaps. Data were gathered using semi-structured interviews and a mail survey based on established socio-psychological scales to measure beliefs, values and attitudes that are expected to shape landholder behaviour.

Most interviewees had observed changes in climatic patterns but expressed a level of scepticism, often expressing concern over the extent of disagreement among scientists and the extent of vested interests by those presenting information. Most rural landholders surveyed were not climate change 'deniers', with 70% agreeing with the statement, 'The climate is changing and that human activity is a major influence'. Landholders were adopting a wide range of tactical and strategic responses to drought and climate change, which is consistent with rainfall as a critical factor in agricultural production. More than half the survey respondents said they were implementing five of the 14 potential adaptations included in the survey. Climate change was nominated as an influence on six adaptive behaviours by 50% or more of survey respondents. Belief in, and knowledge of, climate change were both linked to the adoption of more strategic adaptations. However, most respondents reported having limited knowledge of climate change topics. Engaging landholders in dialogues about climate change could therefore be an effective way to influence on-property management.

INTRODUCTION

Global climate modelling has identified a number of regions where climatic conditions are likely to become more extreme, including south-western and south-eastern Australia (Giorgio 2009; Chandler 2009). Giorgio's (2009) projections indicated a generally more variable climate with more heatwaves, longer drought periods and greater intensity of precipitation. Robinson (2009) suggested that these projections would have serious implications for the Murray-Darling Basin and Australia's future food security.

Australia's rural landholders have a strong record of being responsive to the myriad political, economic and climatic factors influencing their enterprises. This is reflected in a generally optimistic view of their ability to adapt to the challenges of climate change. According to Grothmann and Patt (2005), people have to feel vulnerable before they begin to think about making changes and adapting their behaviour. People systematically underestimate the likelihood of a hazard affecting them (Thornton *et al.* 2002). While awareness of climate change and its associated risks has increased in recent years, there remains a considerable amount of scepticism about the cause and science (Reid *et al.* 2007; Grothmann and Patt 2005; Thornton *et al.* 2002). Recent research in Australia (Thwaites *et al.* 2008; Milne *et al.* 2008; Evans 2009; Reid *et al.* 2007) suggested that rural landholders and communities continue to be less receptive to climate change than the general public. This is not surprising, given significant threat that climate change poses to incomes in the rural sector. However, despite some climate change scepticism among rural landholders, this research found rural landholders were making relevant adaptive changes.

Our motivation for exploring rural landholder responses to climate change is that their collective actions may have implications for others, including those downstream, and because their actions may or may not be sufficient to reduce their vulnerability to the risks posed by climate change. We were also interested in exploring the extent to which climate change, specifically, is influencing rural landholder decisions. This chapter draws on our findings from a case study in two rural communities of north-eastern Victoria (Chiltern and Eskdale) to explore these topics.

FACTORS THAT INFLUENCE LANDHOLDER BEHAVIOUR

This research draws on key frameworks that have been used by social scientists interested in the uptake of sustainable practices in agriculture (Cary *et al.* 2002; Pannell *et al.* 2006; Pickworth *et al.* 2007; Nelson *et al.* 2006). A number of studies have explored the links between adaptation to climate change and the elements of the five capitals (natural, human, social, physical and financial capital). Positive links have been found with education levels, experience

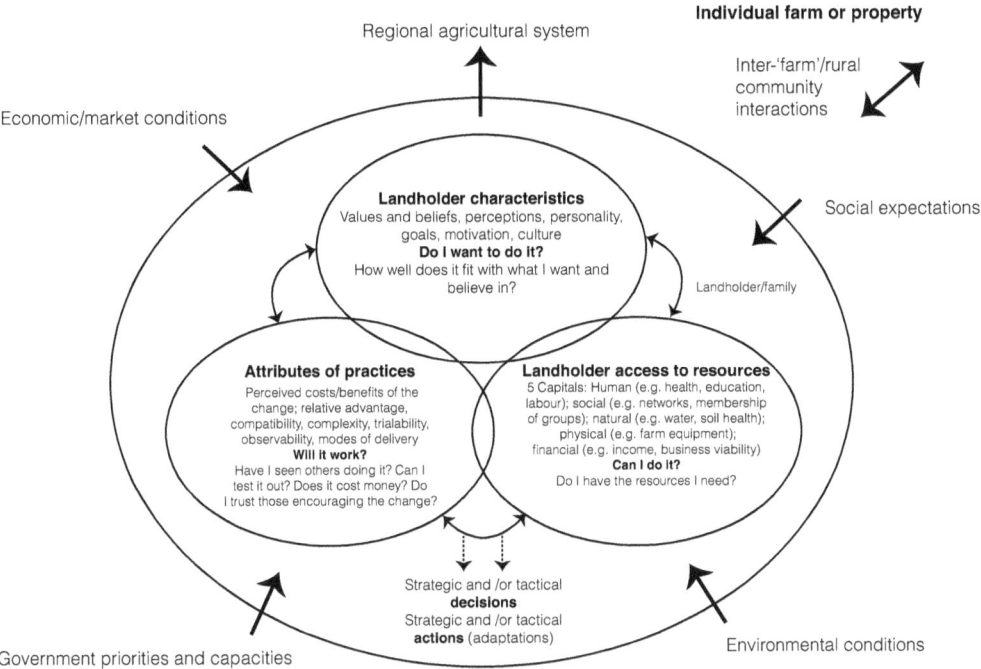

Figure 16.1: Socio-psychological factors likely to influence landholder adaptive behaviour to climate change.

of past conditions, strong community leadership, social norms in favour of environmental (and climate change) action, economical viability of farm businesses, and farm size (deWit 2006; Leith and Haward 2010; Milne *et al.* 2008).

However, Stern *et al.* (1998, 1999), Ajzen (1991) and Grothman and Patt (2005) argued that the cognitive processes of individuals are more important influences on adaptation responses than the socio-economic factors that are more commonly referred to. The former influences include people's value and belief systems, attitudes and perceptions, personalities, motivations, goals and culture. Collectively, they inform landholders' judgements about the extent and nature of their response to an external threat. Positive links to adaptation have included direct experience of climate change (e.g. prolonged drought conditions) (Milne *et al.* 2008; Leith 2006), confidence in your ability to adapt (Milne *et al.* 2008; Grothmann and Patt 2005), trust in science (Leith 2006; O'Connor *et al.* 2002) and an appreciation of the potential benefits of adaptation. Figure 16.1 is our attempt to synthesise these approaches by embracing the full range of socio-psychological factors that are likely to influence why a rural landholder might want to adapt to climate change and believe they are able to adapt to climate change.

The case study research in north-eastern Victoria explored a range of key questions, but the focus of this chapter is on the following questions.

1   Do rural landholders believe in human-induced climate change?
2   What do rural landholders know about climate change?
3   Do rural landholders perceive a risk of personal loss as a consequence of climate change?
4   How are rural landholders responding to the risk of climate change?

5   What are the potential implications of rural landholder actions?
6   To what extent is climate change a key factor in rural landholder decisions?

NORTH-EASTERN VICTORIA: THE STUDY AREA

The research was undertaken in the north-eastern region of Victoria where rainfall and water availability have typically been regarded as very secure, i.e. relatively high and reliable with uninterrupted river and spring water flows. It is an area rich in natural assets from snow-topped mountains to fertile river valleys and open plains. The region covers approximately 1.9 million ha of Victoria, including three major catchments: the Upper Murray, Kiewa and Ovens. The two study sites of Eskdale and Chiltern (Fig. 16.2) embody very different social contexts. Eskdale is remote, with a high proportion of multi-generational farmers. Chiltern is closer to Albury–Wodonga, with a larger proportion of hobby farmers and absentee landholders. The research team relied on regional and local knowledge to ensure that the two districts represent a typical cross-section of the 'amenity' and 'production' landscapes in north-eastern Victoria. Although there are some differences between the responses of rural landholders in Eskdale and Chiltern, these are not the focus of this chapter.

RESEARCH APPROACH

Data were gathered using in-depth semi-structured interviews (qualitative component) and a mail survey (quantitative component).

Figure 16.2: The study areas of Eskdale and Chiltern in north-eastern Victoria.

Table 16.1: Land use on properties owned by the interviewees

Major land use characteristics	Eskdale	Chiltern
Dairy farmers with irrigation	8	1
Sheep/beef growers		3
Beef growers with irrigation	4	
Beef/dairy growers(no irrigation)		2
Non-farming occupations, lifestylers, hobby farmers	2	6
Horticulture		2
Total	**14**	**14**

IN-DEPTH INTERVIEWS

The research team worked with staff from the North East Catchment Management Authority and key landholders within each study site to identify 20 interviewees with diverse farming enterprises and life experiences. Of those contacted, 14 individuals from both study sites were willing to be interviewed. Each informant owned or managed a rural landholding of greater than 10 ha. The research team relied on local knowledge to provide confidence that the inform-ants represented a cross-section of landholders. The land use characteristics of the interview-ees are summarised in Table 16.1.

A semi-structured and open-ended interview schedule (SOEI) (Patton 1990) was used. This technique involves asking informants a similar set of questions, worded in the same way and asked in the same or similar sequence. Using SOEI tends to minimise the variation in the questions asked by interviewers, reducing interviewer bias and eliciting more standardised and comparable interview data. The interview schedule can be found in Thwaites *et al.* (2008). All interviews were taped and key points and themes recorded in handwritten notes, ensuring the anonymity of interviewees was maintained. Content analysis techniques were used to analyse the responses to identified patterns and themes (Berg 2001).

Interviewees were initially asked to talk about their experiences of the 'climate' over the past few years: what they believe is happening, what they think about the idea of climate change, its causes and consequences, how they respond to the regional projections information provided by CSIRO, how likely they think it is that the projections will come to fruition, what actions they have taken or would like to take, the resources needed to take action and what factors have influenced their thoughts about climate change. The information obtained about the actions that people are taking or considering was used to develop the survey topic seeking information about adaptations.

MAIL SURVEY

The mail survey (the quantitative research component) was specifically designed for the purpose of building a statistical model exploring the factors influencing rural landholder responses to climate change, i.e. the relationships between a wide range of socio-psychological factors and a number of adaptive actions landholders are making.

Where possible, the survey was based on established socio-psychological scales that opera-tionalise key aspects of the identified theories. A scale is a series of items (questions or state-ments) that together explore the different dimensions of a theoretical construct. For example, theory suggests that beliefs (what is believed to be true) have three elements which, according to the Value–Belief–Norm theory (Stern *et al.* 1998), include ecological and social world views, awareness of consequences and ascribed responsibility. Accordingly, the research team drew on existing research to identify questions and statements that could be adapted for use in the landholder survey to explore those elements of beliefs about climate change. Some items included in the survey have been employed in other studies by the research team (e.g. Curtis *et al.* 2001, 2008). Other items are new, but in most instances have been adapted from existing scales. For example, Eckersley's (2007) categories of world views were used to develop a scale testing landholders' degree of optimism or pessimism about the state of the world. Five of the 12 items from the New Environmental Paradigm (Dunlap and Van Liere 1978) were used to ascertain the strength of landholders' environmental convictions. Landholders' inclination to adapt to climate change was also examined by testing their personal sense of obligation and moral duty to act (Stern *et al.* 1998). Jorgensen and Stedman's (2001) commitment to place research was used to test the relationship between the extent of people's commitment to where they live, and their readiness to adapt to climate change.

The survey was distributed to respondents as a 12 page booklet. It comprised 10 topics and 136 items. Booklets were hand-delivered to every identifiable resident in each district, with 100 rural landholders in Chiltern and 86 landholders in Eskdale receiving surveys. An overall response rate of 55% was achieved, with 94 usable surveys returned.

The survey data analysis included a range of quantitative techniques, including general inference based on the descriptive statistics for each variable using means, medians, frequencies and percentages where appropriate. Some groups of variables belonged to theoretical scales. The Cronbach's Alpha test (de Vaus 1991), using a threshold of 0.7, was used to see if respondents answered these groups of items consistently and the scale could be considered valid. Each of the possible predictors of behaviour indicated by pairwise analyses was scrutinised for a response rate above 80% to ensure modelling integrity. The final list of predictors was then used in a stepwise linear modelling process that used Akaike's information criterion (Venables and Ripley 2002) as the step criteria. The modelling process was used to identify the set of variables that collectively contributed most to a higher score for each adaptive behaviour or set of behaviours. The amount of variance explained by the model is indicative of the extent to which key independent variables have been included in the study. A model that explains 30% of variance is considered useful in the social sciences (social sciences typically have a large number of potential influential variables).

KEY RESEARCH FINDINGS

In this section we consider the research results according to our six key questions.

DO RURAL LANDHOLDERS BELIEVE IN HUMAN-INDUCED CLIMATE CHANGE?

For the interview process, respondents were asked if they believed in climate change, could describe climate change, what kinds of 'changes' in the climate they had observed, knew about the causes of those changes, had observed any consequences and knew of potential remedial actions and believed there was any relationship between 'climate change' and drought. Content analysis of the interview data enabled a grouping of their responses into three broad categories – non-believers, unsure and believers – in large part reflecting their views about the reality of human-induced climatic change. These groupings are similar to those found by Thwaites *et al.* (2008).

Interviewees who were categorised as 'believers' generally thought climate change was real and induced by human actions. These interviewees typically felt that the worsening drought was a symptom of climate change. Believers tend to find the scientific evidence for climate change credible, although some seemed confused about various technical aspects of the phenomena:

> *I just know the last year in the fire seasons were horrendous and it seems to be every summer it's getting hotter, drier and windier. Especially the last two seasons we seem to be getting more wind and more dry wind, windier and it's turned around and coming from the south. The temperatures has changed, it's simply hotter. Our winter pattern where we used to get good rain early we don't seem to be getting that rain anymore ... definitely climate change, definitely.*

'Non-believer' interviewees tended to question the existence of human-induced climate change more strongly than 'unsure' interviewees. They also referred to themselves as 'sceptics', but were typically very negative about what they felt were the vested interests of climate change believers and the credibility of 'expert' information. Non-believers tended to attribute any observed changes in the climate to a 'natural' climate cycle.

> *I don't believe in it ... it's only nature doing its thing ... Yeah I think there's a shift in our seasons ... our rainfall, [it's] not as much as it should be ... that the human race is affecting climate change and polluting everything ... but I think it's all political and people are making money out of it. To me it's [climate change information] confusing and because it's confusing – it's not convincing.*

Those who were unsure seemed to believe that something was happening but were not clear about the cause, i.e. whether it was just part of the natural cycle or was human-induced climate change.

Survey respondents were presented with a set of statements exploring views about the different dimensions of the climate change phenomenon. They were asked to rate each statement according to a five-point scale ranging from 'strongly disagree' to 'strongly agree' (Table 16.2). The survey data suggests that views held by the unsure and non-believers (interviewees) are not as widely distributed as the views of believers. Only 6% of those surveyed said there was 'No such thing as climate change' (i.e. outright deniers), while 70% agreed with the statement, 'The climate is changing and that human activities are influencing the change'.

WHAT DO RURAL LANDHOLDERS KNOW ABOUT CLIMATE CHANGE?

Survey respondents were presented with a series of locally and globally relevant statements about climate change (Table 16.3) and asked to indicate their level of knowledge about each topic. A clear majority of survey respondents said they had some knowledge of the general notion of climate change and its global implications, i.e. the potentially serious impacts of global climate change, the explanation by scientists of global warming and an understanding of the difference between the greenhouse effect and the ozone hole. None of those interviewed, and only a small proportion of those surveyed (10%), said they had some knowledge of regional climate change projections and their implications.

Interestingly, the explanations by interviewees of what they understood climate change to be often came across as confused, not reflecting basic concepts found in formal and scientific

Table 16.2: Views about climate change, Chiltern and Eskdale survey respondents 2009

Views about climate change	Disagree (%)	Not sure (%)	Agree (%)	N/A (%)	Mean score
Climate is always changing. It is a naturally occurring phenomenon that has been going on forever (n=90)	9	11	80	0	3.99
The climate is changing (n=90)	8	21	70	1	3.9
Human activities are influencing changes in climate (n=90)	11	20	68	1	3.83
It is not too late to take action to address climate change (n=90)	9	27	60	4	3.64
If we do nothing climate change will have dire consequences for all living things, including humans (n=90)	16	34	48	2	3.53
The climate maybe changing but currently we are in a cycle (n=90)	13	33	53	0	3.51
Humans will always adapt to their environment	20	26	52	2	3.39
Climate might be changing but it is not as bad as it is being portrayed (n=90)	32	30	38	0	2.99
Climate change is out of our control (n=90)	42	39	19	0	2.73
There is no such thing as climate change (n=90)	79	14	6	1	1.84

definitions. The non-believers and those who were unsure talked about the information being confusing and therefore not convincing.

Table 16.3: Self-reported knowledge of climate change, Chiltern and Eskdale survey respondents 2009

Climate change topic	Little/no knowledge (%)	Some knowledge (%)	Sound/ very sound knowledge (%)	Not relevant (%)	Mean score
Actions that householders can take to reduce their carbon emissions	10	40	47	2	3.46
Potentially serious impacts of global climate change	22	42	32	3	3.16
The explanation provided by scientists for global warming	29	37	30	4	2.99
The difference between the greenhouse effect and the ozone hole	27	44	23	4	2.94
How carbon can be taken out of the atmosphere and stored	39	29	30	2	2.8
The implications of a 1–2° increase in global temperatures for agriculture in your district	36	47	16	2	2.72
Steps that an agricultural enterprise could take to become carbon neutral	46	37	16	2	2.55
The Federal government's plans for a carbon emissions trading scheme	49	39	7	6	2.5
The broad climate change projections for your region by 2050	60	27	10	3	2.18

HOW ARE RURAL LANDHOLDERS PERCEIVING THE PERSONAL RISK OF LOSS OR DAMAGE AS A RESULT OF CLIMATE CHANGE?

Interviewees were presented with recent climate change projections for north-eastern Victoria produced by CSIRO, and asked what they thought they could lose (or gain) as a consequence. Interviewees typically talked about increased difficulty in farming if the projections were to eventuate. The types of impacts identified included reductions in farm productivity and land values and an increased exit rate from farming, including through the sale of land. Interviewees' perceptions of climate risk seemed influenced by their antipathy towards the way climate change information was being presented. Many interviewees felt that the climate change 'crisis' was being 'peddled' by those who have something to gain.

Those surveyed were presented with a list of potential climate change impacts and asked to indicate the likelihood of their property being affected by each impact. Many respondents believed they 'would be' affected 'if the climate projections were realised', with at least half the respondents identifying nine impacts they thought were 'highly likely' (Table 16.4). Most respondents thought that all impacts would be negative, but relatively small.

HOW HAVE RURAL LANDHOLDERS RESPONDED TO THE RISK OF CLIMATE CHANGE?

Interviewees were asked about management decisions they had made over the past few years in response to climatic conditions. The responses were then included in the survey. Survey respondents were then asked to indicate which of the management options they had taken, or had considered taking, in the past five years.

Australia lacks clear agreed criteria for distinguishing the quality of particular practices (e.g. what distinguishes sound from maladaptive actions). At present it seems that any action in response to drought or climate change is viewed as a positive action and an indicator of capacity to adapt, while not taking direct action is considered a negative response and an indicator of vulnerability. Furthermore, contemporary climate change dialogues tend to assume that

Table 16.4: Perceived risk of climate change impacts to interviewee's property, Chiltern and Eskdale survey respondents 2009

Potential types of impacts	Unlikely (%)	Highly likely (%)	Small negative impact (%)
Increased intensity/frequency of bushfires	6	86	66
Water supply for pasture production	8	74	70
Water supply for stock and domestic	16	74	74
Pasture productivity	12	71	71
House and gardens	15	70	67
Quality/health of native vegetation	12	68	74
Survival of some native vegetation	15	66	72
Attractiveness of property	10	60	61
Pest plants	17	50	59
Family health/wellbeing	20	45	51
Personal health/wellbeing	21	44	52
Insect pests	16	41	53

n=91

anticipatory actions (those taken ahead of symptoms of change) are more desirable because they indicate some level of preparedness by industry, community or individuals to engage with longer-term consequences of climate change. It is possible, as discussed later, that some actions taken by landholders will not be socially or environmentally sustainable in the longer term. It is also possible that landholder actions may not reduce their vulnerability to climate change risks. As Pannell (2010) pointed out, landholders typically have the capacity to respond quickly to threats and it is possible that acting early may lead to poorly informed decisions.

The research team attempted to rate the quality of the identified current and planned adaptations according to a spectrum of 'reactiveness' versus 'anticipatory' or 'tactical' versus 'strategic' actions (1 = reactive/tactical, 10 = anticipatory/strategic). These ratings are presented in Table 16.5.

It is important to point out that the team's rating of adaptations is only a tentative or preliminary assessment. For example, the team thought that 'improving water capture' through the building or extension of dams is often considered a critical and immediate first response to drought but is not a complex or challenging innovation for landholders. The team therefore considered this to be a highly tactical action and allocated a rating of 2. By comparison, actions to improve water use efficiency were seen as reflecting acceptance of the need to do more with less water, but provided no indication that the landholder was fundamentally changing the nature of their enterprise and was given a rating of 4. Value-adding to farm products was considered to reflect a commitment to fundamentally realigning the farm enterprise and was given a rating of 9.

More than half the survey respondents said they were implementing five of the 14 potential adaptations included in the survey (Table 16.6, rows 1–5). All these were categorised as largely tactical in nature. Climate change was nominated as a specific influence by 50% or more of survey respondents for six of the adaptive behaviours (improved water efficiency and water capture, reduced stocking rates, establishment of perennial pastures, a move from annual to perennial pastures, planting trees, shrubs and native grasses). It seems that the vast majority of respondents considered the option to purchase or lease more land locally or elsewhere to be largely irrelevant. The research team had categorised such actions as being highly strategic.

WHAT FACTORS INFLUENCE RURAL LANDHOLDERS TO TAKE ACTION TO ADDRESS CLIMATE CHANGE?

The survey included a large number of items exploring factors identified as possible socio-psychological influences on adaptive behaviours. These included belief in climate change, risk perception, knowledge of climate change, perceived ability to adapt, world views, values, personal norms and commitment to place.

The pairwise analysis and regression modelling produced groupings of influential factors found to be highly correlated to each adaptive behaviour. There were significant relationships (positive and negative) between many of these items and specific adaptations. There was also considerable variation in the sets of factors linked to each of the adaptive behaviours.

Surprisingly, there was little correlation between socio-economic variables such as age, occupation and property size and the identified adaptive behaviours. Some of the more substantive relationships (highly correlated) were between knowledge about climate change, world views and personal values and some of the 14 adaptations (Table 16.7). For example, knowledge about steps that an agricultural enterprise can take to become carbon neutral was found to be an important influence on decisions to improve water capture and water use efficiency, establish perennial pasture, reduce stocking rates and fence off native bush. The

Table 16.5: Rating of adaptive actions – tactical vs strategic

	Rating	Action
Tactical	1	Nothing more we can do
	2	Improve water capture; Improve water use efficiency; Reduce stocking rates; Plant trees, shrubs, natives; Move to low input production system
	3	Increase fodder storage; Change enterprise mix
	4	Move from annual to perennial pasture; Fence off native bush
	5	Establish perennial pasture
	6	Diversity farm enterprise
	7	*(no actions were rated a '7')*
	8	Increase off-farm work; More intensive land use; Purchase/lease more land locally
	9	Value add to farm products; Sell/lease land to others
Strategic (PROPOSED)	10	Acquire more land elsewhere

world view that 'humans must live in harmony with nature in order to survive' and the personal value of 'living a varied life, filled with challenge, novelty and change' were both positively linked to the adaptations of diversifying the farm enterprise, changing the enterprise mix and increasing off-farm work.

CONCLUSIONS

A decade of drought in the study area appears to have alerted rural landholders to what the future might be like if such conditions are going to become more frequent. Given rainfall is critical to agricultural success, it is not surprising that rural landholders adapt quickly to changing uncertainties surrounding this variable. In the absence of a definitive framework for assessing the quality of particular adaptations, it is difficult to know what the off-farm impacts of these adaptive behaviours will be on the wider community and the environment. For

Table 16.6: Changes in management over the past five years, Chiltern and Eskdale survey respondents 2009

Adaptive actions (tactical/strategic rating)	N/A (%)	Plan to/are doing/have done over past 5 years (%)	Influenced by climate change (%)
Improve water use efficiency (2)	9	82	81
Improve water capture (2)	12	77	77
Establish perennial pasture (5)	20	68	67
Reduce stocking rates (2)	22	65	66
Plant trees, shrubs, native grasses (2)	19	67	58
Move from annual to perennial pastures (4)	38	47	53
Increase fodder storage (3)	40	38	44
Fence off native bush to restrict stock access (4)	39	40	34
Increase off-farm work (8)	48	36	31
Move to low-input production system (2)	52	27	29
Change enterprise mix (3)	57	24	27
Diversify farm enterprise (6)	49	23	27
Value-add to farm products (9)	52	20	22
Move to more intensive land use (8)	56	16	20
Purchase or lease more land in local district (8)	67	12	17
Sell or lease land to others (9)	71	8	6
Acquire more land in another district (8)	74	2	5

This table is a compilation from part A and part B of Section 5 in the survey.

n=90

example, if individual property owners tap into groundwater or expand their capacity to capture surface water, there are likely to be implications for downstream landholders and urban populations and environments. Nor do we know the extent to which landholders' risk related to potential impacts of climate change will be reduced through these actions.

Our overall findings, however, indicate that most rural landholders in north-eastern Victoria are not climate change deniers and are responding to climate change. Most perceived a risk to their property but considered the likely impacts to be small. The vast majority have considered or have already adopted a range of responses (improving water capture and water use efficiency, reducing stocking rates, establishing perennial pasture, planting trees, shrubs and native grasses) to recent climatic pressures. Climate change was nominated as an influence on these adaptive behaviours by 50% or more of survey respondents.

Interestingly, belief in climate change did not show in the pairwise analysis as a significant factor in behavioural change. However, locally relevant knowledge about what an agricultural enterprise could do, and views about the need to protect individual rights, were found to be key factors influencing tactical responses. The more strategic adaptations were likely to be made by those who believed 'humans must live in harmony with nature in order to survive' and who seek a 'varied life, filled with challenge, novelty and change'.

Some economists believe that engaging landholders in dialogue is a waste of public resources because they already have a good record of adapting to climatic challenges and have plenty of time to do so (Pannell 2010). Our findings indicate that knowledge about climate change is a key factor influencing a shift in behaviour toward more strategic adaptations. However, most respondents reported having limited knowledge of climate change topics, particularly knowledge about the implications for agriculture in their region. Consequently,

Table 16.7: Relationships between influential factors included in the survey and specific adaptations, Chiltern and Eskdale survey 2009

Influential factors	Improved water capture	Water use efficiency	Est. perennial pasture	Reduced stocking rates	Annual to perennial past	Diversify farm enterprise	Change enterprise mix	Increase fodder storage	Increase off-farm work	Value-add to farm products	More intensive land use	Low input production system	Plant trees etc.	Fence off native bush
Multi R² (Measure of good fit > 15%)	15	17	29	34	18	41	48	28	49	46	19	43	40	36
Knowledge of steps that an agricultural enterprise can take to become carbon neutral	P	P	P	P										P
A worldview that US, India, and China should reduce their emissions before Australia	P	P	N							P			N	N
A worldview that humans must live in harmony with nature in order to survive						P			P			P	P	
A perception of personal/property risk from pest plants		P										P	N	N
Ability to adapt to reduced water supply for stock and domestic	P							N			N			
Importance of protecting individual rights	P		P	P			P			P	P	P	P	P
Importance of protecting the environment						P	P					P	P	P
Importance of living a varied life, filled with challenge, novelty and change						P				P			N	
Importance of ensuring a fair go for all		P		P									P	P
Importance of equal opportunity for all		P		P			P					P	P	P

P = positive relation; N = negative relation.

engaging landholders in dialogues about climate change should still be seen as a cost-effective way to influence on-property management.

REFERENCES

Ajzen I (1991) The theory of planned behaviour. *Organisational Behaviour and Human Decision Processes* **50**, 179–211.

Berg BL (2001) *Qualitative Research Methods for the Social Sciences.* Allyn and Bacon, Boston.

Cary JW, Webb TJ and Barr NF (2002) *Understanding Landholders' Capacity to Change to Sustainable Practices: Insights for Practice Adoption and Social Capacity for Change.* Bureau of Rural Sciences, Canberra.

Chandler R (2009) Adapting to climate change: confronting model uncertainty in future projections. Presented at *Greenhouse 2009: Climate Change and Resources Conference.* 23–26 March, Perth. CSIRO.

Curtis A, Robertson A and Tennant W (2001) 'Understanding landholder management of river frontages: the Goulburn Broken Catchment'. Johnstone Centre, Albury, NSW. <regional.org.au/au/jc/r/p.htm>.

Curtis A, McDonald S, Sample R and Mendham E (2008) 'Understanding the social drivers for natural resource management in the Wimmera region'. Institute for Land, Water and Society report. Charles Sturt University, Albury, NSW. <www.riv.csu.edu.au/research/ilws/research/publications/docs/46%20Wimmera.pdf>.

de Vaus DA (1991) *Surveys in Social Research.* Allen and Unwin, Sydney.

de Wit M (2006) 'Climate change and African agriculture: how farmers perceive and adapt to climate change'. Policy Note no.10. Centre for Environmental Economics and Policy in Africa. <www.ceepa.co.za/docs/POLICY%20NOTE%2010.pdf>.

Dunlap RE and Van Liere K (1978) The new environmental paradigm. *Journal of Environmental Education* **9**, 10–19.

Eckersley R (2007) Nihilism, fundamentalism, or activism: three responses to fears of the apocalypse. *The Futurist* (magazine of the World Future Society), January–February.

Evans C (2009) Rural communities' attitudes and opinions to climate change. Presented at *Greenhouse 2009: Climate Change and Resources Conference.* 23–26 March, Perth. CSIRO.

Giorgio F (2009) Regional climate change projections: what have we learned and where do we go from here? Presented at *Greenhouse 2009: Climate Change and Resources Conference.* 23–26 March, Perth. CSIRO.

Grothmann T and Patt A (2005) Adaptive capacity and human cognition: the process of individual adaptation to climate change. *Global Environmental Change* **15**(3), 199–213.

Jorgensen BS and Stedman RC (2001) Sense of place as an attitude: lakeshore owners' attitudes toward their properties. *Journal of Environmental Psychology* **21**, 233–248.

Leith P (2006) Conversations about climate: seasonal variability and graziers' decisions in the eastern rangelands. School of Geography and Environmental Studies, University of Tasmania, Hobart. pp. i–vii, 1–59, 1 CD-ROM. <http://ecite.utas.edu.au/40494>.

Leith PB and Haward M (2010) Climate change adaptation in the Australian edible oyster industry: an analysis of policy and practice. University of Tasmania, Hobart.

Milne M, Stenekes N and Russell J (2008) *Climate Risk and Industry Adaptation.* Bureau of Rural Sciences, Canberra.

Nelson R, Webb T and Byron I (2006) 'Socio-economic data: prioritising collection to support Australian Government natural resource management programs: principles and priorities'. Prepared by ABARE-BRS for the National Land and Water Resources Audit, Canberra.

O'Connor RE, Bord RJ, Yarnal B and Wiefek N (2002) Who wants to reduce greenhouse gas emissions? *Social Science Quarterly* **83**(1), 1–17.

Pannell DJ (2010) Telling farmers how to adapt to climate change. Pannell Discussion no. 168, 30 August. <cyllene.uwa.edu.au/~dpannell/pd/pd0168.htm>.

Pannell DJ, Marshall GR, Barr N, Curtis A, Vanclay F and Wilkinson R (2006) Understanding and promoting adoption of conservation practices by rural landholders. *Australian Journal of Experimental Agriculture* **46**(11), 1407–1424.

Patton MQ (1990) *Qualitative Evaluation and Research Methods.* 2nd edn. Sage Publications, California.

Pickworth J, Casey AM, Maller C and Stenekes N (2007) 'Adapting to change in fisheries: report to the Seafood Industry Partnership Project'. Bureau of Rural Sciences, Canberra.

Reid S, Smit B, Caldwell W and Belliveau S (2007) Vulnerability and adaptation to climate risks in Ontario agriculture. *Mitigation and Adaptation Strategies for Global Change* **12**(4), 609–637.

Robinson M (2009) Developing and implementing a national climate change research strategy for primary industries. Presented at *Greenhouse 2009: Climate Change and Resources Conference.* 23–26 March, Perth. CSIRO.

Stern PC, Dietz T and Guagnano GA (1998) A brief inventory of values. *Educational and Psychological Measurement* **58**(6), 984–1001.

Stern PC, Dietz T, Abel T, Guagnano GA and Kalof L (1999) A value-belief-norm theory of support for social movements: the case of environmentalism. *Human Ecology Review* **6**(2), 81–97.

Thornton B, Gibbons F and Gerrard M (2002) Risk perception and prototype perception: independent processes predicting risk behavior. *Personality and Social Psychology Bulletin* **28**, 986–999.

Thwaites R, Curtis A, Mazur N and Race D (2008) 'Understanding rural landholder responses to climate change'. Report no. 48. Institute for Land, Water and Society, Charles Sturt University.

Venables WN and Ripley BD (2002) *Modern Applied Statistics with S.* 4th edn. Springer, New York.

What we learned about measuring change in vegetation extent and condition

Ted Lefroy and Geoff Park

It is impractical to try to restore ecosystems to some 'rightful' historical state (Davis et al. 2011).

Hundreds of millions of dollars of public and private funds have been invested in vegetation management programs in Australia over the last 30 years, with a significant proportion in Victoria, the birthplace of the Landcare movement. In just the four years that the Landscape Logic project was active, our three partner catchment management authorities (CMAs) in Victoria invested $260 million in environmental management programs. A major weakness in all these programs, and a weakness in any claim that we really are practising adaptive management, is the lack of data that enables us to determine the effect of this investment. When Landscape Logic formed, our three partner CMAs in Victoria identified this lack of information as a major gap in their understanding and practice.

Our research questions with respect to native vegetation were therefore as follows: whether we could quantify change in vegetation extent and condition, attribute change due to human intervention and other drivers, and distinguish between change from public and private funding sources.

Chapters 10–16 in Part II have reflected the way those questions were broken down into tractable pieces of research. It soon became apparent that we needed to tackle the questions of extent change (Chapters 10 and 11) and condition change (Chapter 12) separately. Extent change is covered in two chapters, reflecting the need to combine ecological knowledge and spatial analysis (Chapter 10) with an understanding of the social and economic drivers of environmental change and local knowledge to construct landscape histories (Chapter 11). Chapter 10 also combines social and ecological research at the human scale of individual landholdings, to identify what sort of intervention produced the change apparent at larger scales and how it was funded. Vegetation condition change is handled as a separate question (Chapter 12) because of the absence of agreed and measurable definitions of desired condition states applicable to specific vegetation communities and the lack of metrics capable of detecting change, particularly of distinguishing between human and other influences. Chapter 13 arose in response to the need for better information about where to invest. It aims to link high-quality spatial data on vegetation types with a searchable database capable of representing vegetation type and composition by tenure, watershed and connectedness and to identify features of individual patches such as shape, size, perimeter length and proximity to other patches. Chapter 14 presents the results of three PhD theses targeted at improving the application of remotely sensed data to the measurement of vegetation extent and structural aspects of condition. Chapters 15 and 16 investigate social aspects of vegetation management. Chapter 15 examines how a community influences the behaviour of its members, especially new arrivals, and the extent to which this is affected by demographic change as rural areas experience a decline in the number of full-time farmers. Chapter 16 was instigated by our partner CMAs, who were interested in knowing what sort of strategies and tactics landholders are adopting in response to climate change.

OUR FINDINGS

The new knowledge that emerged from the seven chapters on vegetation change is summarised below.

VEGETATION EXTENT CHANGE

The combination of aerial photo interpretation and landscape history workshops found that vegetation extent across the three study areas in northern Victoria declined from 1946 to the late 1980s and early 1990s, after which decline in vegetation cover levelled out. Regulation appears to have played a major role in arresting that decline through the introduction of grazing controls on public lands in the 1980s and clearing legislation in the late 1980s, but social and economic factors were also found to be major influences. Interactions between environmental and socio-economic factors produced a range of drivers varying in scale from local (fire) to regional (drought) and from fast and immediate (land clearing and land use change) to slow and incremental (declining wool prices, improved rabbit control and demographic change).

Immature vegetation evident in the landscape as regeneration and revegetation, the product of Landcare, other environmental programs and natural regeneration, has produced a modest 1–2% increase in cover over the last 20 years and appears set to restore mature vegetation cover to 1946 levels within a further 20 years, assuming it survives to maturity. Regeneration accounted for twice the area of the more expensive and labour-intensive revegetation, which has implications for extension programs and incentive schemes. It also raises questions about the relative habitat and conservation value of these two forms of novel communities.

PUBLIC AND PRIVATE INVESTMENT

Joint public–private funding of vegetation management was found to have exceeded solely private investment by a ratio of 4:1 over the period of study. This imbalance appears to have been influenced more by the appearance of large publicly funded programs in the 1990s, notably the National Heritage Trust, than by any absolute decline in private investment. This finding challenged the assumption used in state and regional reporting that there has been one privately funded hectare of vegetation management for every hectare receiving some public funding, i.e. a 1:1 ratio, or twice the area of managed vegetation than appears in publicly funded records, sometimes referred to as the 'two times' assumption. The study also found that the conditions attached to the public funding meant that it was twice as likely to occur in areas of high conservation value. That is, publicly funded investment was more likely to occur in areas where native vegetation is underrepresented due to high rates of clearing, commonly because it was more desirable for agricultural production, such as valley floors. The Patch Data Viewer tool was developed to aid such investment decisions by identifying patches by vegetation community, catchment, management region, proximity to similar patches and other contextual information.

VEGETATION CONDITION CHANGE

Assessment of vegetation condition change faces two problems: the lack of clear and measurable vegetation condition targets or endpoints, and the lack of metrics capable of detecting change in the desirable attributes within reasonable time-frames. The development of a quantitative state-and-transition model (Chapter 12) provides an ecologically defensible process for arriving at achievable and measurable targets. It avoids the current assumption implicit in many vegetation management goals and targets of a linear successional trajectory from some current state to an ideal or reference state, by acknowledging that a particular vegetation community can exist in multiple states at any one time depending on land use history, ecological processes and current disturbance regime. It also recognises that the effort required to move between these states varies considerably, and in some instances the transition may not be possible. This raises an important question about the relevance of historical reference or benchmark states, still the most widely used approach in Australia. This concept has been challenged in the literature for some years (Oliver *et al.* 2002; Hobbs 2007; Curtis and Lefroy 2010; Davis *et al.* 2011) but the debate has yet to have a major influence on policy and management in Australia, although there are some positive signs (Zammit *et al.* 2010).

REMOTELY SENSED MEASUREMENT OF VEGETATION CHANGE

Aerial photography was the method chosen for measuring extent change in this study because it remains more accurate and more easily interpreted than remotely sensed imagery despite the latter's ready availability and growing sophistication. The three PhD studies undertaken as part of Landscape Logic each added insights that potentially improve the application of remote sensing to measurement of vegetation change. One study found that the grid spatial resolution of remotely sensed imagery needs to be several times finer than the size of the target, to achieve 75% detection of square patches and even more for elongated patches. Another showed that measures of the texture of multi-spectral satellite imagery could be used as correlates of some attributes of vegetation structure. The third demonstrated that airborne laser scanning (LiDAR) showed strong correlation with field measurements of vegetation structure in stands of native forest, notably canopy cover, tall tree height and understorey vegetation

SOCIAL DRIVERS OF ENVIRONMENTAL CHANGE

Chapter 15 identified the importance of social norms in environmental management, both strong injunctive norms that justify boundary-crossing behaviour (spraying weeds on a neighbour's property) and weaker but significant descriptive norms that help to define a community and its important values. This was revealing research, as it provides policy-makers and environmental managers some insight into the behaviours most likely to be reinforced by public intervention. As the authors observed, this is potentially a low-input/high-output area for public policy, and program evaluations would be well advised to include consideration of the extent to which they have established or reinforced positive social norms. The climate change study (Chapter 16) found that over half the respondents in a survey of 94 landholders had implemented practices designed to adapt to climate change (improved water harvesting, better water use efficiency, reduced stocking rates, establishing perennial and native pastures) although the majority acknowledged that they had limited knowledge of the subject. This indicates both motivation and an opportunity for well-targeted dialogue and extension but also raises questions about downstream impacts.

RESEARCH METHODS AND FURTHER QUESTIONS

One observation emerging from the research described in Part II (particularly Chapters 10, 11, 12, 15 and 16) has strong parallels with conclusions reached from the water quality research described in Part 1 (Chapter 9). This is that inquiry at three scales (landscape, property and site), using a combination of ecological and social methods, was necessary to detect change in environmental condition in landscapes dominated by private ownership and to infer cause.

Landscape-scale analysis of historic aerial photos plus landscape history workshops with local residents and experts were required to validate mapped change in vegetation extent at regional scale and to infer the most likely cause. Property-scale surveys and interviews were required to establish the types of management that had been undertaken, when they occurred and how they were funded. Site-level analysis was required to establish quantitative descriptors for the various condition states for woodland vegetation and identify the attributes that could be tracked through time as measured changes in structure and composition. Social research was also necessary to understand the values, incentives and social norms that influenced people to undertake vegetation management.

IMPLICATIONS

These findings have several implications for vegetation management and research. First, that public policy has had a positive impact on vegetation extent in our study area by helping to arrest decline through regulation, and is set to achieve a modest gain one to two decades into the future through a range of incentive programs. However, patience and long-term commitment is required of policy-makers and managers given that it takes 40 years to achieve mature vegetation cover from regeneration or revegetation. Second, we need agreed and measurable goals for vegetation condition expressed in terms of desirable states for each major vegetation type, and these will be more ecologically defensible and economically achievable if they avoid reference to some ideal historical state which may no longer be achievable. Third, assessment of progress requires inquiry at multiple scales (landscape, property and site) using a combination of ecological and social inquiry.

REFERENCES

Curtis A and Lefroy EC (2010) Beyond threat- and asset-based approaches to natural resource management in Australia. *Australian Journal of Environmental Management* **17**, 6–13.

Davis MA, Chew MK, Hobbs RJ, Lugo AE, Ewel JJ, Vermeij GJ, Brown JH, Rosenwig ML, Gardener MR, Carroll SP, Thompson K, Pickett STA, Stromberg JC, Del Tredici P, Suding KN, Ehrenfeld JG, Grime JP, Mascaro J and Briggs JC (2011) Don't judge species on their origins. *Nature* **474**, 153–154.

Hobbs RJ (2007) Setting effective and realistic restoration goals: key directions for research. *Restoration Ecology* **15**, 354–357.

Oliver I, Smith PL, Lunt I and Parkes D (2002) Pre-1750 vegetation, naturalness and vegetation condition: what are the implications for biodiversity conservation? *Ecological Management and Restoration* **3**, 176–178.

Zammit C, Attwood A and Burns E (2010) Using markets for woodland conservation on private land: lessons from the policy–research interface. In *Temperate Woodland Conservation and Management.* (Eds DL Lindenmayer, A Bennett and RJ Hobbs) pp. 297–308. CSIRO Publishing, Melbourne.

Part III

INTEGRATING SCIENCE AND PRACTICE

18

Bayesian networks as integration tools in collaborative research

Jenifer Ticehurst, Carmel Pollino and Wendy Merritt

SUMMARY

This chapter describes the experience of a team of modellers and decision support specialists tasked with integrating the research components and results of different disciplinary teams within the Landscape Logic project. One of the aims of Landscape Logic was to develop models and embed these into decision support systems (DSS). The development of these tools was a collaborative exercise with natural resource management (NRM) decision-makers and disciplinary experts contributing their knowledge of natural resource planning and investment to the models. The modelling team proposed using Bayesian network (BN) models as the integration tool. These models provide a framework for:

- integrating knowledge from different biophysical and social research components;
- readily incorporating qualitative and quantitative information of varying quality;
- explicitly accommodating the uncertainty inherent in addressing NRM issues;
- conceptualising cause and effect in graphical form that can be readily understood by a range of audiences through the model development stages.

The graphical structure of BNs enhances hypothesis generation, knowledge elicitation and information-sharing and promotes trust and openness between participants in the model development process. Accordingly, research teams and collaborating bodies from regional NRM organisations and government environmental agencies were introduced to the basic functions and development of BNs through training workshops, with the intention that the modelling team would then develop BNs in consultation with these groups. The interest and capacity of some of these groups led them to undertake the development of their own BNs, while others preferred to await the outcomes of the Landscape Logic project. We found that partners with greater capacity (staffing and funding) were more willing to pursue the approach independently by funding or developing their own models. For those with fewer resources, this was not feasible. The involvement of different Landscape Logic project partners in the development of BNs enabled us to investigate the applicability of BNs as an integrative tool within the project.

Over the life of the project, at least 14 NRM regional staff received formal introductory training in the development of BNs and at least that many again were exposed to and/or involved in developing the final models. At least six researchers developed their own BNs to completion, while an estimated 20 more participated directly in the development of a BN. In total, well over 50 researchers and resource managers have increased skills in building BNs and, for some, model development in general.

This chapter documents our experiences in developing BNs within a collaborative environment and documents feedback undertaken to gain an appreciation of how project partners viewed their experiences of being trained in and using BNs. Feedback indicated that BNs were perceived as useful tools for packaging, integrating and communicating project research outcomes to resource managers, facilitating integration between researchers and the regional partners and providing a common modelling technique that could be applied across different disciplines to enable integration of models into a DSS. The conceptual simplicity of BNs was seen as promoting a common language for communication across Landscape Logic teams, disciplines and partners. The non-BN and BN-based research products were regarded as equally beneficial, and indeed as complementary, by the NRM organisations. The outcomes of Landscape Logic reinforce the usefulness of BNs as a collaborative modelling tool, particularly where there is an emphasis on integration.

INTRODUCTION

The integration team within the Landscape Logic project had four functions:

- to integrate, communicate and package research outcomes for the project's NRM partners and other end users;
- to facilitate integration between researchers and the regional partners, in cooperation with the knowledge broking team (Chapter 19, this volume);
- to develop and apply purpose-built models as components of and inputs to integrated DSSs;
- to build the capacity and skills of Landscape Logic researchers and NRM partners.

BNs were selected as the appropriate integration tool to fulfil these needs. BNs are a modelling approach that is increasingly used to assist research and planning in a diversity of fields. Their popularity in ecology and in NRM has grown since Cain (2001) published protocols on the use of BNs to support water planning and management. BNs are a type of integration model (Letcher and Bromley 2006; Jakeman et al. 2007) with features consistent with integration needs in NRM (Ticehurst et al. 2008). A comparison of integration models is shown in Table 18.1. BNs were of particular relevance to Landscape Logic because they:

Table 18.1: Review of the strengths and capabilities of various integration methods

Model property		System dynamics	Bayesian networks	Meta models	Coupled complex models	Agent-based models	Expert systems
Model purpose	Prediction		✓	✓	✓		✓
	Forecasting			✓	✓		✓
	Decision-making	✓	✓	✓	✓		✓
	System understanding	✓			✓	✓	✓
	Social learning	✓			✓	✓	✓
Input data type	Qualitative and quantitative		✓				✓
	Quantitative only	✓		✓	✓	✓	
Focal range	Focused and in-depth				✓		
	General and broad	✓		✓			
	Compromise			✓			✓
	Both					✓	
Explicit uncertainty	Yes		✓				✓
	No	✓		✓	✓	✓	
Model output	Individual impact					✓	
	Aggregated impact	✓	✓	✓	✓		✓

Adapted from Jakeman *et al.* (2007).

- can incorporate quantitative and qualitative data. This supports seamless integration of social, ecological and economic factors (Bromley *et al.* 2006) and reduces uncertainties associated with the impact of variables on one another through the use of probability distributions (Varis and Kuikka 1997);
- lend themselves to development, testing and application in participation with users due to their graphical nature;
- specify the states, or classes, used to define each variable in the BN in accordance with important and suitable thresholds that are relevant to the decision-makers (e.g. a threshold in annual income, below which people will no longer invest in NRM), while being consistent with the available level of subjective or quantitative knowledge;
- provide a record of corporate knowledge by documenting understanding and assumptions of how a system works in a way that can be reviewed and updated;
- are well-suited to adaptive management where managers are encouraged to plan, implement actions, monitor results and review and update plans (Smith *et al.* 2007).

While reviews of the use of BNs in water resource modelling and management exist (Castelletti and Soncini-Sessa 2007) there are few if any published studies evaluating end user responses. This chapter explores the role that BNs played within the Landscape Logic project and the extent to which the researchers and intended end users found them a useful tool. As it is too early to examine the actual adoption of the Landscape Logic BNs, we discuss the perceived value and challenges of BNs as a measure of their potential for adoption.

METHODS

The typical steps used to develop a BN are summarised in Figure 18.1. Generally, after a focal issue and scale have been defined, a cause and response representation of the issue is summarised in an influence diagram. The strength of the interactions is then defined by populating the model with probability distributions. The BN development approach easily lends itself to stakeholder input, particularly in the development and review of the conceptual diagram (Steps 2 and 3, Fig. 18.1), collating data to populate the model, eliciting expert opinion where required (Step 5) then testing the outputs of various scenarios to assess the accuracy of a model's performance (Step 6). The BNs developed as part of Landscape Logic generally followed these steps, with some variation in stakeholders' participation.

Figure 18.2 provides an example of the different BN development stages, showing:

- an influence diagram (Step 2, Fig. 18.1) developed during a stakeholder workshop, representing the habitat and management activities relevant to the survival and persistence of the giant freshwater crayfish (*Astacopsis gouldi*), a threatened crustacean found in northern Tasmania;
- the framework after it had been amended following expert review (Step 3, Fig. 18.1);
- the final BN (Chen and Pollino 2010) populated with data and used for testing different scenarios (Step 7, Fig. 18.1).

As the Landscape Logic project had an emphasis on active collaboration, the integration team was tasked with developing the BN models with research and NRM partners. After being introduced to BNs through workshops, the NRM regions expressed interest in applying the technique themselves, specifically requesting training in model development to fulfil their planning requirements. Similarly, some of the research teams pursued BNs as a framework for representing their research outputs. These included the river health team (Chapter 3, this

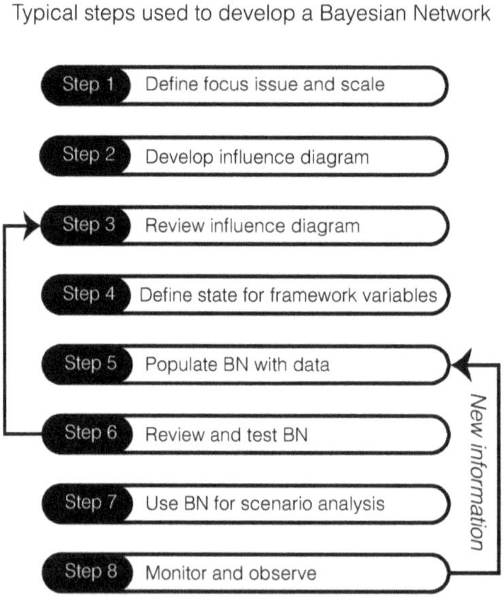

Typical steps used to develop a Bayesian Network

Step 1 — Define focus issue and scale

Step 2 — Develop influence diagram

Step 3 — Review influence diagram

Step 4 — Define state for framework variables

Step 5 — Populate BN with data

Step 6 — Review and test BN

Step 7 — Use BN for scenario analysis

Step 8 — Monitor and observe

New information

Figure 18.1: Typical steps in developing a BN. Influence diagrams developed and reviewed in Steps 2 and 3 represent critical factors affecting an outcome variable (adapted from Ticehurst *et al.* 2011).

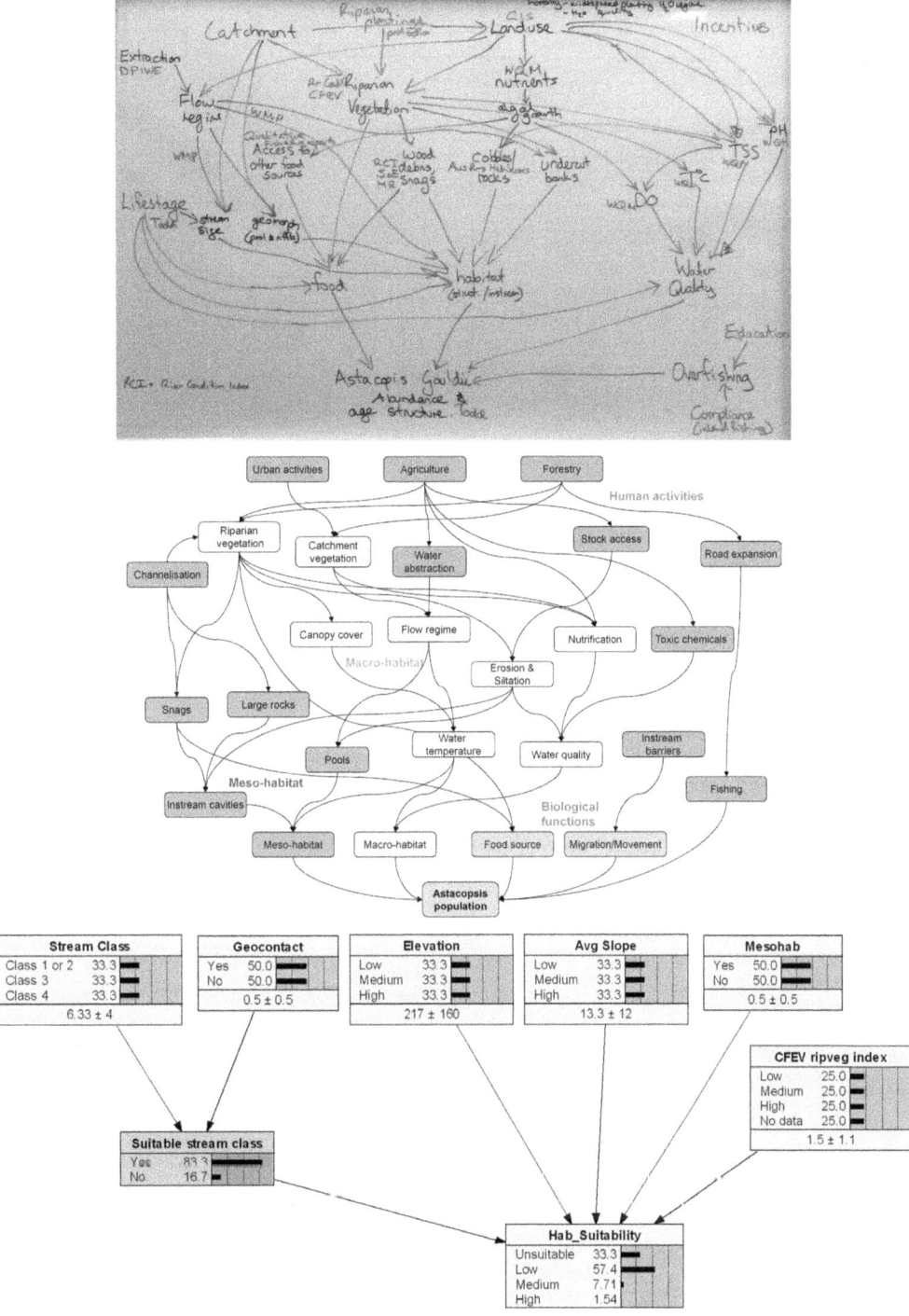

Figure 18.2: Evolution of a BN. a) Initial influence diagram for the *Astacopsis gouldi* BN (Step 2, Fig. 18.1). b) Reviewed influence diagram for the key factors affecting *Astacopsis gouldi* populations (Step 3, Fig. 18.1). c) The final working BN model of juvenile *Astacopsis gouldi* habitat suitability, populated with data, as would be used for running scenarios (Step 7, Fig. 18.1).

Table 18.2: Development details for Landscape Logic BNs

Title	Focus/purpose	Key references
Estuary BN	Predict estuary condition	Pollino (2010)
River Health BN	Predict river condition following various environmental management scenarios	Chapter 3 this volume; Magierowski *et al.* (2010)
Site Native Vegetation Quality BN	Predict change in quality of native vegetation at a particular site following various management activities	Chapter 12 this volume; Rumpff *et al.* (2010)
Landscape Native Vegetation Quality BN	Predict change in the quality of native vegetation in a particular landscape following various management activities	Merritt and Manger (2010)
Adoption of Riparian Management BNs	Four BNs to predict the level of adoption of various riparian management practices, and what factors affected that adoption	Chapter 6 this volume; Ticehurst and Curtis (2010)
Adoption of Native Vegetation Management BN	Predict the level of adoption of various native vegetation management practices, and what factors affected that adoption	Merritt and Manger (2010)

volume) and the native vegetation quality team (Chapter 12, this volume). A list of the BNs developed in Landscape Logic and their intended purposes is given in Table 18.2.

EVALUATING THE DEVELOPMENT AND POTENTIAL UPTAKE OF THE LANDSCAPE LOGIC BNS

This section outlines how information was collected on the attitude of end users to BNs as a management tool and the use of BNs by researchers.

For NRM bodies and other potential end users such as state agencies, the perceived usefulness, benefit and potential for uptake of the Landscape Logic BNs (Table 18.2) were investigated through surveys. Training workshops for the BNs and other Landscape Logic products were held in October 2010 (Tasmania) and December 2010 (Victoria). A summary of these surveys is given in Table 18.3. Through the surveys, participants were asked for their opinion on:

- whether the BNs could have a role in supporting decision-making;
- how valuable the different phases of BN development were (i.e. the development of the influence diagram compared to using a completed model) (Survey 4a, Table 18.3);
- how beneficial they thought each of the products (including the non-BN tools) would be to their work;
- the extent to which the BNs had contributed to improved understanding (Survey 4b, Table 18.3).

It is important to note that attendees at these training workshops had not necessarily attended previous Landscape Logic workshops.

At the end of Landscape Logic, researchers were asked to complete an online survey about the usefulness of BNs in their research (Survey 5, Table 18.3).

CAPACITY-BUILDING FOR MODEL DEVELOPMENT AND USE

This section, largely summarised from Ticehurst and Pollino (2007, 2010), describes the methods we used to train the NRM regions in the development of BNs and how we monitored their progress and views about BN technique.

Table 18.3: Focus questions and number of respondents to the surveys

Survey	Date	Response group	No. of responses	Focus questions
1	March 2007	NRM regions	14	How effective was the training?
				Could BNs assist you in your decision-making processes? If so, how?
2	May 2007	NRM regions and stakeholders	20 (8 trained, 12 untrained)	How could BNs assist you in your decision-making processes?
				Are BNs meeting your expectations?
				What barriers are you finding in developing BNs?
3	September 2009	NRM regions and stakeholders	8 (key contact per region, 3 responses for one region)	When you were first introduced to BNs did you think they would be useful?
				With your experience of BNs now, has your opinion changed?
				Is the development of BNs by regions feasible and/or valuable?
				Which components of BNs are of most use?
				How much time have you/your organisation had to develop BNs?
				What barriers are you finding in developing BNs?
4	a) October 2010 (Tas)	NRM regions and stakeholders	14	Could these BNs have a role in supporting your decision-making?
	b) December 2010 (Vic)		5	How valuable are the different phases of the BN development (i.e. influence diagram vs completed model)?
				How beneficial do you think each of the Landscape Logic products (including the non-BN tools) would be to your work?
				How useful were the products in improving your understanding?
5	January 2011	Researchers	14	Initially, did you think that BNs would be useful for your research and/or for integration within Landscape Logic?
				Were BNs useful for your research and/or for integration with Landscape Logic (i.e. did they meet your expectations)? If so, how?
				What were the major hurdles in developing BNs?

In response to introductory workshops in 2006, the NRM regional bodies requested training and support to build their independent capacity to develop and apply BNs to their work. Consequently, in March 2007 a two-day training workshop was held in Hobart with two or three representatives from each of the six NRM partner regions. This workshop covered the theory and practices of the BN development process (Steps 1–6, Fig. 18.1) through presentations and hands-on tutorials (Survey 1, Table 18.3). Follow-up meetings were held in May 2007 with each region to see how their BN development was progressing and what problems needed to be addressed (Survey 2, Table 18.3), with the intention of holding another follow-up meeting in August. At the May meetings it was apparent that staff from the different regions were working at different paces and had different levels of interest in the approach. The intended August 2007 meetings were replaced with an offer for each region to reschedule a meeting at its convenience. Participation varied between regions over the next two years. A follow-up survey was conducted via email in September 2009 to gauge progress across the regions (Survey 3, Table 18.3).

RESULTS AND DISCUSSION

This section explores how BNs have been viewed by Landscape Logic researchers and NRM partners, and the extent to which BNs helped meet the functions of the integration team.

INTEGRATING AND COMMUNICATING RESEARCH OUTCOMES FROM LANDSCAPE LOGIC TEAMS TO END USERS

After an introduction to BNs, some of the NRM regional staff and researchers could see potential for their application in decision-making and research. The first NRM region survey (Regions Survey 1, Fig. 18.3) indicated 64% of participants believed that BNs would assist decision-making in their organisation. The comparative value for the researchers (Researchers Survey 5, initial, Fig. 18.3) was only 43%. After some experience in developing and/or using BNs, the proportion of regional staff and researchers who saw value in BNs in their work increased to 88% for regional staff (Regions Survey 3, Fig. 18.3) and 57% for researchers (Researchers Survey 5, final, Fig. 18.3). One researcher who was undecided about the usefulness of the technique commented:

> BNs are great at forcing the modeller to conceptualise the whole system and thinking about how variables interact. But when there is poor understanding as well as limited data, it can be impossible to build a working BN. Or the resultant BN can be misleading (overparameterised etc).

When asked whether they thought BNs were a useful tool for adopting science into management, researchers' responses matched those of the NRM regions (86% yes, 14% undecided; Researchers on adoption of science, Fig. 18.3).

Results from the survey of NRM regions in September 2009 indicated that 88% of the regional partners felt the development of influence diagrams was a very useful process (Fig. 18.4). One region reported having changed its planning processes to include the development of influence diagrams as a means of promoting discussion and putting into perspective its relative influence over the condition of natural resources. A total of 64% of researchers found the development of influence diagrams to be useful. Compared with the NRM respondents, the researchers found greater use in populating the BNs and in applying completed models. However, most NRM regional staff had not seen or used a completed BN at the time of that survey (September 2009). Following the training workshops in October 2010, the responses of NRM staff suggested that the completed BNs were significantly more useful than the influence diagrams (Fig. 18.5).

These results indicated that BNs were generally considered useful by NRM regions and researchers and assisted in their work. For the NRM regions, key advantages of BNs were:

- their facilitation of discussions to clarify the issues, goals, definitions and assumptions among staff within a region;

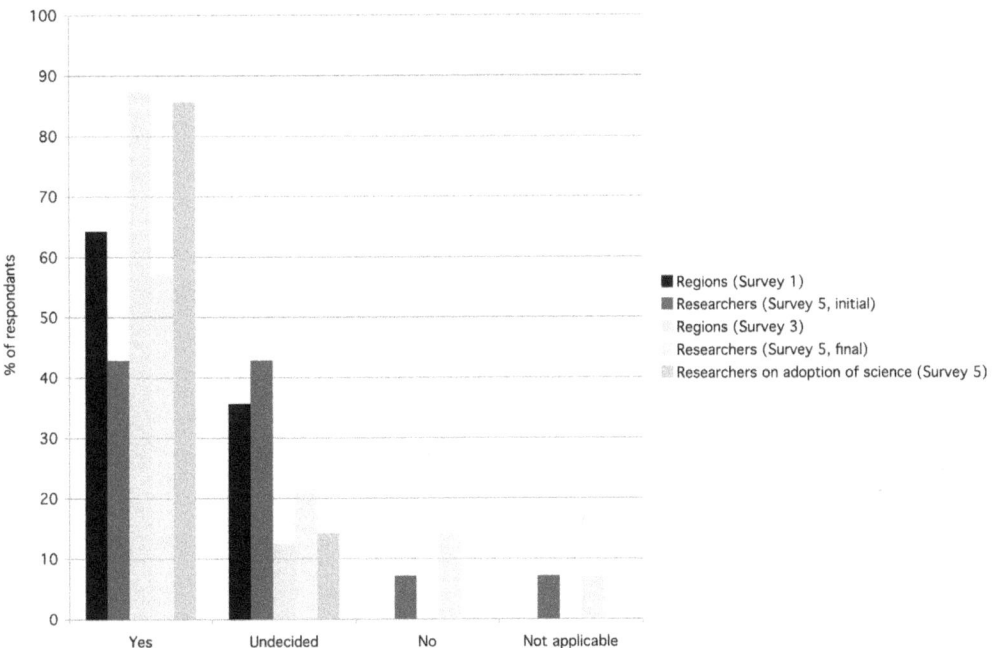

Figure 18.3: Regions' views on whether BNs could assist their decision-making (March 2007, Survey 1, Table 18.2; September 2007, Survey 3, Table 18.2) compared with researchers' views on whether they expected that BNs would be useful (initial) and whether they actually found them to be useful (final) for their research (Survey 5, Table 18.2). Researchers were also asked their opinion on whether BNs were a suitable tool to aid the adoption of science into management (Survey 5, Table 18.2).

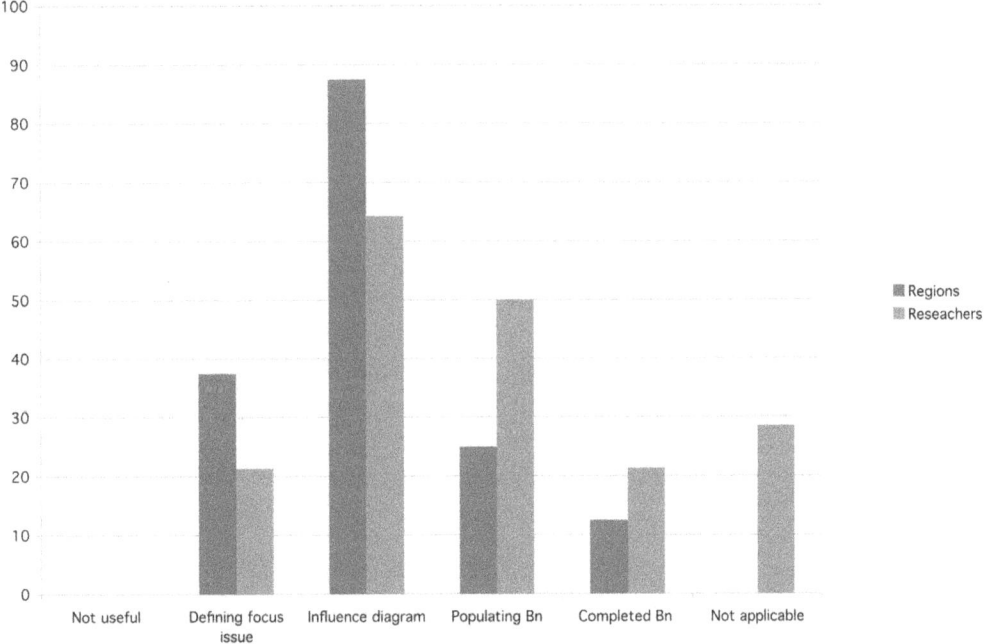

Figure 18.4: The steps that were most useful in developing BNs. Regional responses, September 2009. Researchers' responses, January 2011.

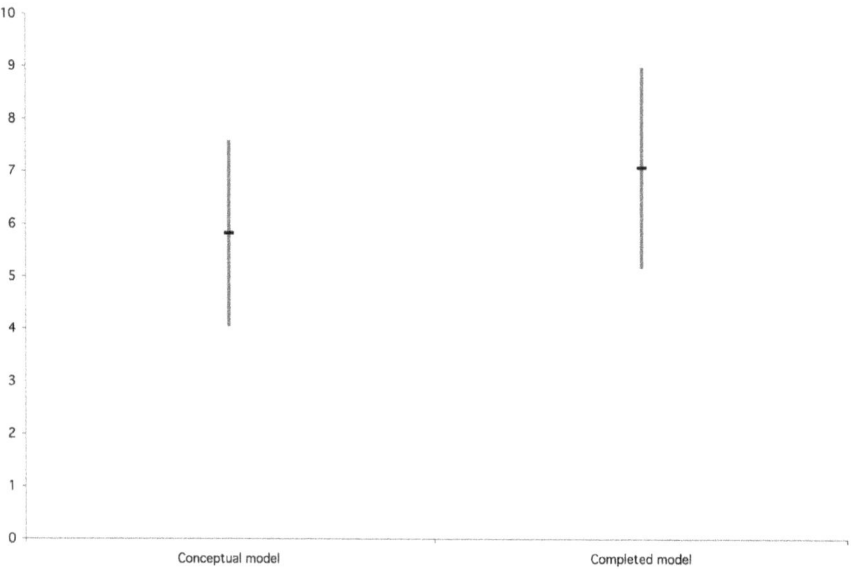

Figure 18.5: The mean +- 1 standard deviation of participants' views on the usefulness of the BN influence diagram and completed model (1 = least beneficial, 10 = most beneficial) following training workshops in Tasmania and Victoria.

- a common framework for staff to document current understanding, assumptions and decision-making processes (this corporate information is otherwise often contained only within the heads of individual staff members);
- a process where the questions that need to be asked and answered are the same as those required in planning and management (this builds on the existing processes rather than adding a new process or task);
- the capacity to graphically illustrate the complexity of the NRM operating context to its investors and stakeholders (responses from Survey 2, Table 18.3).

BN attributes that researchers viewed most favourably were:

- exploration of causality to explain and interpret findings with NRM policy-makers and managers;
- a systemic conceptualisation process that assisted modellers by requiring definitions of focal issues and scales;
- a basis for communication between different researchers and research disciplines, thereby supporting a common understanding of the research being conducted;
- clarification of research questions and influencing factors;
- interaction of probabilities and use of mixed data types;
- capacity to run BNs backwards to allow investigation of system sensitivity and discover previously unknown parameter sets that can deliver desired outcomes.

Limitations of the BN technique identified by staff from NRM regions included the trade-off between complexity and usefulness, the level of work required to complete a BN and uncertainty of whether BNs will improve the quality of their current decision-making processes. Some researchers identified limited relevance to some areas, with one researcher noting:

I found that BNs weren't really suitable for the type of research I was doing (Survey 5, Table 18.3).

Evidence suggested that the benefits gained from the BNs did not vary significantly from benefits from other products such as DSSs (which had a user-friendly software interface over BNs) or other Landscape Logic products that did not use BNs such as the estuarine decision tree (Chapter 4) and the Patch Data Viewer (Chapter 13). Figure 18.6 shows that, following various product training workshops, participants ranked the benefits of BNs as very similar to those of the other Landscape Logic products. Some cautionary comments were made by project researchers:

BNs were definitely useful, (but) I would say they were also definitely over-emphasised as research end products by some.

We need to be careful not to overstate the usefulness of BNs. For example, there may have been other ways to facilitate interaction and collaboration across disciplines; and existing techniques can explore causality, some better than BNs (e.g. structural equation modelling).

… should never be the sole integration tool.

Despite these notes of caution, BNs were overall regarded as at least equally useful as other more standard modelling approaches.

FACILITATING INTEGRATION BETWEEN RESEARCH, RESEARCHERS AND REGIONAL PARTNERS

Survey responses from researchers (Survey 5, Table 18.3) highlighted the value of using the BNs for integration and engagement between Landscape Logic researchers, external

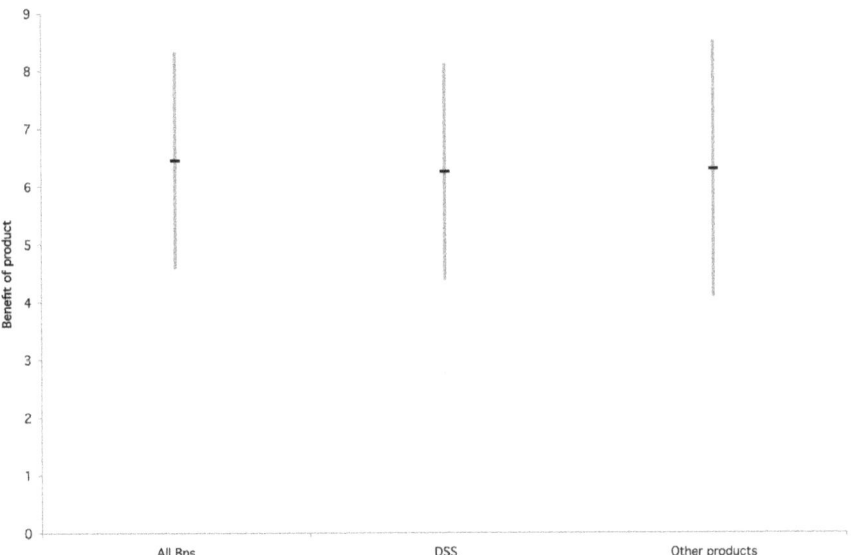

Figure 18.6: The expected benefits of products to NRM regions, other intended end users and their organisations (1 = least beneficial, 10 = most beneficial).

researchers and the regional partners. Six researchers in Landscape Logic attempted to develop their own BN through engagement with other project researchers. Of the six, three respondents felt that they could collaborate cooperatively with other researchers, while the remainder of BN developers experienced difficulty with engagement. Four of the six researchers also engaged end users in the BN development, and only one had not taken the BN to the end users for training (Survey 5, Table 18.3).

While there have been some success stories and indeed lasting collaborations established out of Landscape Logic, there were challenges through the process. The development of integration tools for the NRM regions was complicated in some instances by changing research priorities and questions in response to changes in the NRM institutional environment. In one case, it became clear that the data needed to populate a particular BN was not going to be collected by one of the research teams, and consequently the integration team needed to make use of existing sources of data to populate the BN. In other cases, a lack of clarity concerning the initial research question meant that research findings were not in the exact form or at appropriate scales to directly populate the model, necessitating patching different sources and scales of data. Other difficulties were late changes in the model structure and delay in the delivery of outputs from the researchers. These experiences reflect the challenges of working collaboratively within a large project that tackles complex problems in a real world environment.

DEVELOPING AND APPLYING INNOVATIVE MODELS TO DIFFERENT DISCIPLINES: THE INTEGRATED DSS

The collaborations between the various disciplinary scientists and the integration team presented several opportunities for sharing knowledge, methodologies and expertise. BNs played a key role in this process. For example, the social research team had considerable experience in conducting and analysing social surveys using traditional statistics such as paired analysis and linear regression, but had not previously used BNs. Rafferty (1995) outlined some limitations of using straight statistical techniques in social research and, as early as 1986, Bayesian approaches were being used to overcome these limitations. However, most of the earlier work using Bayesian techniques in social research focused upon Bayesian statistics and there were only a few examples where BNs had been applied (e.g. Nedevschi *et al.* 2006; Whitney and Walsh 2010). Ticehurst *et al.* (2011) describe an approach were BNs were used to supplement a conventional social statistical analysis. They were found to be a beneficial addition to the conventional statistical analysis because they provided a useful tool for structuring, clarifying and communicating to stakeholders the results of the investigation at a level of complexity beyond that possible with conventional analysis. Consequently, BNs were used to enhance further social studies within Landscape Logic (Ticehurst and Curtis 2010).

BN models developed for specific disciplines lent themselves to integration within two DSSs developed as part of Landscape Logic. The capacity to define particular nodes as common links between component models allowed different teams to work independently on their models, knowing that they could be readily linked within a DSS. With the exception of the Adoption of Riparian Management BNs, all the BNs outlined in Table 18.3 were eventually incorporated within one of the DSSs. The Tasmanian Aquatic Condition DSS (Merritt *et al.* 2010) links a catchment water quality simulator to river health and estuary BNs, and allows modelling of land use change on exports of nitrogen and phosphorus and the likely impacts of these nutrient levels on condition indices in Tasmanian rivers and estuaries. The Victorian Native Vegetation Condition DSS (Merritt and Manger 2010) enables users to explore the likely change in native vegetation quality in response to a range of landholder actions and environmental conditions. This DSS uses a series of nested BNs to explore the likely impacts of land management on vegetation quality. It uses the landscape-scale Adoption of Native Vegetation Management BN to explore the likely adoption of various management actions in the light of the social, economic and environmental characteristics of landholders and the structure of assistance programs. The predicted mix of landholder actions can then be used to determine the likely change in native vegetation quality across the landscape using the Landscape Native Vegetation Quality BN. Finally, the Site Native Vegetation Quality BN estimates how native vegetation is likely to respond to management interventions given its current condition and its past and present management.

CAPACITY- AND SKILL-BUILDING

The approach used to develop BNs (Fig. 18.1) is intuitive and complements decision-making processes and research projects as reflected by the NRM regions' interest in the modelling approach. Over the life of the Landscape Logic project, at least 14 NRM regional staff received formal introductory training in the development of BNs, with at least that many again being exposed to and/or involved in developing the final models. At least six researchers have developed their own BNs to completion, and it is estimated that over 20 more participated directly in the development of a BN. In total, well over 50 researchers and resource managers have an increased skill in building BNs and, for some, model development in general.

Although recognising the value of BNs, the views of regional staff on the value and feasibility of developing BNs for their own organisation changed over time. In May 2007, 100% of respondents indicated that it was both valuable and feasible for them to develop their own BNs. Two years later, with little additional development of their BNs, fewer regional staff thought that it was valuable (63%) and feasible (50%) to develop these models in-house. This reflects the increased understanding of the resources required to develop a robust BN populated with data, as well as recognition of the considerable barriers to development.

In total, 62% of all researcher respondents answered that they would continue to use or develop BNs after the completion of Landscape Logic, indicating that they had found significant value in the contribution that BNs could make to their work and felt that they had the capacity to use and develop BNs further.

Although many researchers and NRM regional staff were trained in the use and development of BNs and were offered follow-up support, particular barriers needed to be overcome for people to develop a complete BN. After initial training (March 2007, Survey 1, Table 18.3), the NRM regional staff realised that having the time to sit down and develop the BNs would be a substantial barrier to developing their own models. Two months later, after they had been given some time to progress their model development, NRM staff (including those not engaged in formal training) were asked to rank the barriers to progressing their BNs from greatest to smallest (Table 18.4). Regardless of previous training, both regional staff and stakeholders considered time to be the top two barriers to BN development. A lack of data resources was noted as the next greatest barrier by trained participants, and as fourth by the untrained. While BNs readily allowed the use of expert and stakeholder elicitation to populate the model in the absence of other data, elicitation is a time-consuming process – moving from an influence diagram to a populated BN in the absence of data was a major hurdle for the NRM regions.

Bearing in mind the small survey sizes, there is evidence that regional NRMs have enthusiasm for BNs but that regional staff and stakeholder lack of time and limited data resources

Table 18.4: Barriers to the development of BNs by NRM regions (May 2007 survey) and researchers (January 2011)

Barrier rank	Trained NRM participants		Untrained NRM participants*		Researchers	
	Issue	Score**	Issue	Score**	Issue	Score**
1	Region's time	1.1	Region's time	2.1	Data resources	2.3
2	Stakeholder time	2.2	Stakeholder time	2.5	Stakeholder time	3.2
3	Data resources	3.3	Technical understanding of BNs	2.6	Researchers' time	3.8
4	Technical understanding of focus issue	3.6	Data resources	4.5	Additional support	4.0
5	Technical understanding of BNs	4.0	Technical understanding of focus issue	4.7	Technical understanding of BNs	4.3
6	Additional support	5.8	Additional support	5.1	Technical understanding of focus issue	4.4
7	Equipment and software	7	Equipment and software	5.5	Equipment and software	6.2

* These participants did not attend the March 2007 training and may not have been previously exposed to BNs.
** Weighted average of individual participant scores.

create significant barriers to in-house development. There is a risk of loss of skills in BN development if there is slow progress and high staff turnover. Two of the regions overcame this barrier by dedicating time and money through specific project funding or by initiating a new collaboration with researchers external to the Landscape Logic project, to complete the BN development. One NRM regional staff member commented in the September 2009 survey that 'we are not funded to undertake activities such as DSS and strategic planning'. Our regional partners required support – financial, institutional or through embedded researchers or model developers – to participate fully in the use of this modelling approach.

Interestingly, the researchers also found that their time, their stakeholders' time and data resources were the top three barriers to BN development (Table 18.4), with data resources the greatest barrier. However, one researcher noted that:

> We had to address all the above issues [listed in Table 18.4] but I don't see any of these issues as 'barriers to development': rather, they are stages to be worked through.

These comments highlight how BN development was prioritised within their work requirements, as reflected by their funding objectives, compared to the priorities of the NRM regions. One comment from a regional respondent to the September 2009 survey neatly summed up the requirements for collaborative modelling and successful completion of a BN as 'the projects, ideas, personnel, data and partnerships need to be aligned'.

CONCLUSIONS

BNs were considered useful by the environmental managers in NRM regions and by researchers in assisting the integration aspects of the Landscape Logic project, as measured by the four key roles of the integration team (integrating research outputs, integrating researchers and end users, building DSSs and building capacity of researchers and managers). The NRM regions and researchers saw BNs as a useful means of packaging, integrating and communicating research outcomes into a form useful for management. BNs were effective tools for facilitating integration between researchers and the regional partners, although it was found that alignment of issues and goals and the ability of partner organisations to allocate sufficient staff time and other resources played a significant role in the effectiveness of those interactions. A particular strength was that BNs aided the development and application of models that incorporated knowledge from different disciplines such as social science and ecology. Six researchers led development of separate BNs, 14 managers received formal training and over 50 researchers and resource managers increased their skills and capacity in BN modelling over the life of the project. This development of capacity and skills has given the regional organisations, specifically through the use of influence diagrams, a new way to view the landscapes and systems they are managing, to formalise current knowledge and to engage with research providers in the future. The increased capacity for managers and researchers to collaboratively define problems and develop solutions is perhaps the most significant legacy of this exercise in using BNs in NRM.

ACKNOWLEDGEMENTS

This work was funded through the Commonwealth Environment Research Facilities (CERF) program, an Australian government initiative supporting public good environmental research. The time and feedback from staff of the participating NRM regions (Goulburn Broken, North East and North Central Catchment Management Authorities in Victoria and NRM South, NRM North and Cradle Coast NRM in Tasmania) and the research partners in the Landscape Logic project (Dept of Primary Industry and Water, Forestry Tasmania, Dept of Sustainability and Environment, CSIRO, ANU, RMIT and UTAS) is greatly appreciated.

REFERENCES

Bromley J, Jackson NA, Clymer OJ, Giacomello AM and Jensen FV (2006) The use of Hugin(R) to develop Bayesian networks as an aid to integrated water resource planning. *Environmental Modelling and Software* **20**(2), 231–242.

Cain J (2001) 'Planning improvements in natural resources management: guidelines for using Bayesian networks to support the planning and management of development programmes in the water sector and beyond'. Centre for Ecology and Hydrology, Wallingford, UK. <http://www.norsys.com/downloads/BBN%20Guidelines%20-%20Cain.pdf>.

Castelletti A and Soncini-Sessa R (2007) Bayesian networks in water resource modelling and management. *Environmental Modelling and Software* **22**(8), 1073–1074.

Chen SH and Pollino CA (2010) Bayesian networks for modelling habitat suitability of an endangered species. In *Proceedings of the 2010 International Congress on Environmental Modelling and Software Modelling for Environment's Sake*. (Eds DA Swayne, W Yang, AA Voinov, A Rizzoli and T Filatova). International Environmental Modelling and Software Society (iEMSs) 5th Biennial Meeting, Ottawa, Canada.

Jakeman A, Letcher RA and Chen S (2007) Integrated assessment of impacts of policy and water allocation changes across social, economic and environmental dimensions. In *Managing Water for Australia: Social and Institutional Challenges*. (Eds K Hussey and S Dovers) pp. 97–112. CSIRO Publishing, Melbourne.

Letcher RA and Bromley J (2006) Typology of models and methods of integration. In *Sustainable Management of Water Resources: An Integrated Approach*. (Eds C Giupponi, A Jakeman, D Karssenberg and M Hare) pp. 287–323. Edward Elgar Publishing, Cheltenham, UK.

Magierowski RH, Davies PE and Read SM (2010) 'The Tasmanian River Condition Bayesian Network'. Technical Report no. 26. Landscape Logic, Hobart.

Merritt WS and Manger P (2010) 'Victorian Native Vegetation Condition Decision Support System: User Guide (Version 1.0)'. Technical Report no. 30. Landscape Logic, Hobart.

Merritt WS, Kelly R and Manger P (2010) 'Tasmanian Aquatic Condition Decision Support System: User Guide (Version 1.0)'. Technical Report no. 22. Landscape Logic, Hobart.

Nedevschi S, Sandhu JS, Pal J, Fonseca R and Toyama K (2006) Bayesian networks: an exploratory tool for understanding ICT adoption. *International Conference on International Communication and Technologies Development*. 25–26 May 2006, Berkeley, California. Submission 236.

Pollino CA (2010) 'Predicting change in condition of estuaries in Tasmania: a description of the Landscape Logic Bayesian network'. Technical Report no. 27. Landscape Logic, Hobart.

Rafferty AE (1995) Bayesian model selection in social research. *Sociological Methodology* **25**, 111–163.

Rumpff L, Duncan DH, Vesk PA, Keith DA and Wintle BA (2010) State-and-transition modelling for adaptive management of native woodlands. *Biological Conservation*. doi:10.1016. In press.

Smith C, Felderhor L and Bosch OJH (2007) Adaptive management: making it happen through participatory systems analysis. *Systems Research and Behavioural Science* **24**, 567–587.

Ticehurst JL and Curtis A (2010) 'The development of Bayesian networks to explore the adoption of riparian management practices in Tasmania'. Technical Report no. 25. Landscape Logic, Hobart. <http://www.landscapelogicproducts.org.au/site/system/files/54/original/Tech_report_25_Jen_BN.pdf?1294893977>.

Ticehurst JL and Pollino C (2007) Build collaborative models or capacity? Comparison of techniques for building Bayesian Networks for the Natural Resource Management Regions of Australia. In *Proceedings of MODSIM Conference*. 10–14 December, Christchurch, New Zealand.

Ticehurst JL and Pollino CA (2010) Build collaborative models or capacity? Reflections from two years on. In David A. Swayne, Wanhong Yang, A. A. Voinov, A. Rizzoli, T. Filatova (Eds.) *Proceedings of iEMSs 2010 International Congress on Environmental Modelling and Software Modelling for Environment's Sake*. International Environmental Modelling and Software Society 5th Biennial Meeting. Ottawa, Canada. (Eds DA Swayne, W Yang, AA Voinov, A Rizzoli and T Filatova). <http://www.iemss.org/iemss2010/index.php?n=Main.Proceedings>.

Ticehurst JL, Letcher RA and Rissik D (2008) Integration modelling and decision support: a case study of the Coastal Lake Assessment and Management (CLAM) tool. *Mathematics and Computers in Simulation* **78**, 435–449.

Ticehurst JL, Curtis A and Merritt WS (2011) Using Bayesian Networks to complement conventional analyses to explore landholder management of native vegetation. *Environmental Modelling and Software* **26**, 52–65.

Varis O and Kuikka S (1997) Joint use of multiple environmental assessment models by a Bayesian meta-model: the Baltic salmon case. *Ecological Modelling* **102**, 341–351.

Whitney P and Walsh S (2010) Calibrating Bayesian network representations of social behavioural models. *Advances in Social Computing* **6007**, 338–345.

19

Research to adoption: the role of the knowledge broker in participatory research

Geoff Park, Greg Pinkard and Rod McLennan

SUMMARY

This chapter describes the experience of knowledge brokers in the design, delivery and evaluation of Landscape Logic as a participatory landscape research project. Across the five-year time-frame of Landscape Logic, the role of knowledge broking evolved from establishing the collaboration between six regional natural resource management (NRM) organisations and eight research institutions, to eliciting and refining the research questions, establishing and maintaining networks between researchers and natural resource managers, facilitating communication of results, to finally negotiating the meaning and implications of research findings with researchers, environmental managers and policy-makers.

Knowledge broking is a people-centred process and the primary task was developing and maintaining communication networks between all involved parties to ensure the research focus was mutually beneficial and could evolve in response to new information and end user needs.

Significant differences in history, culture and practice between NRM regions were revealed. These influenced how communication and negotiation were managed throughout the collaboration.

New understandings about the influence of human intervention on native vegetation and water quality were generated through the research partnership and an important role of knowledge broking was to make these available in forms that were understandable, practically relevant and adoptable by natural resource managers. To achieve this, a combination of products and communication methods were used including computer-based decision support systems, technical reports, fact-sheets, training workshops, seminars, information sessions, a blog site, targeted email and small on-demand briefings.

This chapter reflects on the effectiveness of the knowledge broking approaches and draws general conclusions about knowledge broking in large collaborative environmental research development and extension projects. A number of key lessons emerged:

- maintaining networks between stakeholders required understanding both their institutional and biophysical landscapes;
- involving end users and researchers in the formulation of research questions enabled local knowledge to be valued and contributed to the development of products that were tailored and relevant to NRM decision-makers;
- successful knowledge broking needs to be adequately resourced and flexible enough to respond to changes in the timing of research outputs and priorities within partner organisations;
- a combination of approaches including facilitation, relationship management, web technology, mentoring, training, product testing and evaluation is important for interdisciplinary and participatory research.

INTRODUCTION

Knowledge broking involves facilitating the exchange of knowledge so that those who need it can access it in a form that is relevant to their situation and needs (Land and Water Australia 2005). Knowledge brokers have been used in organisations across a range of sectors for several years. Australian NRM organisations such as the Cooperative Research Centre (CRC) for Freshwater Ecology, Australian Biosecurity CRC, Greening Australia and CSIRO have used knowledge brokers to ensure their strategic commitments to knowledge exchange are met. Although their functions varied, knowledge brokers were employed in most research hubs within the Commonwealth Environment Research Facilities program of which Landscape Logic was a part.

Knowledge broking typically refers to processes used to mediate between sources of knowledge (usually science and research) and users of knowledge. Knowledge broking is usually applied to help knowledge exchange work better for the benefit of all parties. Knowledge brokers help people to ask the right questions, identify the best sources of the information they need and work with the knowledge sources to get information into a form that will be most appropriate for research end users. Skilled knowledge brokers facilitate feedback from both parties, feeding information to researchers as much as to the users of research outputs and potentially challenging the way research outputs are presented and research questions are explored (Campbell and Schofield 2007).

In Landscape Logic, knowledge broking was identified as a core activity from the outset. It was established as a discrete team within the overall management structure and involved in key decision-making through the management group over the four-year life of Landscape Logic. The knowledge broking project was awarded a budget similar to that of the core research projects, amounting to 10% of the total project expenditure. This was an important factor in enabling the knowledge broking team to structure resource allocation and effort across activities and jurisdictions as the program evolved. While the knowledge broking team was able to draw on previous experience in other programs and share ideas with other CERF hubs, the context of Landscape Logic meant that the project had to forge its own path to accomplish its aims.

The task of the knowledge brokers was to build a shared understanding among the partners about Landscape Logic activities and purpose, and to manage an outreach program to other regional organisations and key NRM stakeholders across Australia. Three key elements of the knowledge broking project identified at the outset were:

1   communication and knowledge sharing, focusing on the development of information and knowledge products and services tailored to the needs of end users involved in improved catchment decision-making;
2   supporting adoption through training and allied activities to ensure that new knowledge led to improved NRM decision-making;
3   evaluation and feedback through the development and execution of the program evaluation framework.

The last of these was initiated by the knowledge broking team but carried out independently (see Chapter 20). A fourth element emerged during the project, that of maintaining relationships and networks across the hub.

Knowledge broking activities were designed to synthesise, package, communicate and evaluate material from across the program. This required a strong link with the project's communications manager, the distinction being that the knowledge brokers were involved in the whole chain of events involved in product development, delivery and adoption.

UNDERSTANDING THE CONTEXT

Landscape Logic involved collaboration between researchers and six regional organisations responsible for NRM: three catchment management authorities (CMAs) in Victoria and three NRM regions in Tasmania, collectively referred to as NRM or catchment management organisations (CMOs). The primary end users for the outputs of Landscape Logic were initially identified as these six NRM partner organisations, with the other 50 regional NRM bodies in Australia as secondary audiences (see Chapter 1). As the project unfolded, the nature of the end user audience changed to encompass a broader range of research users (Table 19.1), which differed between Victoria and Tasmania. For example, consultants, industry bodies and local government were a more important target audience in Tasmania where the research focus on water quality was of immediate interest to a wide range of stakeholders.

Knowledge broking activities were designed to support dialogue between the researchers and research end users by:

•   identifying the needs of the end users including personally knowing who they were and understanding their roles, current approaches to decision-making and preferred methods of knowledge acquisition and sharing. This was accomplished through

Table 19.1: Summary of Landscape Logic end user audiences and knowledge needs

Research user audience	Principal needs
Regional NRM organisations	Improved understanding of catchment processes
	Understanding the impact of past NRM investments and likely prospects of future interventions
	Knowledge and tools to assist with NRM decision-making and monitoring design
Commonwealth and state government agencies	Understanding the impact of past NRM investments and likely prospects of future interventions
	Implications of research outcomes for policy and program design
Local government	Knowledge and tools to assist with NRM decision-making and planning
NGOs	Understanding the impact of past NRM investments and likely prospects of future interventions
	Knowledge and tools to assist in decision-making
Industry bodies	Input from research findings to develop recommended practices and industry codes of practice
Consultants	Knowledge and tools to assist them service their clients

regular contact with CMOs, encouraging their involvement in the research focus formulation phase and dialogue at annual meetings and training events;

- understanding and respecting existing networks that operate at the regional scale to facilitate sharing of knowledge generated through Landscape Logic;
- providing regular opportunities for interaction between researchers and regional partners in a range of settings including field days, seminars and forums, electronic discussion groups and meetings.

Many regional NRM bodies, including those involved in Landscape Logic, aspire to practise adaptive management, an approach to managing natural resources where implementation of policies and strategies is treated as an opportunity for learning (Allan and Curtis 2005). The foundational literature on adaptive management (Walters and Holling 1990; Holling 1978) provides insights into the management of complex dynamic systems such as forests and fisheries which exist in a background of shifting societal goals and resource demands.

The knowledge broking approach employed in the establishment and operation of Landscape Logic was well suited to supporting adaptive management as it drew on the context and needs of our partner CMOs, was flexible in timing and degree of engagement and aimed to generate products such as models and decision support systems that could be directly applied to support management and decision-making. NRM is complex and there are significant differences between states in terms of regional capacity, statutory responsibility and financial resources. For example, a typical Victorian CMA has a staff of 50 and an annual budget in excess of $20 million, while a typical regional NRM organisation in Tasmania has approximately a quarter of the staff and financial resources. An additional complexity that emerged during the project was the introduction of a major new national funding program, 'Caring for Our Country', which established new priorities and reduced core funding to regions. In Victoria, the adoption of new policy on land and biodiversity management (DSE 2009) affected both regional and state government engagement – CMAs and state agencies became focused on implementation of reforms associated with new policy imperatives. Regular staff turnover is a feature of many NRM regions and in some cases hindered engagement with Landscape Logic activities.

NATURE OF KNOWLEDGE BROKING IN LANDSCAPE LOGIC

Knowledge broking activities were tailored to the different phases of Landscape Logic as follows:

- formulation – establishing the partnership and developing the funding proposal;
- design – identifying and refining research questions and approach;
- research – investigation, data collection, analysis and model development;
- harvest – interpreting the implications of research, communicating and extending results and products, product testing and training.

The last of these was not planned at the outset; it was the subject of an additional funding application to extend the project by six months, during which we would concentrate on communication and training. There was significant overlap between these phases, and the knowledge broking activities evolved and adapted to match the needs of the research teams and our partners in each phase.

In the project formulation phase, the knowledge brokers worked closely with the project director to develop the funding proposal and to secure collaboration from partner organisations including CMOs and researchers. The knowledge brokers were able to draw on existing networks and relationships with key staff and directors of CMOs in securing their formal involvement as partners. Likewise, existing relationships with some of the key researchers and their organisations were useful in gaining their participation.

Although the high-level aims of Landscape Logic were expressed in the successful research proposal, the specific research questions were not established until project inception. In many ways this was a crucial aspect of Landscape Logic, enabling the primary users of the research (CMOs and state government agencies) to negotiate the research focus and specific research questions with the project scientists.

In the project design phase, the knowledge brokers were responsible for coordinating discussions between researchers and research users that enabled both groups to better understand their respective skills, needs and contexts. Key activities in this phase included field trips, providing access to information and datasets, facilitating meetings and workshops and relationship management. This phase extended for nearly 12 months, resulting in a focus on two domains particular of interest to our CMO partners: vegetation extent and condition change in Victoria and water quality in Tasmania. The specific research questions, methodologies and study areas were then shaped by interaction between scientists and research users. During this time the knowledge brokers were also active contributors to the development and implementation of the Landscape Logic website, communications strategy and evaluation plan.

In the research phase, the emphasis of Landscape Logic shifted to investigation where the scientific teams focused on traditional approaches (e.g. data mining, experimental design, data collection and analysis, model development) to address the research questions. In keeping with the participatory research approach of Landscape Logic, there was regular interaction between scientists and research users. A good example was active collaboration between the knowledge brokers and the integration team in the application of Bayesian networks (BNs) by staff from CMOs and state agencies (Chapter 18, this volume). A feature of the investigatory mode was the pivotal role of the knowledge brokers in regular feedback, reporting and communication between the Landscape Logic partners. Annual meetings of all Landscape Logic researchers and representatives of our partner organisations were an important mechanism for sharing ideas and progress.

Table 19.2: Summary of activities during the harvest and adoption phases of Landscape Logic

Event type	Target audience	No. of events	No. of participants
Information seminar	Regional CMOs, state government agencies, consultants, local government, NGOs and industry bodies	12	316
Training workshop	Regional CMOs, consultants, NGOs, local government, state government agencies	5	94
Policy briefing	State and Commonwealth government agencies	4	40

In the harvest phase, the emphasis shifted to interpretation and communication of results, training and product development and testing. This involved a series of tailored events designed for research users with a focus on the overall findings from specific areas of Landscape Logic research, their implications for natural resource managers and policy-makers, and training in the use of models and decision support systems (Table 19.2). The knowledge brokers played a key role in the organisation and facilitation of these events, which included:

- developing event programs;
- liaising with individual researchers and research teams;
- assisting and mentoring researchers in development of content and presentations;
- promoting events to end users and encouraging participation by individuals and organisations;
- ensuring there was useful feedback and evaluation.

The harvest phase of Landscape Logic extended over six months. It included a total of 21 events, 14 in Tasmania and seven in Victoria, attended by 450 participants.

In the second year of Landscape Logic it became apparent that additional resourcing, expertise and effort would be required to optimise adoption by end users. The knowledge broking team was expanded to include the Landscape Logic director, one person responsible for science integration and evaluation and the communications manager in addition to the two knowledge brokers. While there was no explicit delineation of roles, members of the expanded team collaborated to meet the variety of project demands. For example, the knowledge brokers supported the director in the development of synthesis products such as fact-sheets, and supported the science integrator and integration team in delivering training in the use of models and decision support tools. The intention to engage with NRM organisations beyond our NRM partners was prevented by the late maturation of products and the adoption support required by our NRM partners.

KEY CHALLENGES AND LESSONS

This section highlights what we learned from the application of knowledge broking in Landscape Logic and the wider implications for this activity in participatory and interdisciplinary research more broadly.

UNDERSTANDING LANDSCAPE HISTORY AND INSTITUTIONAL CONTEXT

A key lesson from Landscape Logic was that the significant differences in history, culture and operation between regions had a major influence on how communication, consultation, negotiation and engagement needed to be conducted. In Tasmania, the regional NRM organisations were relatively new and evolving institutions. At the inception of Landscape Logic the regions had been in place for only three years and stakeholder relationships were just being formed. By comparison, Victorian CMAs had been operating for almost 10 years, extending a legacy of catchment management that began in the early 1980s. The Victorian regions had more-established stakeholder networks, significantly larger budgets and staff numbers and clearly identified investment priorities. These differences shaped the nature of the research focus in each state. Water quality was seen as the highest priority issue in Tasmania while in Victoria, where there were well-established river health and water quality programs, understanding native vegetation change following decades of investment was identified as a critical knowledge gap. These differences have a significant bearing on the way in which the knowledge brokers interacted with key stakeholders.

It was clear that the knowledge and experience held by Tasmanian and Victorian knowledge brokers about NRM programs, operating environment and landscape processes was critical in working with our regional partners in establishing a relevant research focus. The initial three-day field trips through the partner regions were extremely useful in building understanding of the landscapes and an appreciation of the NRM context.

> Visiting partner regions early in the program was extremely useful ... It helped us understand the complexity of the issues and how NRM organisations were endeavouring to do a good job with limited resources (Landscape Logic researcher response to online evaluation; see Chapter 20).

UNDERSTANDING AND VALUING LOCAL KNOWLEDGE

While Landscape Logic was essentially a scientific endeavour, it drew heavily on the local knowledge of landholders, extension staff, consultants and advisors in specific landscapes. The knowledge broking team was instrumental in identifying key contacts with relevant established networks, which facilitated input to the research investigations and product development. The knowledge brokers were strong advocates of local input at all stages of the collaboration from selection of study areas to data collection, model design and validation and communication of key findings. Key lessons were that effective incorporation of local knowledge was important for successful participatory landscape research (Chapter 11), that this needs to occur throughout the project and that knowledge brokers were important intermediaries in this process.

USING EXISTING NETWORKS

Over the life of Landscape Logic, collaboration with multiple regional and state-wide networks ensured ample opportunities for information sharing, input, training and feedback. It is estimated that over 500 different individuals were exposed to the project. It is important to identify and utilise established formal and informal networks when undertaking projects such as Landscape Logic, that face discontinuity between available funding and the long-term nature of the processes under study and the long-term requirement for implementation and management.

An important outcome has been the establishment of relationships between key individual researchers and specific research users, which will continue beyond the end of the project.

RECOGNISING THE CAPABILITIES AND LEARNING STYLES OF RESEARCH USERS

Within Landscape Logic we were dealing with a wide range of research users, often with broad NRM knowledge but generally limited detailed technical understanding and skills. For example, Bayesian networks were chosen as the primary framework for expressing the outputs of a number of the research projects, but very few of the research users had been exposed to graphical modelling in any form. Most research users were keen to be involved, but the amount of time that they could commit varied between partner organisations over the life of the program. This was due more to staff turnover, workloads and changing organisational priorities than to lack of interest but it had a major impact on the operation of the knowledge brokers and some impact on the uptake of research products.

The needs of research users were strongly dependent on their context. For example, staff in CMOs were seeking practical information and products to assist them in regional NRM decision-making. State government officers were focused on key findings that had implications for policies and meeting legislative responsibilities. NGOs and consultants were seeking information and tools to assist them better deliver useful knowledge to their clients. As the preferred language of researchers is not always understood by non-scientists, a key role of the knowledge brokers was to act as translators for the different user groups.

The knowledge brokers were instrumental in identifying the needs, capabilities and learning styles of the different end user groups to assist with the design and delivery of research outputs. Our experience showed that it is important to offer a range of approaches including information sessions, field days, hands-on training and individual briefings, and that these need to occur throughout the program life to be successful. Product design and delivery needs to be carefully considered and flexible, to accommodate changing circumstances.

INVOLVING END USERS AND RESEARCHERS IN RESEARCH DESIGN

Interaction between researchers and research users was used to identify research issues, objectives, study areas and, to some extent, methods. This process required approximately 12 months, with the advantage of greater joint ownership. While Landscape Logic identified potential key research users at inception, reviews of research project design, methods and data saw the target audience expand as the project progressed. Despite our best efforts, it became apparent that some questions relating to water quality research in Tasmania could have been more clearly and precisely defined at the outset; they were revisited during the course of the research. In the case of measuring change in vegetation condition and quality, this was an area that was poorly defined with no generally agreed metrics capable of detecting change over time. Discussions on this lasted the life of the project and were instrumental in prompting original research that could lead to resolution of this issue (Chapter 12).

> *Researchers listened to regional needs. This hasn't happened much in the past, especially for biodiversity (Victorian CMA staff member response to online evaluation; see Chapter 20).*

KNOWLEDGE BROKING IS MORE THAN COMMUNICATION AND INFORMATION TRANSFER

'Communication' is frequently used as a euphemism for 'marketing'. While marketing is essentially a one-way process aimed at transferring knowledge to a range of audiences, participatory research allows exchange of knowledge between researchers and managers. The partnership between researcher and research user knowledge is important if end users are to own the project and its results to the extent necessary for adoption. This ownership needs to be built from the outset of program design and sustained through involvement in research activities. A flexible array of strategies is required to meet this need. This is a role typically above and beyond that which researchers can usually undertake within the scope of their investigations.

In Landscape Logic, we recognised early that a suite of approaches was required for knowledge broking, including networking, mentoring, training, relationship management and involvement in specialist workshops designed to incorporate local knowledge (Chapter 11). The knowledge brokers played an important role in ensuring that potential research users were kept informed of progress. During the last 12 months of the project, the knowledge brokers worked with different researchers in the joint delivery of training and communication of products, findings and key messages.

As the project progressed it became clear that we needed to manage the project's legacy in terms of data and product custodianship and accessibility. The knowledge brokers explored options and facilitated arrangements with more permanent programs to ensure the availability of datasets and tools through existing mechanisms, such as the Tasmanian NRM Portal maintained by the state government.

ESTABLISHING AND MAINTAINING EFFECTIVE RELATIONSHIPS WITHIN THE RESEARCH PARTNERSHIPS

The research teams in Landscape Logic involved a range of disciplines (e.g. ecology, hydrology, social science, modelling) and were geographically spread across a number of locations. This presented challenges to building effective relationships between teams and sustaining focus on the overall objectives. The knowledge brokers' role was to ensure, as much as possible, that functional and positive relationships were maintained between research teams. They acted as 'honest brokers', encouraging interaction, building trust and facilitating communication and information sharing. They also ensured that the scientific outputs of research teams were available in a consistent, understandable and user-friendly form.

KNOWLEDGE BROKING NEEDS TO BE APPROPRIATELY RESOURCED AND STRUCTURED

Throughout the life of Landscape Logic there was a clear need for dedicated resourcing of knowledge broker activities within the key partner organisations. Effective engagement takes time, and feedback from some partner CMOs indicated that directly resourcing staff within their organisations would have improved the effectiveness of their participation. Future programs of this nature could consider a more distributed resourcing model, and introduce greater flexibility so that knowledge broking activities can adapt to changing needs as the program evolves and institutions undergo change.

PLAN FOR A DESIGNATED 'HARVEST AND DELIVERY' PHASE

It is extremely difficult to meet deadlines for the delivery of research outputs when some research teams are dependent on the outputs of others. Our experience was that some key findings and intermediate products became available too late in the project to be adequately incorporated or tested. This created challenges for the research teams, who needed to identify secondary and usually inferior sources of data or information, and for the knowledge brokers, especially as other demands on end users at this time were also significant. A large number of information and training events were held during the last three months of the project, and their effectiveness would have been enhanced if this period had been longer. We recommend that the project plan incorporate a dedicated 'harvest and delivery' phase of up to a year, to be undertaken after the core research activities have been completed. This has as many potential benefits for researchers as it does for research users, through first-hand experience of the application of their work.

BE PATIENT AND PERSISTENT AND FOSTER GOODWILL

In a project involving many different people, disciplines, cultures and organisations it is impossible to meet all needs at all times. Effort needs to be targeted through links with the people who are best placed in organisations to judge when and how best to engage. Researchers need the freedom to pursue their investigations; equally, environmental managers have competing priorities and need adequate notice if they are to provide feedback on research outcomes and products. A key role of the knowledge brokers was to provide continued support and encouragement to researchers and project managers during challenging stages of the project. Another key role was regular communication with partners to update them on progress, forecast future opportunities for engagement and elicit feedback. They needed to be persistent in working with researchers to ensure that the outputs were relevant, user-friendly and delivered in a timely manner. Through regular contact and updates with research users, the knowledge brokers were able to engender a sense of understanding and goodwill as well as continued interest in the project.

> The efforts of individuals [knowledge brokers] in making the partnership work and engaging stakeholders were also recognised (response to online evaluation; see Chapter 20).

CONCLUSIONS

Landscape Logic was an ambitious and challenging undertaking. Like the complexity of the landscape systems and processes that were the focus of the science, the social landscape of participants, institutions and networks was also complex. Knowledge broking played a key role in maintaining the overall research partnership through the development and support of the networks of researchers and end user partners. A focus on fostering and managing productive relationships from inception and a commitment from the leadership of Landscape Logic to support this role was crucial. Knowledge broking is very contextually sensitive and there is no single best model. It will be influenced by the needs and expectations of the knowledge users, the experience of researchers and their familiarity with one another, the experience of the knowledge brokers, the resources at their disposal and the attitudes of managers to their role. Ultimately, however, in a participatory and interdisciplinary research project such as Landscape Logic, effective knowledge brokers need to be well-known and trusted by those involved and have a strong knowledge base that encompasses both technical and institutional domains. The role of dedicated knowledge brokers was influential in Landscape Logic, and was made possible only through the active support of others in key roles, such as the project director, science integrator, communication manager and research team leaders.

REFERENCES

Allan C and Curtis A (2005) Nipped in the bud: why regional scale adaptive management is not blooming. *Environmental Management* **36**(3), 414–425.

Campbell A and Schofield N (2007) *The Getting of Knowledge: A Guide to Funding and Managing Applied Research*. Land and Water Australia, Canberra.

Dept of Sustainability and Environment (DSE) (2009) *Securing Our Natural Future: A White Paper for Land and Biodiversity at a Time of Climate Change*. Dept of Sustainability and Environment, Victoria.

Holling CS (1978) *Adaptive Environmental Assessment and Management*. John Wiley and Sons, New York.

Land and Water Australia (2005) *Knowledge and Adoption Strategy: Managing Information and Knowledge for Adoption Outcomes*. Land and Water Australia, Canberra.

Walters CJ and Holling CS (1990) Large-scale management experiments and learning by doing. *Ecology* **71**, 2060–2068.

Evaluating collaborative landscape research: views of participants and end users

Rebecca Kelly

SUMMARY

This chapter draws on information gathered from team members and partners involved in the Landscape Logic project to evaluate progress against seven key result areas (KRAs) identified in the project's evaluation framework. This framework was developed in the second year of the project by research team leaders and placed particular emphasis on the project's effectiveness at integrating research across disciplines and knowledge types. Four main opportunities for data collection were identified during the development of the evaluation framework. These were milestone reports to funders, mid-term project reviews by external experts, feedback from potential end users at workshops, seminars and information days, and surveys of the experiences and perceptions of the Landscape Logic team, partners and potential research end users. The data presented in this chapter are drawn primarily from the surveys. Survey respondents were asked to provide an assessment of the overall success of the project as well as each KRA. They were also asked to identify what worked well and what could have been improved, and the extent to which we were successful in integrating research from multiple disciplines and sources. Three key findings emerged. First, that integration, a focus of several

KRAs, was interpreted in three different ways – as participation (bringing researchers and managers together), as interdisciplinarity (bringing different disciplines together to solve problems) and as development of decision support systems and tools (bringing different knowledge types together to produce aids to decision-making). Second, that developing an evaluation framework at inception would have been more useful by providing feedback during the life of the project and that the summative evaluation carried out at the end of the project was overly ambitious in that many of the KRAs were not achievable within the project's life. Third, that structuring an evaluation process around KRAs proved to be an effective approach, but more value could have been derived if all team members had been involved rather than just the team leaders, there were greater consistency between data collection methods and evaluation results were communicated to the research teams and partners during the project's life.

EVALUATION FRAMEWORK

An evaluation framework is a structured process for clarifying goals and developing the means by which their achievement can be measured. Evaluation can contribute to both retrospective assessment of research programs (summative evaluation) and the process of development (formative evaluation). While it is important to make summative assessments of the achievements of research programs, it is increasingly accepted that more effort should be invested in formative assessments that can lead to program improvement (Curtis *et al.* 1998). Sound, honest and timely evaluation throughout the life of a program can help establish realistic goals and objectives, identify and refine appropriate research questions and provide direction on collaboration, end user needs and adjustments to the organisation and management of a research program.

Evaluations of research programs typically focus on effectiveness, efficiency and appropriateness. Assessments of effectiveness typically involve judgements about the achievement of goals, typically assigned by program leaders and funding bodies, and the extent to which these goals address the needs and interests of key stakeholders (Scriven 1993). Roux *et al.* (2010) identified three main stakeholder groups – funders, end users and researchers – and suggested that their primary interests were respectively efficiency, relevance and rigour. As it is in the long-term interests of all stakeholders that all three criteria be met, our evaluation targeted all three stakeholder groups.

Unravelling the logic of a program is an important first step in identifying objectives used to assess program effectiveness (Chen 1990; Rossi and Freeman 1985). The Landscape Logic evaluation framework was developed in the second year of the project by research team leaders, the project's two knowledge brokers (see Chapter 19) and the author, who was employed part-time to oversee project reporting and evaluation. A consultant specialising in evaluation guided the group through an analysis of the project's objectives and how they might be achieved. This resulted in a hierarchical framework where each level identified what would be required to achieve the next level:

1 higher-level outcomes (longer-term impacts flowing from the achievements of the KRAs);
2 KRAs (the specific outcomes or objectives);
3 uptake strategies (including integration, communication and extension);

4   groups that could directly use and benefit from outputs and activities;
5   specific outputs and new knowledge products;
6   activities required to produce the outputs;
7   staff development activities;
8   formal engagement of key stakeholders;
9   staffing and resources;
10   context and scope of the project.

Ideally, reporting and evaluation would occur at all levels of this framework. In our case, the evaluation framework was centred on level 2, the seven KRAs. It involved identifying the audiences for each KRA, the indicators that could be used to measure the extent to which they were achieved and the methods used to monitor these indicators (Figure 20.1).

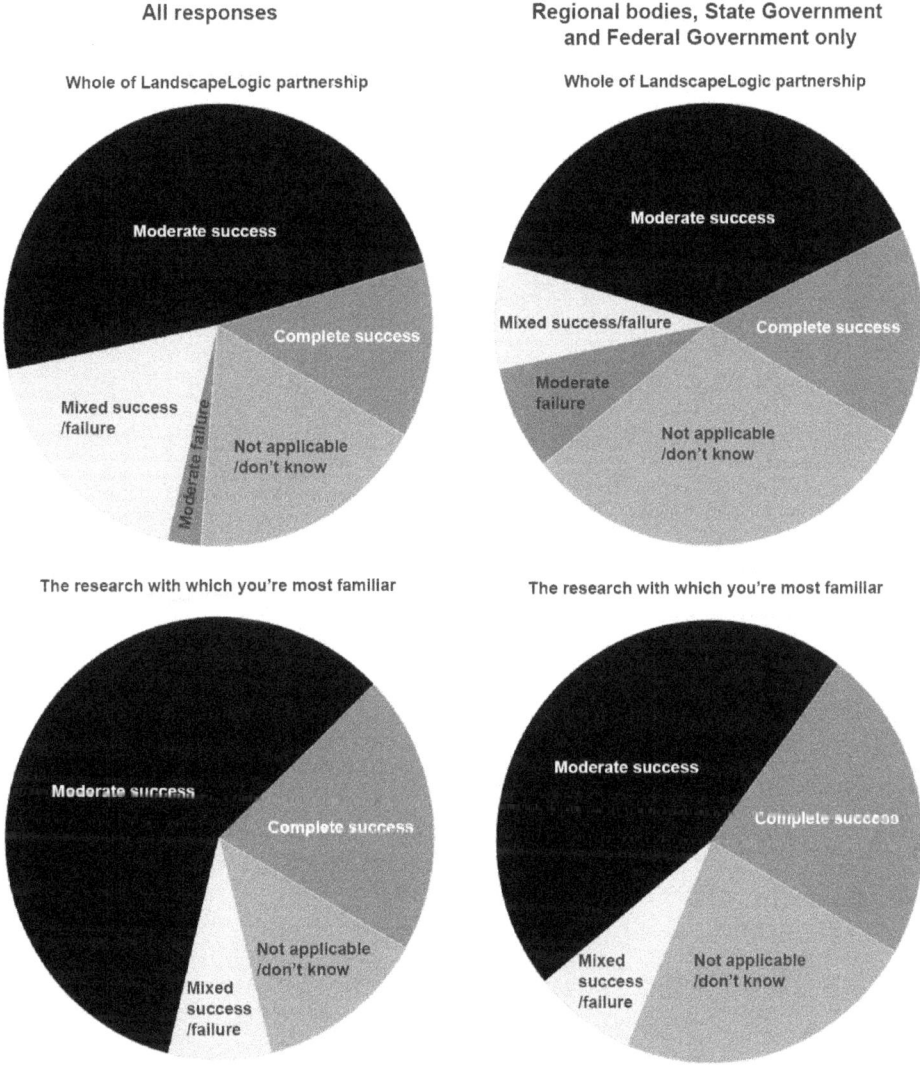

Figure 20.1: Research team and partners' perceptions of the success of Landscape Logic.

Four opportunities were used to collect data to measure progress against the KRAs: twice-yearly project milestone reports, external mid-term project reviews, feedback from potential end users at workshops, seminars and information days, and surveys designed to explore the experiences and perceptions of the Landscape Logic team and partners. This chapter draws mostly on the information gathered through three surveys, one of research team members and partners carried out at the conclusion of the project and two carried out at training and information sessions designed to evaluate the effectiveness of specific decision support tools.

The team and partners online survey invited respondents to provide an assessment of overall project success and the extent to which each KRA was achieved, their views about what worked well and what could have been improved, and their views about the extent to which we were successful in our goal of integrating different disciplines and sources of knowledge to improve understanding of landscape change. There were 41 responses to the team and partner survey, from 89 invitations. These comprised seven project leaders, 10 staff, nine students, seven representatives of regional natural resource management (NRM) bodies, two state government agency staff, five federal agency staff and one 'other'.

ASSESSMENT OF OVERALL SUCCESS

The majority of respondents rated Landscape Logic as a success overall but there was less certainty among regional, state and federal agency partners (i.e. funders and end users), with a higher proportion indicating 'Don't know' (Fig. 20.1). To both the research team and the partners (the latter combining funders and end users), success of the research areas with which respondents were most familiar (Fig. 20.2) was higher than the perceived success of the project overall. Written responses suggested the reasons for success were the partnership approach between regional NRM bodies and researchers, the volume and quality of research completed, and the large number of successful professional partnerships that had been established.

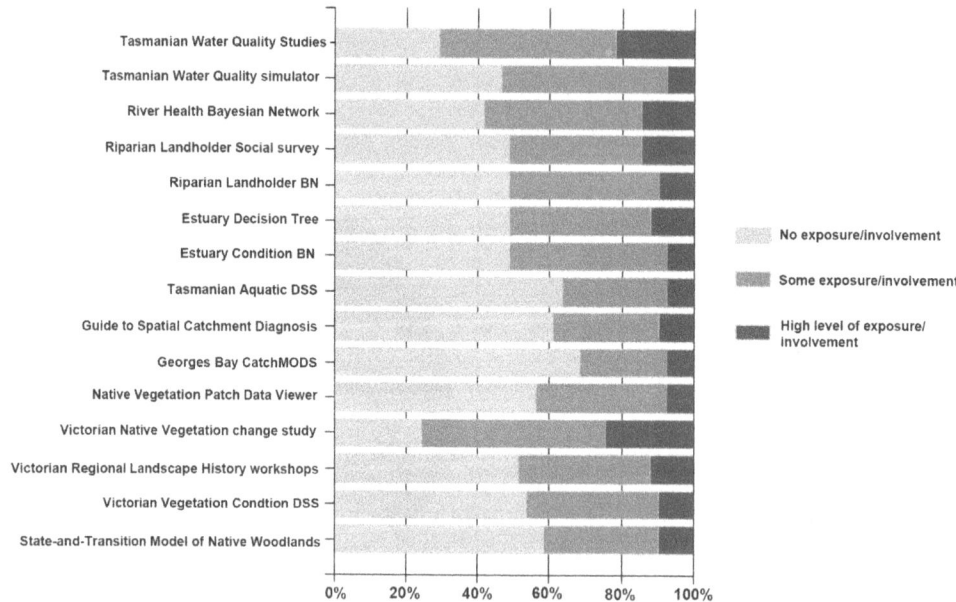

Figure 20.2: Level of exposure to Landscape Logic research outputs among the research team and partners.

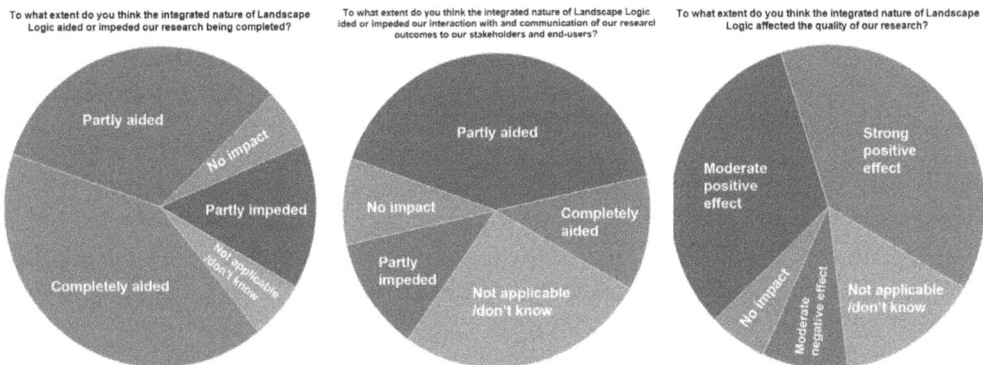

Figure 20.3: Perceptions of the team and partners about the extent to which integration affected completion, communication and quality of the research.

Issues that respondents identified as limiting the success of the project were more numerous and varied than those identified as influencing success, and included:

- underestimating the time required to reach agreement on focus research questions and form functioning teams;
- patchy communication between team members and the management committee (made up of project leaders), indicating a lack of communication within teams as well as between;
- not ensuring that all end users had the capacity to adopt and use the research products;
- initially identifying 'service' teams with a primary role of providing inputs (particularly spatial data and social research) to other research teams. This undervalued their research role and created some tension, which influenced the efficiency of integration;
- the failure of some research teams to engage with other teams and end users early in the project, an issue that was seen to be rectified later in the project;
- occasional conflict between personal goals and team goals, such as completion of a PhD thesis versus providing data and knowledge for integrated outputs;
- poorly communicated and managed delivery of outputs on which other teams depended, which had the effect of transferring stress to other teams and compromising their ability to deliver.

Some of the issues were common to the findings of previous evaluations of integrated landscape research projects, including the conflict of interest between team and personal or career goals (Tress and Tress 2005).

While many respondents were enthusiastic about the level of integration achieved between disciplines, research teams and partners, others identified impediments, missed opportunities and instances where this could have been improved (Fig. 20.3). The general perception emerging from written responses was that integration is often aspired to but rarely achieved by research groups and that there were clearly identified areas for improvement (listed above), but that Landscape Logic was one of the most successful attempts at integrated research in the experience of both research team members and partners.

DELIVERY AGAINST THE SEVEN KRAS

The extent to which team members and partners believed which Landscape Logic had achieved its KRAs as of September 2010, or was likely to within the next six to 12 months, are described below.

KRA 1: Developing new understanding of landscape change and response to intervention based on strong, well-founded science

Approximately 9% of respondents indicated that they did not know or that the question was not applicable to them. All other respondents indicated that this KRA was partly or completely achieved, with the most common response being 'mostly achieved' (47%). In one instance a respondent indicated that the KRA was not appropriately worded: 'If you replace "change" by "dynamics", I'd tick "Mostly achieved" or even "Completely achieved"'. Others indicated that the task was not possible due to insufficient data or the difficulties of working at large scales.

When asked to comment, some respondents explained partial success was due to the lack of available data and the short duration of the project compared to the long time-scales involved. Overall responses indicated that new understanding had emerged from the research.

Responses from potential end users collected at information and training sessions were more mixed. Early feedback from Tasmanian stakeholders attending an information session in late 2009 indicated that they thought the social, water quality and native vegetation research added substantially to new understanding, giving these areas an average of 8/10 for their contribution to KRA 1. Later feedback from four training workshops (July–December 2010) indicated that potential end users in Tasmania still felt strongly positive about new under-standing captured in models and decision support systems, giving an average score of 7/10 to the Tasmanian water quality products (see Chapters 2–8) with the highest ratings given by catchment management organisations (CMOs).

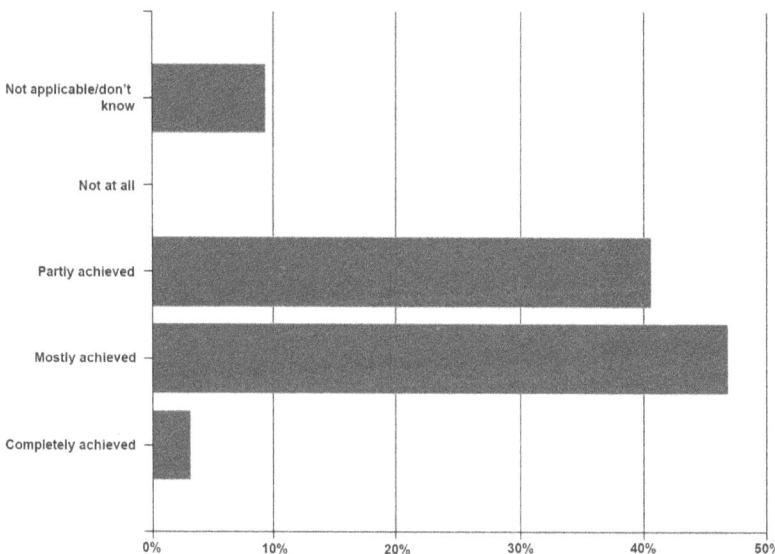

Figure 20.4: The extent to which team and partners perceived achievement of KRA 1: Developing new understanding of landscape change and response to intervention based on strong, well founded science.

In Victoria, feedback from a workshop in December 2010 produced an average score for all research products of 6/10, less than that for the comparable workshop held in Tasmania. Possible reasons for this are that more work was required after that date to fully explore the implications of that research for management and investment (see Chapters 10–12), that a particular interest of this audience was vegetation condition, where well-established goals and metrics do not yet exist, and that this audience of end users have had a longer experience of commissioning and using research outputs.

KRA 2: Incorporating a systems perspective into resource condition planning and investment decisions

Compared with KRA 1, respondents were a lot less certain about whether this KRA had been achieved – nearly 40% of both audiences (regional partners and others) indicated they did not know (or 'Not applicable'). Approximately 15% of the broader audience indicated that the KRA had not been achieved at all, with the majority of both audiences indicating it was partly achieved.

Among potential end users, the main evidence that a systems perspective was being incorporated into their planning and investment decisions came from the adoption of graphical modelling using Bayesian networks, particularly the first stage of conceptual modelling or influence diagrams. Capacity-building workshops were held in the first year of Landscape Logic to train CMO staff in Bayesian network (BN) modelling. Feedback from these workshops indicated that all participants gained new knowledge and most expressed aspirations to develop their own BNs (see Chapter 18). Later follow-up showed that respondents remained very positive and still had aspirations to use BNs to inform their decisions but that none had the staff capacity to warrant full model development. Nearly half indicated that they had sufficient knowledge to carry out the first stage of conceptual modelling. Evidence from the team and partner survey plus anecdotal evidence from CMO staff indicated that some were using

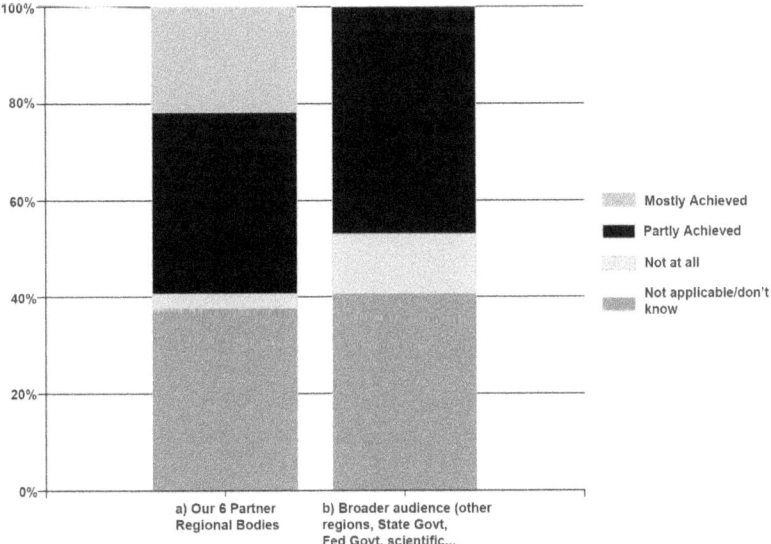

Figure 20.5: The extent to which team and partners erceived achievement of KRA 2: Incorporating a systems perspective into resource condition planning and investment decision.

conceptual modelling to assist in their decision-making. Few had adopted the full BN approach, but this is not seen as necessary to incorporating a systems perspective into decision-making. Issues identified by CMO staff as limiting their uptake of this approach have been staff turnover and the lack of time to invest in developing BNs. There was evidence from written responses that BNs and other integrated models and decision support systems developed by Landscape Logic were seen by CMOs and state agencies in Tasmania and Victoria as potentially useful in planning and investment.

KRA 3: Adoption of specific and generic tools, models, datasets and other knowledge developed by Landscape Logic to aid decision-making

A large proportion of respondents indicated that KRA 3 was not applicable or that they didn't know (>40% for both audiences). Some responses from both audiences indicated 'Not at all', with more of the broader audience (9%) indicating this. Most responses from our partner NRM regions indicated that this KRA will be partly (47%) or mostly (9%) achieved, while 40% of the broader audience thought it would be achieved to some degree.

The most common reason cited for low levels of achievement was that the time-frame was too short to enable adoption. In other words, it was unrealistic to expect significant uptake during the life of the project. Some respondents observed that the BN modelling approach is now more widely known and used due to Landscape Logic. One observed that some tools appeared likely to be used but that others would 'end up on the shelf'.

Feedback from end users in Tasmania indicated a high level of potential adoption of tools and other research outputs, among CMOs and state government agency staff.

In Victoria, there was interest in the native vegetation products but the training was generally perceived as being inadequate to ensure uptake. Perceived impediments to adoption included institutional support, legacy issues such as the need for ongoing support and training,

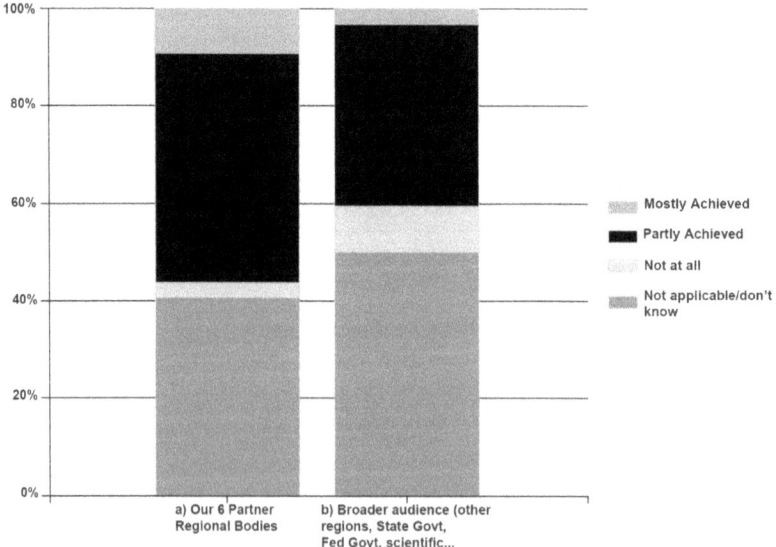

Figure 20.6: The extent to which team and partners perceived achievement of KRA 3: Adoption of specific and generic tools, models, datasets and other knowledge developed by Landscape Logic to aid decision-making.

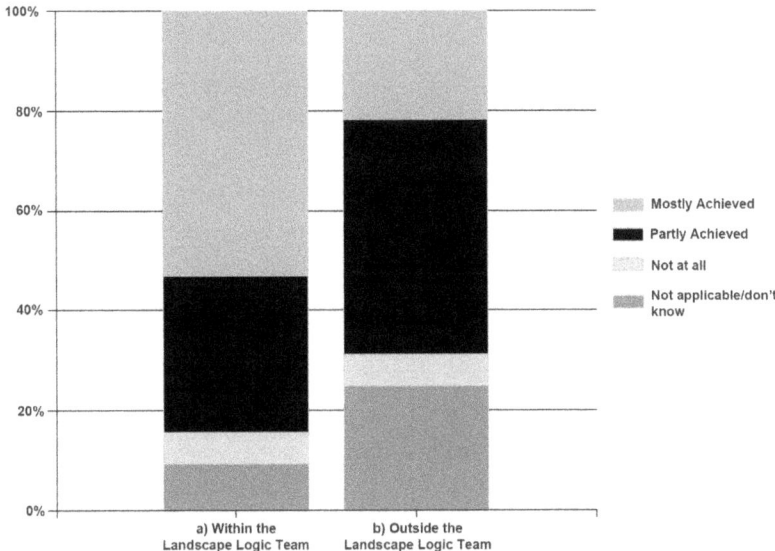

Figure 20.7: The extent to which team and partners perceived achievement of KRA 4: Demonstrated benefits from integrated approaches to landscape science.

and having sufficient time and resources to invest in applying the tools, which may have been a reflection on their complexity.

KRA 4: Demonstrated benefits from integrated approaches to landscape science

Respondents were more certain about the extent to which Landscape Logic will achieve this KRA. Only 6% of respondents indicated they did not think this KRA would be achieved at all, with 9% of the researchers and 25% of others indicating they did not know (or 'Not applicable'). Within the Landscape Logic team, the most common response was that it would be mostly achieved (53%) and another 31% indicated it would be partly achieved. Outside the team, respondents were more inclined to believe it would be partly achieved (47%), with only 22% indicating it would be mostly achieved.

In written responses, some respondents questioned whether there would be any impact outside the team, while others provided anecdotal evidence that this was already occurring. Some respondents acknowledged that there will always be some who would prefer to work in isolation but that there would now be a greater understanding of what could be achieved from integration. Respondents acknowledged that integration of research disciplines and knowledge sources was an ambitious objective of the project but one of its notable successes.

There was positive feedback from potential end users on the extent to which this KRA was achieved, at workshops on the Victorian native vegetation research where vegetation ecology, spatial analysis and social research were combined to quantify change in vegetation extent and attribute cause. Interestingly, this was rated more highly by Tasmanian (7.4/10) than by Victorian participants (6.1/10).

KRA 5: Increased skills and capacity for integrated landscape science within the Landscape Logic team

Overall, 72% of respondents thought that this KRA had been achieved to some degree, with the most common response being 'Partially achieved' (38%). Just over a quarter (28%) of

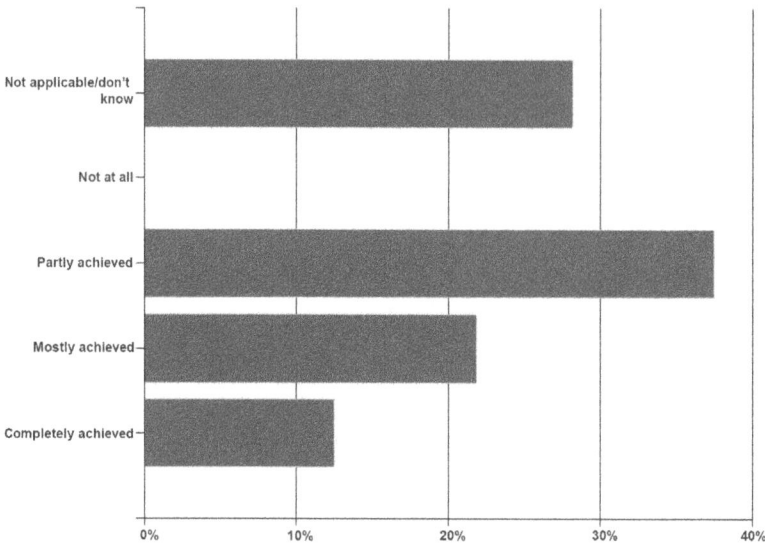

Figure 20.8: The extent to which team and partners perceived achievement of KRA 5: Increased skills and capacity for integrated landscape science within the Landscape Logic team.

respondents indicated that they did not know or that the question was not applicable to them. While some respondents indicated that this was an area of success for Landscape Logic, others expressed the view that there were lost opportunities and that the skills capacity that had been developed might be lost once the project ended, as there were few opportunities for their application.

KRA 6: Enduring collaborations between CMOs, researchers and state agency staff

Over 65% indicated that this KRA was likely to be achieved to some degree, with most indicating that it would be partially achieved (38%). A third (34%) indicated that they did not know or that the question was not applicable. Some indicated that the collaborations that would endure beyond Landscape Logic would involve only a small number of managers and researchers, and that they were dependent on ongoing funding. Several respondents questioned whether collaborations would be 'enduring'.

KRA 7: Improved design of monitoring frameworks for natural resource condition

This KRA elicited a large degree of uncertainty among respondents, with 38% of those from the six partner NRM regions and 50% from the broader audience indicating they did not know (or 'Not applicable'). Most respondents from our six partner regions felt this KRA would be partially (53%) or mostly (9%) achieved, while only 44% of the broader audience believed it would be achieved (38% partially and 6% mostly achieved). Written responses indicated that, while the project's research had contributed to the scientific basis for improved monitoring frameworks and raised awareness that improvement was needed, the major source of uncertainty related to the extent to which this would be acted on in the short term.

Evidence that this KRA was at least partially met was provided by invitations to Landscape Logic team members to participate in workshops to design catchment monitoring and reporting frameworks, with the potential to influence adoption more widely across this region.

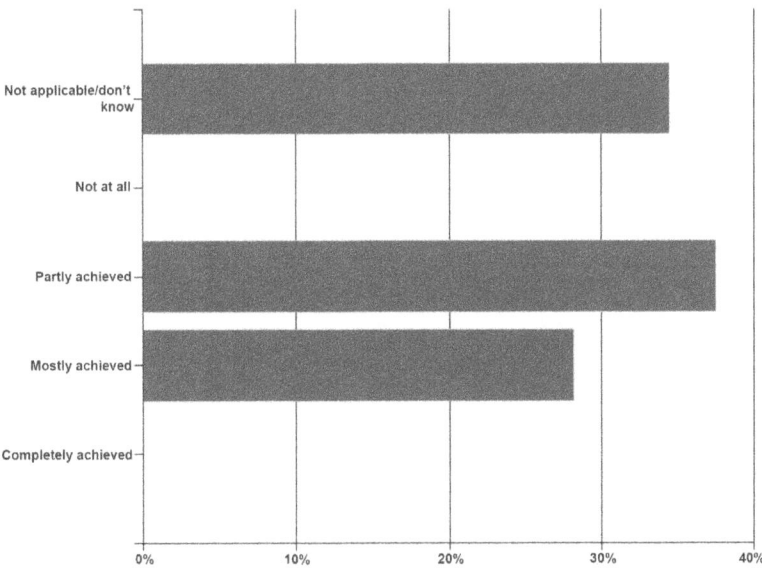

Figure 20.9: The extent to which team and partners perceived achievement of KRA 6: Enduring collaborations between CMOs, researchers and state agency staff.

HOW INTEGRATED WERE WE?

One issue that emerged in responses to questions about the overall success of Landscape Logic was the perceived disconnection between some research teams, particularly between the two domains of water quality and vegetation research. This is reflected in the responses to knowledge about specific products (Fig. 20.2).

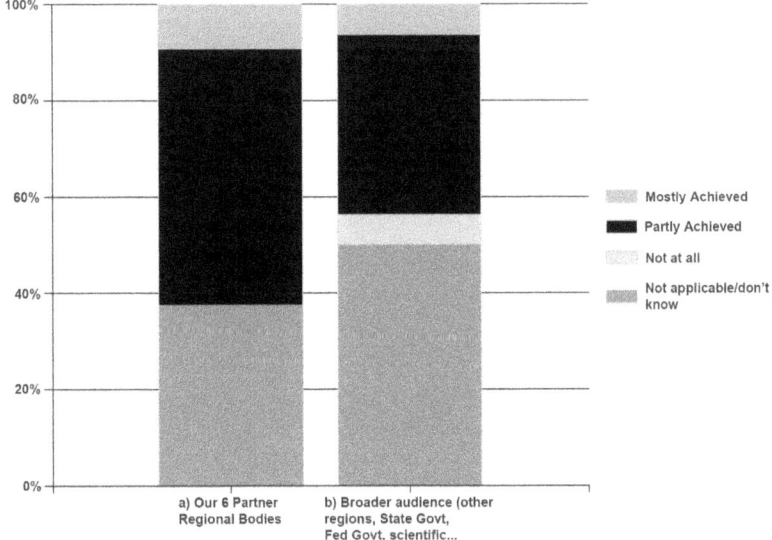

Figure 20.10: The extent to which team and partners perceived achievement of KRA 7: Improved design of monitoring frameworks for natural resource condition.

All research products returned some responses of 'No exposure or involvement in the research' from at least 20% of respondents. For some products this was well over 50%, with the highest at 60%. This reflects the focus of some research teams on very specific stakeholder audiences, but it also suggests there may have been missed opportunities for greater sharing of methods, datasets and understanding about landscape processes. Team members and partners have a key role in creating linkages, opening up new networks and expanding the audiences for research findings and these results suggest there was untapped potential for greater learning.

The survey included three questions intended to elicit the benefits and drawbacks of our approach to integration across disciplines and types of knowledge and between researchers and end users:

1   to what extent integration aided or impeded completion of our research;
2   to what extent it helped us communicate our results to end users and stakeholders;
3   to what extent it affected the quality of our research.

In all cases, integration was considered by the majority of respondents to have made a positive contribution to achieving research goals (Fig. 20.3). No respondents indicated that integration had a strongly negative effect or completely impeded achievement of goals.

Responses seemed to acknowledge that the research objectives intrinsically required an integrated approach, and that delivering science that is useful to NRM could not be achieved otherwise. While some acknowledged that this made the research process more difficult, most indicated that this cost was necessary and that better products were delivered as a result. Some respondents observed that a few teams largely ignored attempts at integration and went their own way, and one researcher indicated that integration diverted them from achieving their research outputs. Others questioned the level of true integration achieved, particularly between the two domains of water quality and vegetation research which have obvious interconnections at multiple scales.

In terms of communicating research results and their implications, some respondents indicated that integration made it harder to draw out key messages and findings. Where effort was made to integrate the outputs from a number of teams, such as the research described in Chapter 3 on ecological river condition and Chapter 6 on attitudes and behaviour to riparian management, this was seen to be more useful for managers. Some respondents indicated that there was still some way to go to fully explore the implications of findings from different research teams.

Some respondents observed that in some cases integration meant that the implications were too general to be of practical use, and that greater value would have been derived from more focused research.

It was clear from the responses that integration meant different things to different people. For many, integration was interpreted as participation between researchers, managers and stakeholders in defining questions, examining preliminary results and interpreting findings and their implications. To some researchers, integration was interpreted as working with fellow researchers from different disciplines and locations. For others, it was about building and delivering 'integrated tools' and decision support systems that brought together knowledge from different sources and different forms such as empirical data, model outputs and expert elicitation. The first two interpretations (participation and interdisciplinary research) in combination are described by European researchers as transdisciplinary research (Tress and

Tress 2005) and by North American researchers as team science (Klein 2008). The different views of respondents regarding the success of integration appeared to depend on their interpretation of integration. The range of responses suggested that Landscape Logic had been very successful in achieving the first definition and moderately successful at the second (with a few notable exceptions), but had missed many opportunities under the third definition.

RESEARCH PROCESS

The survey also asked respondents to identify what worked and what they would do differently, and invited suggestions on improving the research process. Responses indicated particular strengths and areas for improvement that were specific to different stages in the research process.

During the development phase:

- conducting tours of the six study regions with researchers and staff from all partner organisations was regarded as very useful;
- Bayesian networks, especially the conceptual modelling phase, proved to be a very effective platform for integrating knowledge but could have been used more effectively at the outset to clarify the key issues and researchable questions;
- having the time to establish collaborations and jointly identify research questions before detailed research plans were developed was a strength;
- although essential, the time devoted to negotiating and framing research questions with end user partners was cut short in some instances;
- stronger commitment of staff time by end user partners was needed from the outset. It was suggested that this should have been included in contractual agreements.

During the active research phase:

- a more carefully planned staged approach would have enabled some research to be analysed, interpreted and used to better inform the research of related teams, which would have improved research outputs and decision support tools;
- less pressure to integrate research activity early in the process would have enabled team formation and integration to occur in response to evidence as it emerged, rather than having effort partitioned according to anticipated need.

During the delivery phase:

- having more time for integration, either through earlier delivery and exchange of outputs by research teams or having a longer time-frame for research, would be desirable;
- project planning should have included at least 12 months for dissemination of research outputs and working with managers on its application;
- delivery of milestones to dependent research teams by the contracted dates should have been more firmly managed;
- more one-to-one visits between researchers to discuss research methods and results were needed.

Throughout the project:

- teams should be composed of those with a mix of knowledge discovery, knowledge broking and knowledge integration skills rather than the roles being separated;
- there should be more recognition of the roles of people skilled in the use of integration approaches, including use of their particular disciplinary skills;
- better use of tactics are required to engage end user partners at project inception, particularly approaches to overcome the issue of staff turnover in their organisations;
- flexibility, including budgetary, should be retained to adjust to needs as they arise by holding back up to 30% of funds to allow for integrated projects as they emerge;
- more effort should be made to involve students so they know what's going on and feel more part of the whole project;
- there should be fewer PhD projects included, as by their nature they have limited capacity to contribute to integrated outcomes;
- there should be less reliance on one approach (e.g. BN modelling) to accommodate all integration issues.

CONCLUSIONS

The Landscape Logic project set out to bring the expertise of ecologists, social scientists, spatial analysts and modellers into a partnership with environmental managers to document the effectiveness of past attempts at environmental management. Evaluation was a relatively minor activity within the project, involving just under 3% of the total budget, but useful information was collected relatively easily providing feedback from research teams and partners on the extent to which our goals had been achieved. This chapter provides some encouragement that valuable insights into research and adoption can be gained from a relatively simple evaluation framework. Two major findings related to timing of evaluation and interpretation of integration.

The evaluation process used in Landscape Logic was developed halfway through the project. The information that emerged was used by the project's knowledge brokers to confirm our target audiences and design training programs (Chapter 19). However, it became apparent from the evaluation that opportunities to respond to feedback were missed as a consequence of the late start and that using an evidence-based evaluation process from inception could have helped resolve issues as they arose and have better focused the project's activities. Ideally, the process of identifying research questions at inception would have been the most appropriate time to develop the evaluation framework and produce more realistic expectations of what could be achieved within the budget and time available. More effort could then have been placed on communicating evaluation results to team members and partners to let them know how their work was being appreciated, motivate the project teams and provide direction to some projects earlier in their life. In terms of external communication, the evaluation results would have provided good material for a newsletter that could have reached the project's extended audience of some 100 people and more widely communicated the project's activities and the extent to which the research was achieving its aims.

A significant finding of the evaluation was that three different interpretations of integration emerged – participation, interdisciplinarity and development of decision support systems. The evaluation indicated that the first was regarded as having been done well, with the limitation that time commitment by partners could have been greater and that it warranted more consideration, negotiation and possibly formal support through contractual arrangements. The value of the second interpretation, interdisciplinarity, was clearly demonstrated in several instances such as the new understanding that emerged from the synthesis of ecological and social knowledge in Chapters 3 and 6 (see summary, Chapter 9) and in Chapters 10–12 (see summary, Chapter 17). The third was achieved to the least extent and was the area where the evaluation results were most equivocal.

The failure to satisfactorily achieve this third interpretation of integration, development of decision support tools or systems that brought together different types of knowledge was essentially an issue of managing interdependencies between teams, i.e. where the output of one team is a necessary input for another. This was a significant issue that had implications for project performance and team member satisfaction over the project life. One implication from the evaluation is that rather than structure teams on the basis of knowledge discovery, knowledge integration and knowledge broking skills, a solution may have been to integrate the three skills within separate problem-focused teams.

Survey responses also suggested that the evaluation framework would have been more effective if it had been owned by the whole team, not just team leaders. While it was positive that the project leaders were involved in development of the evaluation framework, it would have been more useful to have students, other staff and regional partners also involved in developing the framework. This would have given them a greater sense of ownership and better reflected their goals and aspirations. Despite this, the evaluation yielded very significant information not only on what was achieved but on the actual process of carrying out integrated research, which may be of value to future participants in large-scale integrated research projects.

ACKNOWEDGEMENTS

The evaluation described here could not have been carried out without the contribution of the team leaders in developing the key result areas, the sound advice of Gordon Stone and Jeff Couts in structuring the evaluation framework and the participation of our partners and research teams in the surveys.

REFERENCES

Chen HT (1990) *Theory-driven Evaluations*. Sage, California.

Chen HT and Rossi PH (1983) Evaluating with sense: the theory-driven approach. *Evaluation Review* 7(3), 283–302.

Curtis A, Race D and Robertson A (1998) Lessons from recent evaluations of natural resource management programs in Australia. *Australian Journal of Environmental Management* 5(2), 109–119.

Klein JT (2008) Evaluation of interdisciplinary and transdisciplinary research: a literature review. *American Journal of Preventative Medicine* **35**, 116–123.

Prosavac EJ and Carey RG (1992) *Program Evaluation: Methods and Case Studies*. Prentice Hall, New Jersey.

Rossi PH and Freeman HE (1985) *Evaluation: A Systematic Approach*. Sage, California.

Roux DJ, Stirzaker RJ, Breen CM, Lefroy EC and Cresswell HP (2010) Framework for participative reflection on the accomplishment of transdisciplinary research programs. *Environmental Science and Policy* 13, 733–741.

Scriven M (1993) *Hard-won Lessons in Program Evaluation*. Jossey-Bass, San Francisco.

Tress and Tress (2005) Researchers' experiences, positive and negative, in integrative landscape projects. *Environmental Management* 36, 792–807.

21

Integrating science for landscape management

Ted Lefroy, Allan Curtis, Anthony Jakeman and

James McKee

In the last decade, owing partly to advances in graphical models, causality has undergone a major transformation: from a concept shrouded in mystery into a mathematical object with well-defined semantics and well founded logic ... Put simply, causality has been mathematized (Pearl 1999).

The Landscape Logic project set out with three ambitious aims as reflected in its programs: Knowledge Discovery, Knowledge Integration and Knowledge Broking. The Knowledge Discovery program had the aim of identifying causal links between past interventions in environmental management and the state of the environment in two specific areas: water quality and native vegetation extent and condition in south-eastern Australia. The intent was to improve the information available to environmental managers about the success or otherwise of past interventions. It was an attempt to close the adaptive management loop, much invoked but rarely practised (Allan and Curtis 2005; Allan *et al.* 2008; Lefroy 2008; Lindenmayer and Likens 2010). We knew this was an ambitious aim and were warned it was not only ambitious but naïve to assume we could shed light on cause and effect relationships in such a complex area as the state of the environment. Some of our studies did shed direct light on causal links, as outlined below, while others were less successful. Lack of success was primarily due to the lack of information available to disentangle the drivers from the observed responses. Like most scientific studies, new questions were raised worthy of further work. The quote from Pearl (1999) above is a reminder that failure to unambiguously identify causal relationships is not due to any fundamental flaws in semantics or logic. Philosophical frameworks and mathematical tools exist to elucidate causal links in principle, but as the adage goes, 'In theory, there is no difference between theory and practice. But, in practice, there is' (variously attributed to Yogi Berra and Jan van de Snepscheut).

The Knowledge Integration and Knowledge Broking programs were designed to complement knowledge discovery by performing the transdisciplinary functions of:

- exploring management needs, in particular the desired outcomes of natural resource managers;
- mapping the knowledge base required to elucidate how climate, management interventions and other drivers affect those natural resource management (NRM) outcomes;
- identifying the appropriate level of information required to connect the essential variables in this mapping exercise, including the scale and complexity of relationships to be represented, taking into account the data, information and knowledge available and, as far as possible, their uncertainty;
- undertaking these tasks in a participatory and iterative fashion that included the relevant scientists in the Knowledge Discovery program and elsewhere, NRM agency collaborators, industry groups and landholders.

The strength of the Knowledge Integration and Knowledge Broking programs was their ability to incorporate and synthesise many forms of identified knowledge, not just that obtained from the Knowledge Discovery program. This knowledge existed in many forms including disciplinary socio-economic and biophysical knowledge as well as the perspectives and aspirations of NRM specialists, industry representatives and landholders.

Along the way we identified a number of pre-conditions that have a major influence on inquiry into the causes of environmental condition. This final chapter identifies a number of these pre-conditions, suggests possible remedies that may be useful to others undertaking similar large-scale environmental research, and summarises what we learned about practising interdisciplinary and participatory landscape research.

INVESTIGATING CAUSAL RELATIONSHIPS IN ENVIRONMENTAL MANAGEMENT

Before causality can be formalised, the experience described in the preceding chapters suggests there are some basic requirements, consistent with systems thinking, that need to be met. This is the case especially when dealing with environmental processes occurring over large spatial scales and long time-lags that are responding to an interacting mixture of natural and human influences.

1   There should be clearly defined, measurable and widely accepted end points that include environmental goals, and socio-economic goals where relevant.

2   The boundaries of consideration for the systems under study must be sufficiently large to incorporate the boundaries of the major factors that influence the end points, without attempting to encompassing all external drivers.

3   Each of the major influences acting on the variables of interest, and their interactions, needs to be articulated – graphical mapping worked well in our context.

4   The factors that would provide appropriate or sufficient measurement of the effect of one major variable on another need to be identified (Pearl 2000 p. 425 shows an example of a graphical solution to the problem of co-variate selection).

5   Data must be available or collectable at sufficient spatial and temporal resolution to test and evaluate the strength of relationships articulated in the graphical model; if not, the model should be simplified or changed.

In many cases it is the last condition that prevents exploration of unambiguous causal relationships, and this is a commonly cited reason for the general inability to assess the effectiveness of environmental intervention (Lindenmayer and Likens 2010). But even where that condition is met to a useful extent, it will be useless in shedding light on cause and effect relationships without systematic attention to the other four. Conditions 2, 3 and 4 rely on the collective understanding of experts as well as established empirical evidence to map the likely influences on the desired end points, however, the subjective element of this neither negates nor diminishes the value of the modelling process. As pointed out by the mid-term review of the integration team (Chapter 18), the graphical modelling method chosen for studying the systems of interest to Landscape Logic needed to:

- be clear and transparent so participants could understand each other's knowledge domains, meaning no 'black boxes' (most important property);
- show how a decision or intervention was arrived at even if it later proved to be incorrect;
- have the capacity to be changed and added to as experience grows;
- give the correct answer (the least important property).

The last point reinforces the case made by Kay *et al.* (1999) that conventional scientific approaches to modelling and forecasting are inappropriate when dealing with complex socio-ecological systems, as are prevailing explanations in terms of linear causality. The authors suggest instead that we require narratives in the form of scenarios that can depict causal loops and multiple possible pathways. Chapter 18 reflects on the approach that was adopted across Landscape Logic of using graphical models to help us meet conditions 2, 3 and 4, and concludes that their graphical nature made it easier for managers and researchers to collaboratively formulate problem definitions and explore potential solutions.

Of the individual studies in this volume, Chapter 3 came closest to satisfying the five conditions above. While it stopped short of establishing causal relationships between land use and ecological river condition, it did establish strong inference that grazing above a threshold of 40–60% of the area of a catchment or subcatchment was the main driver of declining condition. A simple and distinct advantage this study had over others was that prior effort had gone into establishing an agreed measurable target for ecological river condition in the form of the Tasmanian River Condition Index, from which this study took its specific end point of the diversity and abundance of selected taxa of macro-invertebrates (NRM South 2009). Overall, the study provides a good example of the application of probabilistic graphical methods to elucidating and systematically examining the drivers of environmental condition.

Chapters 10, 11 and 12 provide good illustrations of the first condition not being met, when the goal is not defined in sufficient detail to enable measurement. The original question of measuring change in native vegetation extent and condition had to be broken into two tasks, measuring extent change and defining condition change. The goal or end point originally adopted for vegetation condition had been developed as a guide to prioritising investment in vegetation. Being an observational measure made relative to a desired baseline or historic state, it was found to be incapable of quantifying change over appropriate time-scales for monitoring (Chapter 12).

Chapters 2 and 4 are examples of condition 5 not being met, where data quality was not adequate for the questions of interest. Questions about the influence of land use on nutrient and sediment in rivers were beyond the spatial and temporal resolution of the available historic water quality data, particularly the lack of event-based monitoring, which was noted as a potential obstacle from the beginning. The two teams responded differently to this problem. Having explored the large-scale data to the fullest extent possible, one (Chapter 2) went down in scale to investigate the impact of selected land management practices (see also Chapter 5). Chapter 4 illustrates a different response, of stepping back to look at broader generalisations about Tasmania's estuaries to help to focus future research and monitoring effort.

Chapter 7 took a different approach again to the question of understanding the processes that drive water quality in agricultural catchments. That study provides an example of the forensic application of diagnostic tools to pin down the likely source of contemporary pollutants, using a combination of first principles and local knowledge within a clearly defined geographic boundary. It attempted to move from the particular to the general rather than the other way round.

Chapters 6, 15 and 16 illustrate how targeted social inquiry can provide data on the economic, social and institutional influences on land use and land management, contributing to conditions 3 and 4 above. While these three chapters describe studies that for convenience were conducted separately from their related biophysical inquiries, some had overlapping geographic boundaries (Chapter 6, Chapter 3 and parts of Chapter 2), meaning that their findings can be used to improve their respective graphical models. Chapter 11, on the other hand, illustrates how a careful marriage of social and ecological methods can be used simultaneously to compile and verify landscape histories.

LESSONS ABOUT MANAGING NATURAL RESOURCES AND PRACTISING TRANSDISCIPLINARY RESEARCH

The five conditions for establishing causality, while necessary, are not sufficient for the results of research to be adopted and applied to environmental management. Some additional factors identified through the course of this research, including a survey of participants (Chapter 20), are summarised below.

MANAGING NATURAL RESOURCES

1   *Have clear end points or goals*. It is not possible to track the state of the environment or the effects of intervention without the objectives being translated into clear goals

and metrics. This is not an issue for researchers alone, as these are questions where fact and value converge. The objectives, subsequent goals and indeed metrics need to be agreed and endorsed by representatives of those who own and manage the resources involved, those affected by its management and those responsible for reporting on their condition (Chapter 3). There is a role for researchers in clarifying the goals that are likely to be technically feasible, scientifically defensible and socially and economically acceptable. But this is not a task for the research community to carry out in isolation; it is one of the areas that demands participatory research (Chapter 12).

2   *Develop conceptual models.* A model of how a system works, particularly graphical models developed in collaboration with managers, is not only an accepted way to initiate any causal modelling exercise. It is also an effective way to reach agreement on the key processes and variables to represent and the interventions likely to be the most effective. No models are perfect, but through experience models can be improved and serve as hypotheses for researchers, as monitoring designs for managers and as a business case for governments and other investors. The advantage of probabilistic approaches such as Bayesian networks is that they can accommodate the complexity and uncertainty encountered in natural systems operating at various scales, and through sensitivity and other analyses be simplified into forms readily accessible to managers and policy-makers (Fig. 21.1; Chapters 3, 12 and 18).

3   *Investigate at multiple scales.* We found that in order to measure response to intervention and attribute change in our studies, it was necessary to investigate at three scales. Landscape-scale analysis was necessary to establish patterns of change, such as correlations between ecosystem response variables and human and natural drivers of change. Property-scale analysis was required to distinguish between human and other types of drivers by determining who did what, when, why and with whose

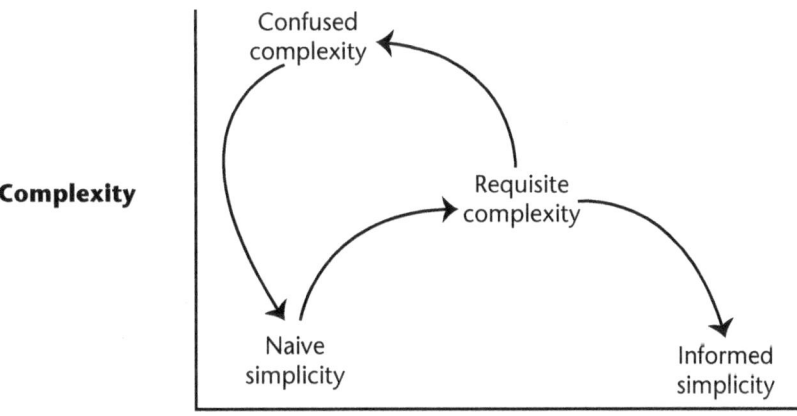

Figure 21.1: Using the simplicity cycle in model-building starts from a position of low utility and low complexity with naively simple conceptual models, then moves through several iterations of high complexity and low utility (confused complexity) until it arrives at a level of complexity sufficient to adequately represent the system (requisite complexity). The ultimate aim is arriving at a state of low complexity and high utility (informed simplicity) by progressively discarding drivers, nodes and links to which the system is found to be least sensitive (from Lefroy *et al.* 2009).

money. Site-scale investigation was necessary to understand the ecological processes that had produced the observed change (Fig. 21.2) (Chapters 3, 6, 10, 11 and 12).

4   *Understand the social context.* Being aware of the social context in which natural resources are used and managed is essential for identifying cost-effective pathways for change. In our studies, it was apparent that demographic change is resulting in a substantial cohort of land managers who are not farmers by occupation and who have largely been unreached by management agencies and programs. This cohort has high conservation values in general but relatively low levels of conservation knowledge or activity. Their activity is low partly because they are poorly connected to conservation networks, funding and sources of knowledge. Connecting these people to existing networks in order to translate their high conservation values into action could be more effective than investing directly in achieving major behavioural change in other groups. In another example, costly interventions to improve water quality by removing willows (*Salix* spp.) appear to have led to perverse outcomes, in that half the property owners surveyed in Tasmania who removed willows did not replace them with native vegetation as intended, thereby increasing year-round light levels and the potential for algal growth (Chapters 6, 11, 15 and 16).

5   *Manage expectations.* A major challenge for researchers and environmental managers is managing the expectations of government at all levels regarding the time-scales

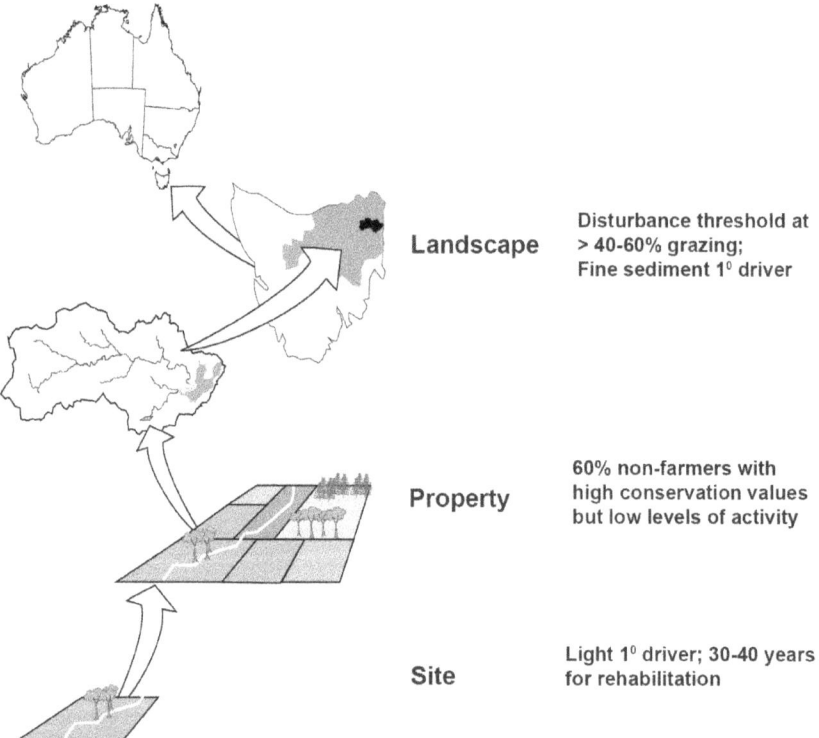

Figure 21.2: Schematic representation of the different levels of inquiry required to infer relationships between land use, land management and river ecological health (landscape, property and site, Ch. 3) showing different drivers dominating at different levels (modified from Lowrance *et al.* 1986).

within which they are likely to see response to intervention. Our studies provide evidence of 30–40-year scales required to achieve measurable change in indicators of vegetation change (recruitment opportunities in a variable climate, survival to maturity; Chapters 10, 11 and 12) and water quality (revegetation to filter sediment and shade streams and thereby shut down algal production and restore trophic pathways; Chapters 2, 3 and 5).

PRACTISING TRANSDISCIPLINARY RESEARCH

1 *Allow time for teams to develop*. Acknowledging the sequences involved in group development (storming, forming, norming and performing; Tuckman 1965) and allowing sufficient time for their expression proved to be an important contributor to a collaborative culture. Our experience was that the length of these phases varied with different teams, which required more flexible time-lines for problem definition, scoping research questions and planning research than we had envisaged.

2 *Get the questions right*. This point is closely related to the previous one, and centres on allowing sufficient time for the all-important processes of problem definition and identification of researchable questions. In our case, the six months allocated to the 'storming and forming' stages which included defining research questions was not sufficient for all areas of research or all teams. Getting the questions right has great bearing on the effectiveness of collaborative research, and in hindsight this could have been more flexibly managed to ensure a well-planned start to all projects.

3 *Collaborative model development*. Developing conceptual models or influence diagrams with environmental managers proved to be a very effective tool for involving managers in hypothesis setting and very useful for researchers in gaining a better understanding of the systems under study. A major factor in their success was their graphical structure and the effort that was put into training by the integration team, which introduced the language and associated software of network modelling through 13 workshops with researchers and managers in the first 18 months of the project (Chapter 18).

4 *Avoid identifying 'service' roles for research teams*. The spatial analysis, social research and integration teams were originally conceived as providing a service role to what were essentially seen as biophysically driven research questions. This proved to be a mistake, and was acknowledged during the course of the project to underrepresent the primary research contribution of these teams to the collaboration. This distinction influenced relationships between teams and presented obstacles to progress, which had negative implications evident throughout the course of the project.

5 *Recognise the roles of 'technical' and 'social' integration*. Two different aspects of integration were recognised as contributing to a large collaborative project such as this (Chapter 9). As well as having the mechanics of integration such as modelling frameworks, software and personnel skilled in integration methods (technical integration), it was important to have processes to overcome the geographic, institutional and disciplinary distances between researchers and partners (social integration). While there is a growing array of useful technology that can help to break down geographic constraints, such as internet meeting and file-sharing facilities, there proved to be no real substitute for regular meetings of team leaders (monthly), the advisory board (three-monthly), related research teams (six-monthly) and all researchers and

partners across the project (annually). Breaking down boundaries and ensuring communication between disparate groups required constant attention from team and project leaders and was an important aspect of fostering a collegial culture within and between groups. Social integration essentially means investing a great deal of time in problem framing, relationship managing and stakeholder engagement. The single most challenging issue in our experience was achieving the relevant level of commitment from all participants.

6 *Include dedicated knowledge brokers*. Having skilled communicators with established networks across research institutions, government agencies and NRM managers proved to be very valuable in breaking down cultural, institutional and language barriers between researchers and managers at all levels. The knowledge brokers helped to foster a shared understanding between partners from the first stages of scoping questions, to exchanging information and new knowledge during the course of the research, to negotiating the meaning and implications of findings in the final stages.

Since its emergence as a discipline, the goal of landscape ecology has been to understand the relationships between pattern and process. The factors listed above, particularly the multiple scales of inquiry (Fig. 21.2), emphasise that natural resource management is fundamentally concerned with the role of people in the landscape and the importance of local knowledge in linking pattern and process.

REFERENCES

Allan C and Curtis A (2005) Nipped in the bud: why regional scale adaptive management is not blooming. *Environmental Management* **36**(3), 414–425.

Allan C, Curtis A, Stankey G and Shindler B (2008) Adaptive management and watersheds: a social science perspective. *Journal of the American Water Resources Association* **44**(1), 166–174.

Kay JJ, Reiger JA, Boyle M and Francis G (1999) An ecosystem approach for sustainability: addressing the challenge of complexity. *Futures* **31**, 721–742.

Lefroy EC (2008) Closing the adaptive management loop: why practical experience is necessary but not sufficient and science essential but not always right. In *Biodiversity: Integrating Conservation and Production*. (Eds EC Lefroy, K Bailey, G Unwin and TW Norton) pp. 249–259. CSIRO Publishing, Melbourne.

Lefroy EC, Pollino CA and Jakeman AJ (2009) Using the simplicity cycle in model building. In *Proceedings of the 18th World IMACS/MODSIM Congress*. 13–17 July 2009, Cairns, Australia. p. 2384. <http://mssanz.org.au/modsim09/F12/lefroy.pdf>.

Lindenmayer DB and Likens GE (2010) *Effective Ecological Monitoring*. CSIRO Publishing, Melbourne.

Lowrance R, Hendrix PF and Odum EP (1986) A hierarchical approach to sustainable agriculture. *American Journal of Alternative Agriculture* **1**, 169–173.

NRM South (2009) *Tasmanian River Condition Index Reference Manual*. NRM South, Hobart.

Pearl J (1999) cited in Pearl (2009) *Causality: Models, Reasoning and Inference*. Cambridge University Press, New York.

Tuckman B (1965) Developmental sequence in small groups. *Psychological Bulletin* **63**(6), 384–399.

Index

www.ingramcontent.com/pod-product-compliance
Lightning Source LLC
Chambersburg PA
CBHW041133280526
45792CB00014B/2408